BUS

**ACPL ITEM
DISCARDED**

Y0-BVP-982

Contractor's Guide to Change Orders

ANDREW M. CIVITELLO, JR.

With additional Material by
William Locher

Prentice Hall

CIP data is available from the Library of Congress.

Copyright ©2002 BNI Publications Inc.

Printed in the United States of America
10 9 8 7 6 5 4 3 2 1

All rights reserved. No part of this book may be reproduced in any form or by any means, without permission in writing from the publisher.

This publication is designed to provide accurate and authoritative information in regard to the subject matter covered. It is sold with the understanding that the publisher is not engaged in rendering legal, accounting, or other professional service. If legal advice or other expert assistance is required, the services of a competent professional person should be sought.

—*From the Declaration of Principles jointly adopted by a Committee of the American Bar Association and a Committee of Publishers and Associations.*

ISBN 0-13-042595-8

PRENTICE HALL
Paramus, NJ 07652

http://www.phdirect.com

A Note to the User

Contractor's Guide to Change Orders, second edition, is designed and intended to help guide the contractor through the preparation, presentation, negotiation and approval of change orders. The book should be utilized with the understanding that its authors are not attempting to render legal advice or any other type of expert consultation with regard to particular situations. Further, the reader must understand that the law is constantly changing and that all authority mentioned in this book should be confirmed prior to any use to ensure that the law has not changed or been modified since the date of publication.

A Note About This Book

The second edition of *Contractor's Guide to Change Orders* has been designed for fast reference by buy construction professionals. Its format is arranged to identify and clarify the critical issues surrounding the preparation, presentation and finalization (acceptance) of change orders. Although this book was prepared by construction professionals and attorneys, and includes some legal citations, it is not intended to be used as a research guide for attorneys.

The sample letters contained in this book are important examples of how particular situations have been handled effectively by construction professionals. Many routine change order requests and responses can be handled by communications in the form presented in the sample letters. Certain situations, however, may require approaches that differ significantly from the procedures suggested in this book. Additionally, certain situations may require the advice of counsel or consultants with specialized expertise.

It is important to understand that laws vary from state to state and that trade practices may differ depending upon the geographic location of your business and the project. Contractual notification requirements, reservations of rights and claim statements are critical parts of the relationship between owners and contractors and may determine the contractor's rights to pursue certain change order claims. These communications must be considered carefully. Additionally, the responsible contractor should seek the advice of a competent attorney who specializes in construction law once matters proceed to the claim or dispute resolution stage.

If for any reason the reader desires additional clarifications or has specific concerns that are not addressed in this book, we will be happy to answer your written questions concerning changes, scheduling and project management. Address your written questions regarding change orders, project management, scheduling, and subcontractor control to:

Andrew Civitello, Jr.
P.O. Box 88
Valrico, FL 33594

Address your written inquiries concerning legal applications to:

Gibbs, Giden, Locher & Turner LLP
2029 Century Park East, Suite 3400
Los Angeles, California 90067
Attention: William D. Locher and Regina T. Coleman

Your comments, remarks and suggestions pertaining to the content of this book are always welcome.

About the Authors

Andrew Civitello, Jr. is president of Civitello Project Management, Inc., a construction and management consulting firm located in Brandon, Florida. He is a nationally recognized project management specialist and constructions claims consultant skilled in the areas of contract management, claims entitlement and valuation, CPM scheduling, management efficiency, delay and interference determination, and claim presentation and defense. A veteran constructor, his clients include owners, contractors, design professionals, and attorneys on a wide range of public, institutional, and private projects, throughout the country. He has extensive experience in the areas of job cost, estimating and bidding, scheduling, subvendor coordination, changes management, and dispute resolution. Mr. Civitello has been an advocate and expert on claims ranging to $39 million, an active panelist for the American Arbitration Association on cases ranging to $6 million, and is a qualified expert witness in his fields of expertise. He is the author of best-selling construction books with McGraw-Hill, Prentice-Hall, Simon & Schuster, and M. E. Sharpe, including *The Construction Operations Manual of Policies and Procedures, The Construction Manager, Complete Contracting, Contractor's Guide to Change Orders, The Builder's and Contractor's Yearbook, Construction Scheduling Simplified*, and the *Construction Safety & Loss Control Program Manual*, and is an advisor to those publishers on books by other authors.

William D. Locher is a senior partner with the law firm of Gibbs, Giden, Locher & Turner LLP in Los Angeles, California, and Las Vegas, Nevada. He has represented owners, contractors and material suppliers in construction-related matters and disputes for over twenty years. Mr. Locher has been involved in all aspects of the construction project, from site acquisition and design development to litigating construction contract claims. Mr. Locher lectures regularly on various construction law topics, including contract administration, change orders and construction contract claims. He also serves as a mediator for the resolution of public and private contract disputes. Mr. Locher received a Bachelor of Science degree from Arizona State University and a Juris Doctorate from Pepperdine University School of Law.

Regina T. Coleman is an associate with the law firm of Gibbs, Giden, Locher & Turner LLP. Ms. Coleman's practice also focuses on the representation of persons involved in the construction industry. Ms. Coleman has a bachelor of arts degree from the University of California, Los Angeles and received her Juris Doctorate degree from Loyola University.

Acknowledgments

The authors would like to express their sincere thanks to William Mahoney of BNi Books for giving them the opportunity and encouragement to prepare this book. Mr. Locher would also like to thank his family, Cindie, Alyson, Stephanie and Lauren, for their patience while he carried this manuscript to soccer games, volleyball tournaments and other family outings. Mr. Locher and Ms. Coleman would also like to thank the staff at Gibbs, Giden, Locher & Turner LLP who provided hours of invaluable assistance during the preparation of this manuscript.

Preface

The business of construction contracting has always been a very dynamic and exciting environment. On all construction projects, persons with differing skills join together to transform words and lines on paper into a permanent work of improvement. Successful construction projects require the combined effort of owners, financiers, designers, contractors, and manufacturers. Each of these entities is bound to the other through a variety of contract documents and legal principles. This book discusses these contracts and specifically the change order process, which is an integral part of these contractual relationships.

Changes have always been a part of the construction industry. However, the issues surrounding change orders have become increasingly complex in today's construction industry. This complexity is the result of many factors, including the fact that contractors have become more aware, even savvy, regarding change orders and their contractual rights. At the same time, there has been a marked increase in the cost of labor, materials and administrative overhead, the time pressure on performance, and the volume of construction law. Finally, the level of sophistication found in owners, architects, contractors, and material suppliers has also increased.

Even during the current construction boom, the industry has been highly competitive. To be profitable, every contractor must approach contract documents and contract performance in a lean and advantageous manner. The contract must be completed as expeditiously as possible and without significant interruption, or else the contractor must seek compensation for delays and interruptions. In today's environment, a contractor can no longer afford to include contingencies in bids or to absorb any additional costs that result from extra work, interruptions or schedule changes. "Time is money" is as true today as ever.

The convergence of all of these factors, along with the continuing education of the parties involved in the construction industry, have fueled the growth of construction claims and construction claims litigation. Furthermore, most industry professionals believe that this trend will continue.

One method for avoiding construction claims at the end of the construction project is to resolve the issues that develop during construction through the effective use of a "change" clause or the change order procedures set forth in the contract. The change order process, when used properly and efficiently by both parties to the contract, should reduce the number of construction contract disputes.

Change orders are a normal part of the construction process, and they occur on every construction project, irrespective of size. Owners and architects who acknowledge this fact will be more effective in putting their emotions aside, dealing with changes quickly and effectively, and concluding projects with a minimum of claims.

Contractor's Guide to Change Orders, second edition, is a comprehensive workbook designed to be an invaluable reference for contractors who wish to possess a complete working knowledge of the following matters:

- Why change order arise
- How changes are discovered
- Where to find hidden changes
- Who has responsibility for changes
- How to consolidate all relevant cost, time, and liability components into effective change order requests
- How to evaluate and quantify the impact of changes on the construction schedule (time, interference, delay, and acceleration)
- How to maximize the price for extra work
- How to prepare complete and forceful change order requests
- How to protect *all* contractual rights
- How to get paid for consequential costs, even after "final" payment
- How to develop and use sound, innovative, tested, and proven negotiating strategies, tactics, and techniques
- What powerful records are required for maximum protection
- How to effectively prepare and present a claim for monetary damages
- How to interpret construction contracts
- How to find attorneys and key consultants to maximize advantages and opportunities in arbitration, mediation and litigation

This book contains step-by-step instructions for the most common situations giving rise to changes. By using the procedures described in this book, the contractor can develop his or her own system for the discovery, preparation, presentation and finalization of change orders. With a little time and effort, the contractor can cultivate an environment in his or her organization that:

1. Identifies potential changes quickly;
2. Implements whatever actions are required to preserve contractual rights;
3. Produces complete, professional change order proposals that are viewed as credible by owners and architects; and
4. Is able to maximize the recovery on change orders through negotiation or dispute resolution procedures.

What This Workbook Will Do for You

This workbook is a comprehensive, practical guide to handling one of the most critical areas of construction contracting: discovering (uncovering), preparing, pricing, and presenting, negotiating, completing, and getting paid for *all* changes to construction projects.

Contractor's Guide to Change Orders is designed to give you valuable guidance and working tools for increasing profits. Everything you need for lean, advantageous contract interpretation is in this single-volume workbook. You will be able to uncover all the "hidden" cost items for which you are entitled to receive compensation. What the architect would have previously "clarified," for example, the owner will *now* process and pay for as a change order.

This book stimulates proactive thinking and foresight among professional contractors. It details dozens of money-making and money-saving principles and practices that have been successfully used by many contractors in the change order process.

Each chapter contains step-by-step procedures, checklists, full-sized forms, and word-for-word letters that combine to make up the blueprint for the successful submittal of change orders. You will get all the information you need to increase your effectiveness in requesting change orders, while simplifying your work and boosting your change order approval rate and recovery.

Whether you're the owner, president, executive, project manager, estimator or superintendent in a large contracting company, or all these things in a small or mid-sized firm, you'll be given:

Step-by-Step Instruction

- How to determine contract responsibilities of the owner, architect, engineers, prime contractors, and construction managers.
- How to interpret construction contracts in the most advantageous manner.
- How to get paid for a time delay rather than winding up with a "granted" noncompensable extension of time.
- How to dig into the contract, plans, specifications, and the site to uncover all potential change orders and interferences *before* they impact the job—and your profits.
- How to identify all potential items of change order to maximize your position with respect to costs, time, and profits.

WHAT THIS BOOK WILL DO FOR YOU

- How to prepare complete, effective change order proposals that satisfy all contract requirements while maintaining your most advantageous position.
- How to secure subcontract proposals that are properly substantiated, complete, and correct—the first time, while preserving *your* maximum protections against errors, omissions, and oversights.
- How to calculate and support charges for delays, interruptions, and interferences.
- How to keep pressure on the architect and the owner to reduce response and approval times.
- How to substantiate each component of your change order pricing to reduce or eliminate questions and objections.
- How to easily keep and use powerful records that will discover, define, support, and track change orders, claims, and their precise effects.
- How to negotiate change orders through each step in the process.
- How to administer multiple change orders and the project under control.
- How to prepare for arbitration, mediation or litigation.

Sample Forms and Form Letters

- Direct Project Management and Administrative Costs Form
- Daily Field Report Form
- Field Payroll Report
- Monthly Administrative Time Sheet
- Photograph Layout Form
- Job Meeting and Minutes Form
- Sample Time and Material Form
- Change Order Summary Sheet Form
- Form Letter to Subcontractor Regarding Shop Drawing Submittal Requirements
- Form Letter to Subcontractor Regarding Shop Drawing Resubmission Requirements
- Shop Drawing Submittal Summary Record Form
- Form Letter of Transmittal

Checklists

- Change Order Research Checklist
- Change Order Discovery Checklist
- Change Order General Conditions Checklist and Estimate Sheet
- Checklist for Meeting with Decision Makers
- Construction Claims Checklist

WHAT THIS BOOK WILL DO FOR YOU

Sample Word-for-Word Letters

To Subcontractors:
- Owner's Decision Directing Work
- Coordination of Work in Ceiling Spaces
- Material and Equipment Coordination
- Request for Change Quotation
- Change Order Price by Default
- Confirmation of Telephone Quotation
- Improper Proposal Submission
- T & M Submission Requirements
- Mandatory Job Meeting Attendance
- Lack of Job Meeting Attendance

To Owners:
- General Letter on Changes
- Determination of Responsibility for Questionable Work
- Duplications of Design (Nos. 1, 2, 3)
- Coordination of Not in Contract Equipment
- Equal to Proprietary Item
- Rejection of Equal to Proprietary Item
- Changed Site Conditions
- Pending Change Order
- Sample Change Order Proposal Cover Letter
- Cost Escalation Due to Untimely Action
- Acknowledgment of Actual Work Performed
- Transmittal
- Confirming Special Meeting

To Architects:
- Contract Equipment Coordination
- Transmittal

Sample Change Orders
- AIA Form G701
- Alternative Change Order Format

Relevant, Informed Discussion

Historical background.

The most favorable interpretation of the significant construction contract clauses.

Detailed analysis of change order components and processes.

Where to look on the project site and in contract documents to discover all those hidden changes.

Explanation of *all* compensable change order components, to make sure that your change order prices include all those items that are too easily overlooked and "given" away.

Tested and proven change order negotiating strategies, tactics, and countermeasures.

How to Use This Book

Contractor's Guide to Change Orders is a complete presentation of the issues surrounding the change order process. This book incorporates the knowledge, philosophies, procedures, and techniques that are helping contractors to succeed *now,* across the country. Each chapter is designed to address a specific area of change order consideration, and all are arranged in logical sequence for efficient, systematic use.

Follow these *five steps* toward increasing your effectiveness in requesting change orders.

STEP 1: *Read through the book in sequence.* Although the book is designed to allow any chapter to be reviewed independently, following the prearranged order will help to clarify the interrelationships of the subjects. The concepts set forth herein provide actual methods used to resolve problems successfully. They are easily tailored to fit your own business style and objectives.

STEP 2: *Establish change order files.* Review the recommended administrative procedures for organizing all change order files in Chapter 11, Keeping Change Orders Under Control. The use of these procedures will ensure an organized starting point upon which to build your own effective change order procedures and records.

STEP 3: *Consolidate the sample forms, form letters, and checklists.* Begin with the Change Order Summary Sheet in Chapter 11. From that point, reproduce all sample forms, form letters, and checklists. Modify them in any manner that best suits the way you do business.

STEP 4: *Prepare the sample letters.* The recommended treatment for each letter is outlined in the respective chapters that contain them. Insert the specific information necessary to tailor the letters to your special circumstance. Make any corrections that you feel are necessary to keep consistent with your style. The precepts, forms and letters included in this guide book should become a normal part of your business practice.

STEP 5: *Use this book as your on-the-spot reference.* This book is a contracting management tool that you, your peers and your subordinates will want constantly by your side. Because change orders are a normal part of the construction process, you and your company will be required to deal with many changes simultaneously and at every stage in the change order continuum—on every project that you're involved with. Utilizing this book as a regular resource this workbook will ensure that these new-found ideas, principles and practices do not find their way back on the shelf through disuse. Continue to get the most out of every change order situation on current projects and future projects by referring back to *Contractor's Guide to Change Orders* at every opportunity.

Contents

Preface   *vii*
What This Book Will Do for You   *ix*
How to Use This Book   *xiii*

Part One
INDUSTRY AND CONTRACT ENVIRONMENTS

1
The Contractor's Move to Power   3

1.1   Introduction   4
1.2   The Changing Building Industry   4
       1.2.1   The Contractor in Control   4
       1.2.2   Claims Consciousness   5

2
Contract vs. Contact: Parlaying Subtle Differences into Dramatic Advantages   6

2.1   Introduction   7
2.2   Ending the Confusion About Contract Structures   7
       2.2.1   (Traditional) General Contracting   8
       2.2.2   Design-Build   9
       2.2.3   Construction Management (Pure)   10
       2.2.4   Construction Management with a Guaranteed Maximum Price (GMP)   12
2.3   Clarifying Contract Responsibilities to Guarantee Accurate Assessments   13
       2.3.1   The Owner: 12 Categories of Responsibility   14
       2.3.2   The Architect: 10 Categories of Responsibility   22
       2.3.3   The General Contractor: 15 Categories of Responsibility   29
       2.3.4   A Final Note   35

3
Proven Strategies for Applying Construction Contracts   36

3.1   Strategic Interpretation: Applying Contracts to Secure Power Positions   37
       3.1.1   Contract Law Concepts   37
       3.1.2   Construction Law Concepts   38
3.2   The Contract Documents: Simplified Descriptions to Prevent Oversight   38

3.2.1 Contract Components 38
3.2.2 The Four C's of Contracts 40

3.3 Rules of Contract Interpretation: The Cards Up Your Sleeve 41
3.3.1 Introduction 41
3.3.2 Standard of Interpretation: Reasonable Expectations 41
3.3.3 Ambiguities Resolved Against the Drafter 42
3.3.4 Right to Choose the Interpretation 43
3.3.5 Specific vs. General 43
3.3.6 Usage of Trade Custom 43

3.4 Applying Construction Contracts Without Resistance 44
3.4.1 Introduction 44
3.4.2 Change Clauses 44
3.4.3 The Pass-Through Clause 46
3.4.4 The Dispute Clause 48
3.4.5 Authority (Formal/Constructive) 49
3.4.6 "General Scope" of Work 50
3.4.7 "Reasonable Review" 51
3.4.8 "Intent" vs. "Indication" 51
3.4.9 "Performance" and "Procedure" Specifications 52
3.4.10 Equitable Adjustment 53

Reference 54

Part Two
CHANGE ORDERS EXPOSED

4
Change Order Diagnosis 61

4.1 A Normal Part of the Construction Process 62
4.2 Clarification or *Change*? 62
4.3 Reasons for Change Orders (Additions and Deductions) 63
4.4 Change Order Categories 63
4.4.1 Owner-Acknowledged Changes 64
4.4.2 Constructive Changes 64
4.4.3 Consequential Changes 66

5
Understanding How Change Orders Arise 67

5.1 Introduction 68
5.2 Defective Specifications 68
5.2.1 Cut-and-Paste 68
5.2.2 Silly Specifications 69
5.2.3 Old and Outdated Specifications 70
5.2.4 Inconsistencies 70
5.2.5 Impossibilities 71

CONTENTS xvii

- 5.3 Nondisclosure 72
- 5.4 Lack of Coordination Among Design Disciplines 72
- 5.5 Incomplete Design 73
- 5.6 Latent Conditions (Defects) 73
- 5.7 Owner Changes 74
- 5.8 Improved Information 75
- 5.4 Improvements in Workmanship, Time, or Cost 75
- 5.10 Illegal Restrictions 76
- 5.11 Nonapplicable Boilerplate 77
- 5.12 "Intent" vs. "Included" 77

6
Using the Change Order Process to Your Maximum Advantage 78

- 6.1 The Six P's of Change Orders 79
- 6.2 Prospecting for Change Orders (Discovery) 79
- 6.3 Preparing the Change Order 80
  - 6.3.1 Establishing the Change Order File 80
  - 6.3.2 Change Order Research 81
  - 6.3.3 Change Order Research Checklist 82
  - 6.3.4 Notification 86
  - 6.3.5 Sample Notification Letter to the Owner on Changes 87
  - 6.3.6 Notice Components 89
- 6.4 Pricing the Change Order 89
  - 6.4.1 Now or Later 89
  - 6.4.2 Pricing Methodology 92
  - 6.4.3 Selecting the Proper Tone 92
- 6.5 Presenting the Change Order 92
  - 6.5.1 Proposal Submission 92
- 6.6 Performing the Work 93
  - 6.6.1 Tracking Project Effects 93
- 6.7 Change Order Payment 95
  - 6.7.1 Billing and Payment 95
  - 6.7.2 Claims and Disputes 96

Part Three
PROSPECTING FOR CHANGE ORDERS AND THEIR COMPONENTS

7
Where and How to Find Potential Change Orders 101

- 7.1 Introduction 103
- 7.2 Predesign 103
  - 7.2.1 Adjacent Properties 103

- 7.2.2 Boring (Subsurface) Data   104
- 7.2.3 Building Code Compliance   105
- 7.2.4 Easements/Rights of Way   106
- 7.2.5 Special Agency Approvals   107
- 7.2.6 Interference from Utilities Not Properly Shown   108
- 7.2.7 Plan Approvals (Building Permit)   109
- 7.2.8 Temporary Utilities—Availability Within the Contract Limit Lines   110

7.3 The Contract and Bid Documents   111
- 7.3.1 Award Date   111
- 7.3.2 Named Subcontracts   113
- 7.3.3 Sample Letter to the Owner Regarding Obligation to Determine Responsibility for Questionable Work   115
- 7.3.4 Sample Letter to Subcontractor Regarding Owner's Decision Directing Work   117
- 7.3.5 Price/Bid Allowances   119
- 7.3.6 Contract Time   119

7.4 Plans and Specifications   121
- 7.4.1 "As Indicated"   121
- 7.4.2 Ceiling Spaces (Conflict)   123
- 7.4.3 Sample Letter to Subcontractors Regarding Coordination of Work in Ceiling Spaces   125
- 7.4.4 Changed Existing Conditions   127
- 7.4.5 Column and Beam Locations   128
- 7.4.6 Design Change Telltales   130
- 7.4.7 Design Discipline Interfaces   131
- 7.4.8 Duplication of Design   132
- 7.4.9 Sample Letters to the Owner Regarding Design Duplications   135
- 7.4.10 "Fat" Specifications   139
- 7.4.11 Finish Schedule vs. Specification Index   140
- 7.4.12 Inadequate Level of Detail   140
- 7.4.13 Light Fixture Locations   142
- 7.4.14 Match Lines and Plan Orientations   144
- 7.4.15 Mechanical, Electrical, and N.I.C. Equipment   145
- 7.4.16 Sample Letter to Subcontractors Regarding Material and Equipment Coordination   148
- 7.4.17 Sample Letter to the Architect Regarding Contract Equipment Coordination   150
- 7.4.18 Sample Letter to the Owner Regarding N.I.C. Equipment Coordination   152
- 7.4.19 Numerous Details and Dimension Strings   154
- 7.4.20 Performance and Procedure Specifications   155
- 7.4.21 Proprietary Restrictions (Public)   156
- 7.4.22 Sample Letter to the Owner Regarding Equal toZ Proprietary Item   158

CONTENTS

    7.4.23 Sample Letter to the Owner Regarding Rejection of Equal for Proprietary Item 160
    7.4.24 Specification Section "Scopes" 162
 7.5 Site 162
    7.5.1 Introduction 162
    7.5.2 Grades, Elevations, and Contours 163
    7.5.3 Sample Letters to the Owner Regarding Change Site Conditions 165
 7.6 Change Order Discovery Checklist 169
    7.6.1 Introduction 162

Part Four
CHANGE ORDER PROPOSAL PREPARATION AND PRESENTATION

8
Designing and Constructing Effective Change Order Proposals   *181*

 8.1 Change Order Components 183
    8.1.1 Introduction: The Three Costs 183
    8.1.2 Direct Costs 184
    8.1.3 Indirect Costs 184
    8.1.4 Transforming Indirect Costs into Direct Costs 185
    8.1.5 Direct Project Management and Administrative Cost Form 186
    8.1.6 Consequential Costs (Damages) 188
    8.1.7 Practical Management of the Three-Cost Approach 189
 8.2 Developing the Change Order Proposal 190
    8.2.1 Change Order Identification/Notification 191
    8.2.2 Sample Letter to the Owner Regarding Pending Change Order 193
    8.2.3 Assembling Component Prices 195
    8.2.4 Sample Change Order General Conditions Checklist and Estimate Sheet 196
    8.2.5 Assembling Subcontract Prices 198
    8.2.6 Sample Letter to Subcontractor—Request for Change Order Quotation 199
    8.2.7 Sample Letter to Subcontractor—Change Quotation, Second Request 201
    8.2.8 Sample Letter to Subcontractor Regarding Change Order Price by Default 203
    8.2.9 Sample Change Order Telephone Quotation Form 205
    8.2.10 Sample Letter to Subcontractors Confirming Telephone Quote 207
    8.2.11 Determining Schedule Impact 209
 8.3 Finalizing the Proposal 212
    8.3.1 Introduction 212
    8.3.2 Proposal Format and Timing 213

8.3.3	Sample Change Order Proposal Cover Letter	214
8.3.4	Sample Letter to the Owner Regarding Change Order Cost Escalation Due to Untimely Action	217
8.3.5	Representing Change Order Components	219
8.3.6	Presenting the Total Change Order Price	221
8.3.7	Presenting the Effects on Contract Time	221
8.3.8	Requiring Approval Action	222
8.3.9	Additional Terms and Conditions	223

9
Substantiating Change Order Prices: Settling Arguments Before They Begin   224

9.1 Introduction   225
9.2 Lump-Sum Prices   227
    9.2.1 Sample Letter to Subcontractor Regarding Improper Proposal Submission   228
9.3 Detailed Cost Breakdowns   231
9.4 Time and Material   233
    9.4.1 Sample Letter to Subcontractors Regarding T & M Submission Requirements   235
9.5 Unit Prices   237
9.6 Historical Cost Records   238
9.7 Industry Sources   239
9.8 Invoices—Records of Direct Payment   239
9.9 The Schedule of Values   240

10
Using Project Records to Discover, Define, Support, and Track Change Orders and Claims   241

10.1 Introduction   243
    10.1.1 Active Working Files   243
    10.1.2 Item Completion and Close-Out   244
    10.1.3 Archives   244
10.2 Establishing Dates in the Correspondence   244
10.3 Daily Field Reports   245
    10.3.1 Sample Daily Field Report Form   247
10.4 Payroll Records   251
    10.4.1 Sample Field Payroll Report Form   251
    10.4.2 Sample Monthly Administrative Time Sheet   252
10.5 Photographs—What, When, and How   255
    10.5.1 Introduction   255
    10.5.2 Photograph Layout Requirements   256
    10.5.3 Sample Photograph Layout Form   256
10.6 Construction Schedules   258
    10.6.1 As-Planned, As-Built, and Adjusted Schedules   258

10.6.2 Six Requirements for Presentable Evidence 259

10.7 Using Job Meetings to Establish Dates, Scopes, and Responsibilities 261
   10.7.1 Introduction 261
   10.7.2 Job Meeting and Minutes Guidelines 262
   10.7.3 Sample Letter to Subcontractors Regarding Mandatory Job Meeting Attendance 264
   10.7.4 Sample Letter to Subcontractors Regarding Lack of of Job Meeting Attendance 266
   10.7.5 Sample Job Meeting Form 268

10.8 Shop Drawings and Approval Submittals 271
   10.8.1 Approval Responsibility 271
   10.8.2 Approval Response Time 273
   10.8.3 Treatment of Differing Conditions 273
   10.8.4 Absolute Contractor Responsibility 273

10.9 Time and Material Tickets 274
   10.9.1 Introduction 274
   10.9.2 Sample Letter to the Owner Regarding Acknowledgment of Actual Work Performed 275
   10.9.3 Sample T & M Form 277

Part Five
CHANGE ORDER AND FILE PRESENTATION

11
Keeping Change Orders Under Control: How to Save Time and Improve Records with Administrative Housekeeping 281

11.1 Introduction 282

11.2 Establishing Easy-to-Research Change Order Files 282

11.3 File Content 284

11.4 Correspondence File 287

11.5 Tracking Change Order Trends 288
   11.5.1 Introduction 288
   11.5.2 Evaluating the Change Order Summary Sheet 289
   11.5.3 The Change Order Summary Sheet Procedures 290
   11.5.4 Sample Change Order Summary Sheet Form and Sample Completed Form 291

11.6 Approval Submissions 294
   11.6.1 Introduction 294
   11.6.2 Shop Drawing Review and Coordination 294
   11.6.3 Shop Drawing Submission Requirements 295
   11.6.4 Sample Form Letter to Subcontractors Regarding Shop Drawing Submission Requirements 296
   11.6.5 Submittal Review, Distribution, and Follow-Up 299
   11.6.6 Sample Form Letter to Subcontractors Regarding Shop Drawing Resubmission Requirements 301

11.6.7 Shop Drawing Submittal Summary Record Procedure   304
11.6.8 Shop Drawing Submittal Summary Record Form   305
11.7 Sample Letter of Transmittal   309
11.7.1 Sample Form Letter of Transmittal   309

Part Six
DISPUTE RESOLUTION

12
Winning in Change Order Negotiation   315

12.1 Introduction   317
12.2 Acceptance Time   317
12.3 Agenda   317
12.4 Gentleman's Agreement   318
12.5 Agreement vs. Understanding   318
12.6 Allowances   318
12.7 Alternatives   319
12.8 Arbitration and Mediation   320
12.9 Aspiration Level   321
12.10 Assumptions   321
12.11 Authority   321
12.12 Averages   322
12.13 Boilerplate   322
12.14 Catch-22   323
12.15 Change Clauses   323
12.16 Change the Negotiator   323
12.17 General Contractor as a Conduit   324
12.18 Contingency   325
12.19 "Convenience" Specifications   325
12.20 Concessions   325
12.21 Constructive Clauses   326
12.22 Correlation of Contract Documents   327
12.23 Cost Perceptions   328
12.24 Credits—Turning Them Around   328
12.25 Deadlines   329
12.26 Deadlock   329
12.27 Deliberate Errors   329
12.28 Level of Detail   330
12.29 Discipline   330
12.30 The Eighty–Twenty Rule   331

12.31 Elaboration   331
12.32 Empathy   331
12.33 Designer's Estimates   331
12.34 Equitably Adjustment   332
12.35 Exceptions   333
12.36 Excusable Delays   333
12.37 Use of Experts   334
12.38 Face-Saving   335
12.39 Job Meetings   335
12.40 The Power of Legitimacy   335
12.41 Letter Wars   336
12.42 Lost Notes   337
12.43 "Nonnegotiable" Demands   337
12.44 Objections   338
12.45 Off-the-Record Discussions   338
12.46 Patience   339
12.48 Presentations   339
12.49 Proceed Orders   340
12.50 Promises   341
12.51 Questions   341
12.52 Quick Deals   342
12.53 Reasonable Review   342
12.54 Reopening Change Proposals   342
12.55 Split the Difference   343
12.56 Statistics   343
12.57 Telephone Negotiations   343
12.58 Plain Hard Work   344
12.59 Unit Prices   344
12.60 Value of Work Performed   345
12.61 Conclusion   345

13
Preparing for Winning When Changes Become Claims   *346*

13.1 Introduction   347
13.2 Turning Around Change Order Rejections   348
    13.2.1 Introduction   348
    13.2.2 "Good" or "Bad" Faith Rejections   349
    13.2.3 The "Nothing to Lose" Attitude   350
    13.2.4 Change Amount vs. Litigation Expense   351

   13.2.5 Meetings at the Highest Levels 351
   13.2.6 Checklist for Meetings at the Highest Levels 352
   13.2.7 Sample Letter to the Owner Confirming a Special
      Meeting 353
  13.3 Arbitration/Litigation/Mediation—What Is the Difference? 355
   13.3.1 Introduction 355
   13.3.2 Arbitration 355
   13.3.3 Litigation 358
   13.3.4 Mediation 360
   13.3.5 Conclusions 362
  13.4 Finding an Attorney 362
   13.4.1 Introduction 362
   13.4.2 Characteristics of the Lion 363
   13.4.3 Characteristics of the Pussycat 363
   13.4.4 How to Find Your Lion 363
  13.5 Selecting Consultants 364
  13.6 Construction Claims Checklist 366

Appendix: Sample Contract Change Order *369*

Index *375*

Part One
Industry and Contract Environments

1
The Contractor's Move to Power

1.1	Introduction	4
1.2	The Changing Building Industry	4
	1.2.1 The Contractor in Control	4
	1.2.2 Claims Consciousness	5

1.1 INTRODUCTION

The construction industry is comprised of a large number of small companies operating in one of the most intensely competitive work environments ever to evolve. In this country, all construction firms, large and small, share remarkably similar characteristics: The competitive economic factors affecting resources, time, and management are nearly the same for all businesses.

Moreover, the construction contract agreement on a $1 million job will be very similar to that on a $50 million job. The plans themselves may be more elaborate on the larger project, but the bid documents, General Conditions of the Contract, and the Working Procedure will all be very similar. In addition, the functions, rights, and duties of the owner, design professionals, prime contractors, and subcontractors will likewise be fundamentally the same. So, again, the factors affecting the interpretation of our construction contract documents are nearly the same for *all* businesses, regardless of size or organizational complexity.

As a result of the current competitive environment, many contractors have seen their profitability decline. But some companies continue to record superior rates of growth. These successful firms have recognized that work environments and the nature of the business have changed, and they have moved to accommodate these changes.

In the discussions that follow, we review how the contractor's power and importance have increased as a result of the changes in the contracting environment.

1.2 THE CHANGING BUILDING INDUSTRY

1.2.1 The Contractor in Control

A hard lesson was learned by the building industry as a result of the recessions of the 1970s and early 1980s: Increases in the cost and complexity of doing business dictate that *all* costs must be tightly controlled to preserve profits. This means that if a contractor is going to win a bid and earn a profit, the contract documents must be interpreted in a lean fashion. There is no longer any room in a contractor's vocabulary for the concepts of "contingency" or "absorbed costs." As a matter of survival, contractors have had to become intimately familiar with contracts, their components, structure, meaning, and interpretation.

Contractors have now come to know their contracts better than the designers who produced them. They control their application more effectively. What the architect would have previously "clarified" is now a change order for which the contractor will get paid.

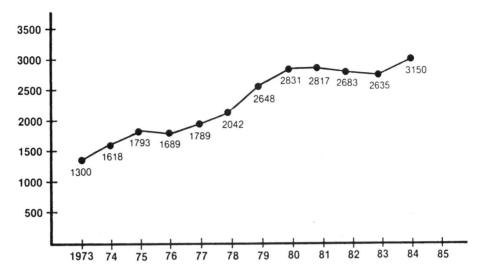

FIGURE 1-1 CASES SUBMITTED TO AAA NATIONALLY
By the American Arbitration Association, 140 W. 51 Street, New York, N.Y.

1.2.2 Claims Consciousness

Many contractors became very good at getting what they were after. When necessary, they recovered damages through a "claims" procedure that introduced damages through the courts or arbitration. These became hot news items, often because the dollars won in many of the awards were very large. The concept of being paid for a time delay, for example, was much more attractive than the historic "granted time extension." Contractors became very interested in developing these techniques. Some even began to look upon the whole idea as a profit center in itself.

The courts became inundated with construction claims. The American Arbitration Association (AAA) became a more visible, prominent force in the contracting legal environment. The AAA began accepting construction cases for arbitration in 1966, but didn't begin to maintain comprehensive records until 1973. From that point, as Figure 1-1 presents, the number of construction arbitration cases submitted per year nationally doubled in the decade that followed, and continues to increase at the same rate.[1]

Even if a contracting company is not particularly interested in pursuing an aggressive claims posture, a responsiveness to the system has now become absolutely necessary to at least maintain an ability to accurately assess day-to-day operating problems as they occur. An intimate awareness of contract structures and resolution options has now become the most fundamental prerequisite for responsible contracting management.

Claims consciousness is a matter of prevention. Many problems in construction, although varying widely in their specific composition, fall within common recurring categories. Errors and omissions, acceleration, delay, interference, and latent conditions will make most lists. It is, therefore, necessary to develop coordinated systems for their prevention, early detection, prompt resolution, and defense *before* all these activities actually become necessary.

---

[1]American Arbitration Association (Commercial), New York City Office, Frank Zotto, associate director, Case Administration.

2

Contract vs. Contact: Parlaying Subtle Differences into Dramatic Advantages

2.1	Introduction	7
2.2	Ending the Confusion About Contract Structures	7
	2.2.1 (Traditional) General Contracting	8
	2.2.2 Design-Build	9
	2.2.3 Construction Management (Pure)	10
	2.2.4 Construction Management with a Guaranteed Maximum Price (GMP)	12
2.3	Clarifying Contract Responsibilities to Guarantee Accurate Assessments	13
	2.3.1 The Owner: 12 Categories of Responsibility	14
	2.3.2 The Architect: 10 Categories of Responsibility	22
	2.3.3 The General Contractor: 15 Categories of Responsibility	29
	2.3.4 A Final Note	35

Note: Numbers in brackets (e.g., [27]) refer to references at the end of Chapter 3.

INDUSTRY AND CONTRACT ENVIRONMENTS

2.1 INTRODUCTION

Familiarity with how the different parties to a given construction contract arrangement are in contact with and responsible to each other will form the basis of solid application of the skills developed throughout this book. While seasoned construction professionals will be familiar with many of the descriptions of contract relationships and contact practicalities that are incorporated in this chapter, a review of the material may offer a fresh perspective that will help you to view an old set of constructs in a new light.

2.2 ENDING THE CONFUSION ABOUT CONTRACT STRUCTURES

There are several common forms of contract and project arrangements that may be selected by an owner for a construction project, depending upon the individual circumstances and preferences. These, of course, can then be modified in any manner and to nearly any extent desired by the drafter. For the most part, however, they will be applied with their familiar formats generally intact. In any event, the strict wording of all contracts must be studied thoroughly and be completely understood. As those who work intimately with contracts should know, the modification of a single word can alter the meaning of the entire document. In addition, historical relationships (the way in which things are "usually" done) may no longer apply. Just because the roofer, for example, *never* cuts masonry reglets as part of his normal operations doesn't mean it's not the roofer's responsibility on *this* job. Or just because the mason *never* provides the portion of the masonry anchor welded to the structural steel doesn't mean that it's not the mason's responsibility on *this* project. Specifications *must* be read *completely*. The contract must be understood in its smallest detail. Assumptions and a blind dependence upon past experiences have absolutely no place in contemporary construction contracting.

Finally, there is an important distinction to be made between *contact* and *contract*. This sounds almost too simple, but the truth is that contact (communication) too often leads to actions, inactions, and assumptions of responsibilities and liabilities due to misunderstanding of *contract* relationships and authorities.

Figure 2-1 shows the communication, or contact, between the functional administrative divisions of every construction project as the owner, the design function, and the construction force. The format occurs (largely due to practicalities) in nearly the same fashion project after project, regardless of technical contract structures. This is an area with the potential for misunderstandings that eventually lead to cost increases and delays. At the site where all the activity is, the job "feels" like most others.

The most common structures of construction contracting arrangements can be classified in their literal compositions as:

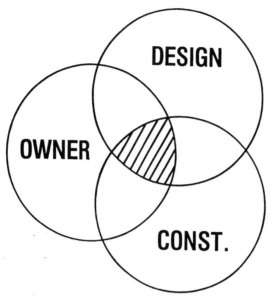

FIGURE 2-1 PROJECT CONTACT (COMMUNICATION)

1. General contractor (traditional).
2. Design-build.
3. "Pure" construction management.
4. Construction management with guaranteed maximum price (GMP).

The concepts introduced throughout the remainder of the book will be discussed in the traditional general contractor-type context. It is important, however, to understand the modifications that have led to the development of the other common structures, so we will begin with their brief descriptions. With a clear understanding of their respective makeups, it will be easy to apply the ideas in this book equally as well, regardless of the specific contract situation in which you may find yourself.

2.2.1 (Traditional) General Contracting

Figure 2-2 represents typical general contracting relationships.

In this structure, the owner contracts directly (and individually) with the contractor and the architect. The architect's general responsibilities beyond production of the design and the general contract documents are to administer the owner-contractor contract, functioning as the owner's agent. The prime, or general, contractor (the GC) enters into individual agreements with and is solely responsible for the complete performance of each of the subcontractors. The owner looks only to the general contractor for satisfaction of performance, and accordingly, the GC's dealings with the subcontractors are, in theory, transparent to the owner. The subcontractors, in turn, look (at least initially) directly to the GC for resolution of difficulties, even if they originate in the owner's contract documents.

Communication between the owner, architect, and general contractor necessarily overlaps to a great extent. This leaves one difficulty during day-

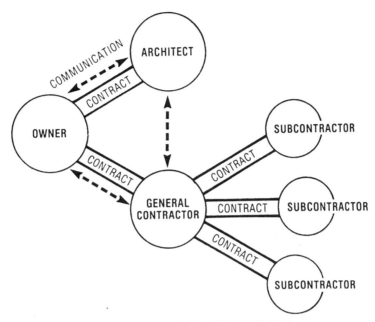

FIGURE 2-2 TRADITIONAL GENERAL CONTRACTOR ARRANGEMENT

to-day operations to be the accurate determination of the proper levels of authority of the design professional versus the owner. The contractor can too easily get caught in a squeeze by having acted on orders from a contractually unauthorized source. In any event, owner- or architect-prepared general conditions take great care in attempts to cleanly separate the contractor and the architect from any implication of a contractual tie. Consider, for example, the disclaimer from the widely used AIA Document A201, *The General Conditions of the Contract for Construction,* prepared by the American Institute of Architects:

> The Contract Documents shall not be construed to create any contractual relationship of any kind between the architect and the contractor ...

2.2.2 Design-Build

Design-build is shown in Figure 2-3. In this form, the owner contracts singularly with the design-build firm. This is usually a general contracting-type company with design capability, whether or not the design function is located wholly or partially in-house. Even if the design force is another company altogether, its connection is just as another subcontractor to the prime: The owner sees only the design-build company contractually. The subcontractor relationships to the construction force of the design-build firm are identical to those of the general contractor–subcontractor relationship just described.

The principal advantage to the owner with this arrangement is simplicity of composition. Only one contract entity need be dealt with, and responsibility for all performance is as clearly pinpointed as possible—directly and to a single company. A serious disadvantage from the owner's point of view is that there is no sensible watchdog mechanism to confirm proper performance of the design-build company. Inspection of the work and evaluation

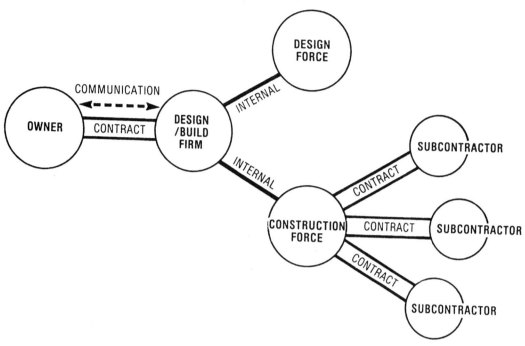

FIGURE 2-3 DESIGN-BUILD

of changes may become a complicated, ineffective operation if the owner cannot have complete confidence in the competence and integrity of the design-build contractor.

It is for these reasons that construction management, both in its pure form and in the many variations that exist, has become increasingly popular. This is particularly true for projects that would normally have a long and/or complex design phase.

2.2.3 Construction Management (Pure)

Construction management means different things to different people. In a real sense, every contractor is and has always been a "construction manager," orchestrating direct hire labor, equipment, subcontractors, suppliers, and money (cash flow). As we're using the term here, however, the functions and services to the various extents as may be defined in the specific construction management agreement are by design being carried out on behalf of the owner. In Figure 2-4, the construction manager (CM) contracts directly with the owner to provide the distinct services in a consultant capacity. There is direct *contact* between the CM and the trade contractors much the same as in the general contracting sense, but there is no direct *contract* between the CM and those trade contractors. There is no direct liability borne by the CM for the performance or nonperformance of those trade contractors. Each of the individual trade contractors is in effect a separate GC, contracting directly with the owner (note the distinction of "trade" versus "sub" contractor).

INDUSTRY AND CONTRACT ENVIRONMENTS

Why then would any owner in his or her right mind even consider a pure construction management arrangement? There are two primary advantages to involving a construction manager as early as possible in the project life cycle.

The first occurs during the design phase. Input of construction expertise as early as the concept stage may significantly improve construction efficiencies. The owner might ultimately obtain the same function while actually reducing the facility cost. Better knowledge of logical construction sequences, for instance, may reduce time, thereby saving everybody money. An owner may benefit from a contractor's superior working knowledge of sensible ways to format the various bid packages. This kind of assistance can make it easier to deal with the trades not only during construction, but through callbacks and guarantee work as well. By being able to take a different look at the same "standard" specification sections and divisions, direct cost savings, greater benefits for the same money, or both can be achieved.

The second reason is to provide the capability for "fast-tracking" the construction sequence without losing control over the individual trade contract structures. In a fast-tracked environment, certain portions of the work can be bid, awarded, and begun at the site well before the entire facility design is complete. The foundation work can start while the structural steel and curtain wall designs are being finalized, or interior partitions can begin before the final wall finishes are selected in total. These individual bid packages are managed and policed by the construction manager as their respective designs become finalized. By being managed by the CM on a fee basis, additional costs borne by the trade contractors as a result of subsequent changes to previously completed designs are passed back to the owner to pay for as

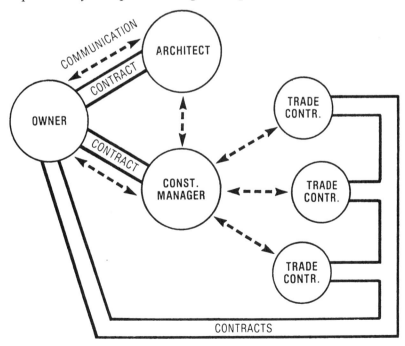

FIGURE 2-4 "PURE" CONSTRUCTION MANAGEMENT

change orders. If, however, the CM is competent and has been given the opportunity to input recommendations through the critical design considerations and their trade contract formations, these instances will be minimized, and the gain in project time will more than offset any increases in direct cost in this category.

2.2.4 Construction Management with a Guaranteed Maximum Price (GMP)

Figure 2-5 illustrates construction management with a guaranteed maximum price. In this format, the relationships described under "pure" construction management exist, except that the construction manager *is* responsible for completing the project for a total sum equal to or less than the guaranteed maximum price. What is more, the CM may be required to sign off to the initial GMP with only 80%, 70%, or even 60% of the drawings complete. The originally quoted GMP may be subject to adjustment based upon changes in scope, but even that knowledge cannot be considered as a blank check. The CM must continually exercise extreme caution in the ultimate application of a firm, superior knowledge of the construction process.

Although the construction manager performs the same functions, the CM now contracts directly with and is ultimately responsible for the performance of each respective subtrade. A review of Figure 2-5 reveals a contract and subtrade relationship nearly identical to that shown in Figure 2-2, the general contractor; the CM in practice appears to again become a GC.

It is this structure, the CM and the GMP, that is becoming increasingly popular in large and/or complex projects. It appears to give the would-be contractor a slightly reduced risk exposure as well as an opportunity to affect the facility design advantageously. It also provides the owner with the ben-

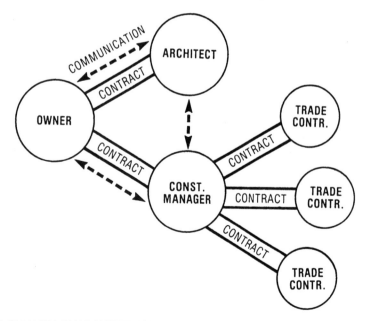

FIGURE 2-5 CM WITH GUARANTEED MAXIMUM PRICE (GMP)

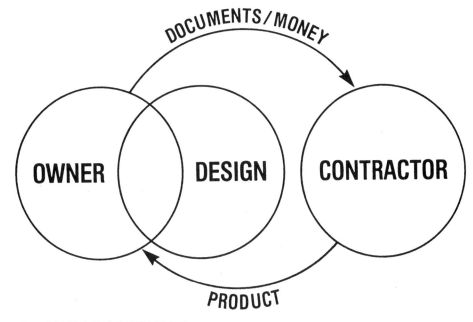

FIGURE 2-6 CONTRACT RELATIONSHIPS

efits of a CM's early input and fast-tracked speed, combined with an added security (however real or perceived) attached to a more tangible "final" project price (at least going into the arrangement).

2.3 CLARIFYING CONTRACT RESPONSIBILITIES TO GUARANTEE ACCURATE ASSESSMENTS

For simplicity, the traditional owner-contractor-architect associations will be assumed in the discussion that follows. Clear lines will be drawn at their respective interfaces insofar as possible. As noted before, it is important to maintain a clear understanding of the specific contract relationships being considered for a particular project to keep the discussion of responsibilities relevant.

The first step in any contract evaluation or performance assessment is to review the documents to confirm just what the relationships are. Any confusions or gray areas should be clarified as early as possible. By doing this at the start, those instances where one either inadvertently or through calculation assumes the role of another will be detected early. The maximum compensation for their effects can accordingly be secured.

In the most basic contracting model, all documents and money flow from the owner to the contractor, and the product moves back from the contractor to the owner. When the system is elaborated upon to include the architect, the design function and its various representations become extensions of the owner. The *contact* shown in Figure 2-1 is accordingly modified in Figure 2-6 to show more accurately the *contract* relationships.

The owner has the right to expect the following:

1. To get what it pays for;
2. Quality construction; and
3. To have the job completed on time.

The contractor has the right to expect:

1. To have all proper documents be complete and correct to the point that they can be depended upon with confidence.
2. To be dealt with fairly.
3. To be paid on time for acceptable performance.

2.3.1 The Owner: 12 Categories of Responsibility

When not barred from a specific contract, the law imposes several warranties, duties, and responsibilities on the owner, whether or not they are specifically highlighted in the contract language.

These responsibilities include the following:

1. Using sound discretion in evaluating the qualifications of low bidders (public bidding);
2. Maintaining and preserving the bidding system integrity (public bidding);
3. Funding the work;
4. Providing all surveys describing physical characteristics of the site;
5. Securing and paying for easements and authorizations;
6. Warranting the adequacy of the plans and specifications; [1]
7. Warranting the suitability of furnished materials; [2]
8. Disclosing superior knowledge;
9. Offering prompt action on clarifications and changes;
10. Providing "final" interpretation of the documents;
11. Cooperating;
12. Assuming ultimate responsibility for the design professionals.

Disclaimers of owner responsibilities and warranties are frequently found in many contracts. While these must be clearly understood if they do exist, they may often be overly ambitious in their attempt to shift the owner's responsibilities onto the contractor. [4] They should, therefore, be scrutinized very closely.

(1) Using sound discretion in evaluating qualifications of low bidders (public bidding)

Public bidding and contracting authorities are charged with the protection of the public trust, i.e., responsibly managing taxpayers' dollars by awarding construction contracts to the lowest qualified bidders. If a contract is to be awarded to someone other than the lowest bidder, the reasons for disqualification of the low bid may be subject to at least criticism and at most legal action. [5] Because the sympathy of the courts usually rests with the public agency in these matters, it is usually very difficult to overturn disqualifications except where collusion, bad faith, fraud, or arbitrary discrimination can be demonstrated.

(2) Maintaining and preserving the bidding system integrity (public bidding)

Public bid requirements are normally set clearly and specifically, and can often be extensive. Bid bonds or other security are needed, various affidavits and statements may be included, and the bid itself may be expanded to detail specific components with unit prices. The respective subbidders may need to be named. Strict conformance to the requirements is required. The bidder who completes the forms and requirements to the last detail as required in the stated bid procedure seals at least those portions of the bid from further consideration and/or competition. Those bidders who do not complete a particular section in a sense maintain the option to keep shopping around for the item. At the extreme, if the improper security for performance has been submitted, the bidder may not even be qualified to perform the work. If the bidder in the second case is accepted, it is clearly unfair to the first bidder who was responsive to the requirements.

The second requirement is that all bids must be submitted *exactly* on time or before. Further, the public entity is required to only accept timely submitted bids. 2:00 P.M. means 2:00 P.M. and not 2:05 P.M. In the words of the Practicing Law Institute:

> If late bids could be accepted, the integrity of the competitive bidding system would be violated because the late bidder would have the opportunity to obtain superior knowledge, and, hence, there would be multiple opportunities for abuse.[2]

In addition, the late bidder had more time to prepare the bid in the first place, compounding the unfair advantage.

Acceptance of late bids at public openings is becoming increasingly rare, but is still possible. Late or otherwise nonresponsive bids should *always* be loudly protested. When nonresponsive bids are at least brought to the attention of the other bidders, protests quickly spread and most often become successful, requiring no further drastic action to correct the situation.

(3) Funding the work

Many would list this as the owner's first requirement. In the case of a public bid, if the funding source is not properly arranged, it is unfair to contractors who expend time, energy and money bidding the project. [10] After the project is under way, the owner's primary responsibility with respect to

---

[2]Kenneth M. Cushman, Bruce W. Ficken, and Richard R. Sneed III, eds., *Construction Litigation* (Practicing Law Institute, 1981).

funding is to secure funds for the payments prescribed by the contract. The relevant considerations include the following:

(A) The timing of payables

Payments to the contractor must be made after receipt of each invoice within the time period specified, or within the requirements of appropriate statutes. The contract language itself will determine which takes precedence. If the contractor has met all procedural requirements and is entitled to payment, the owner may become potentially responsible for subsequent delay and interest charges.

(B) Conformance to rates and amounts corresponding to the actual job progress

If particular quantities of materials can be substantiated to be on the site or installed in place in an acceptable manner, considering all project requirements, the owner becomes responsible for that quantity at the prescribed rate. Any alteration in the contractor's responsibilities made by the owner for its convenience may create a hardship on the general and subcontractors for which they may be entitled compensation. Similarly, if the owner makes nominal deductions from the contractor's scope of work, the owner will be entitled to a credit.

(C) Contingency for changes

Change orders are a normal part of the construction process. Because the fact is that they are integral to *every* construction project, the owner has an obligation to provide for their eventuality. Responsible project funding will incorporate some additional percentage (often 7% to 10%) of the project bid amount to be set aside to be available to accommodate legitimate changes as they occur. Owners may incur additional liability to contractors for unreasonably long response times to change order requests, which cause delay for contractors.

(4) Providing all surveys describing physical characteristics of the site

The contractor is responsible for the correct layout and prosecution of the work. However, it is the owner's responsibility to ensure that all positive assertions it makes regarding site conditions are complete and accurate. [4]

(A) Establishment of property lines

The owner should, at a minimum, establish the property lines, or at least the contract limit lines, within which the contractor has the right to operate.

(B) Site material composition

If significant excavation is part of the project, boring data describing the soil composition that will have to be dealt with are generally a fundamental prerequisite. The boring and soils information should be taken and derived from locations which are relevant to the construction. If, for example, the boring information around the perimeter of a foundation indicates dry gravel with a low water table and no information regarding the soils characteristics is given within the foundation area, the contractor has the right to assume dry gravel with no water in the way of the excavation throughout the entire foundation construction. If actual site conditions differ from those indicated by the owner's data, upon which the contractor was entitled to rely (e.g., when either rock or soup is subsequently encountered while excavating in locations where no specific soils information was provided), the contractor will likely be entitled to compensation. [4] (This idea will come up again in detail in Chapter 7, with specific recommendations for action.)

(C) Baselines and benchmarks

The contractor is responsible to lay out the work, and a typical clause puts it as follows:

> All work shall conform to the lines, elevations, and grades shown on the plans. [47][3]

These data must be physically located on the site; unless disclaimed, the owner assumes responsibility for and warrants the accuracy of the baselines and benchmarks. The contractor will lay out the project based upon the baselines and benchmarks. The cost of errors in the project layout arising from incorrect baseline or benchmark information should be borne by the owner.

(D) Utility locations

The owner is responsible for providing accurate locations of all existing utilities. [6] Locations of telephone lines and so on may be necessary to tie into new building services or may be necessary only to prevent accidental interruption. Correct sanitary and storm line locations and elevations are critical to the design of their underground systems' correct drainage and tie-ins. Invert elevations are also necessary to allow proper estimates of the amount of excavation and backfill for the respective lines. If, for example, the invert elevation indicated on the drawings as 48.0' is actually 43.0', the additional 5' of excavation depth may now require shoring or greater trench width—all at substantially increased cost. If it is due to an improper representation on the drawings of the subject utility location, the owner may be required to pay for the increased costs. On the other hand, it is typical on some projects to encounter underground structures that are not shown on the project plans or drawings. If the owner can demonstrate that it conducted a diligent investigation of the location of substructures and, under the contract, disclaimed responsibility for unknown substructures, the owner should not be responsible for costs arising from discovery of unknown substructures. This is especially true where the number of unknown substructures encountered on the project is typical for that type of project.

(5) Securing and paying for easements and authorizations

(A) Site access

The contractor must have access to the site. [3] If such access is limited or restricted, as may be the case when working within a military complex, a

---

[3]Public Works Standards, Inc., *"Green Book" Standard Specifications for Public Works Construction* (2000 ed.), Surveying, § 2-9.2, p. 14.

security area of a manufacturing facility, or some similar environment, such restrictions and limitations should be made known to the contractor prior to the bid, so that the contractor may adjust its bid accordingly. If these or any other types of access restriction have not been made clear prior to the commencement of the work and are not apparent from a reasonable prebid site investigation, the contractor should not be expected to shoulder the extra burden. [7]

On an open construction site, vehicle, material, and equipment movement in and out of the area must be provided for and maintained. As construction sites grow physically tighter, many projects may have an access picture very closely tied to the sequence of construction. For example, one area may require completion before another to allow a crane to "back out" of the construction, leaving no subsequent crane access to the previous portions of the structure. If changes or other job problems then significantly alter the construction sequence, the issue of site access (getting the crane back into the first portion of the work) may actually become more expensive than all other effects combined. These kinds of problems and related costs may manifest themselves in the form of temporary roads, additional easements, exotic equipment rental costs, and even helicopter lifts.

Space should be provided for workers to park, enter, and leave the site. If it is either expressed or implied in the contract documents or in a reasonable prebid site inspection that space is available for material storage but that space becomes unavailable, the contractor may not be required to bear the resulting extra costs of material storage and carrying charges.

Generally if the owner has knowledge of any conditions which may affect the contractor's access to the site, the owner has an obligation to disclose such conditions to the contractor.

(B) Agency approvals

The Inland Wetlands Commission, the Environmental Protection Agency (EPA), the Occupational Safety and Health Administration (OSHA), or any number of other federal, state and local agencies may have jurisdiction over certain areas of a construction project. The agencies' approval may be necessary at either the design development stage or prior to the Certificate of Occupancy. It is the owner's responsibility to secure all agency approvals prior to the start of work, which can be obtained at that time. In the absence of noted exceptions, the contractor has the right to expect that all such approvals have been obtained.

However, the contractor is typically responsible for securing and paying for the building permit. By the time the building permit is applied for, the owner should have complied with all building code requirements. If delays in securing the building permit arise due to code noncompliance of the design, the owner should be responsible for the resulting costs, including start-up delays. If problems are encountered during the course of construction due to failure by the owner to secure proper agency approvals, such as those necessary when encroaching on inland wetlands, the owner may again be held accountable.

(6) Warranting the adequacy of plans and specifications

The owner warrants the adequacy of the plans and specifications and, therefore, bears the responsibility for any defects in them. [1] Defects can exist in many forms (detailed extensively in Part Three) but can be thought of as boiling down into the two distinct categories of product and time. Most defective specifications problems involve the accuracy of the technical specifications.

In fact, the owner warrants each of the following:

1. The facility and its various components *can* be built as designed.
2. All the components *will* fit together as indicated in the plans (at least fairly closely).
3. It is physically possible to complete the project.

It is the owner's responsibility to ensure that the plans and specifications are clear, complete, comprehensive, and comprehensible (unambiguous—the four C's that will come up again in Chapter 3. However, in the event an alleged ambiguity is encountered, it is generally the owner's responsibility to interpret the contract.

The owner warrants that adequate time has been provided to complete the project. If the contract allows 300 working days to complete the project, the owner warrants that the project *can* be built in those 300 working days. The bidder has a right to expect that by employing reasonable methods and care, it will complete the project within the contract period. If it is discovered, for example, that by employing reasonable schedule logic the job will overrun, it is likely that the owner will be responsible for granting the contractor the additional time needed to complete the job, costs to accelerate the construction (by employing excessive means) to meet the original end date, or both.

(7) Warranting the suitability of furnished materials

Where the owner furnishes material or equipment to the contractor for use in the work, the owner warrants that the material and equipment will be suitable for their intended purpose.[4] [2] In this case, the owner is also responsible for the timing and coordination of the respective items. Material deliveries must be made within the current requirements of the progress schedule. Shop drawings and other coordination information must be submitted and distributed correctly and in a timely manner.

(8) Disclosing superior knowledge

The owner has a duty to disclose to the contractor superior knowledge of any item that may directly or indirectly relate to the work, where that knowledge is either unknown by or has not been available to the contractor.[5] [8] Referring back to the boring data example (point 4), assume that the bor-

---

[4]*Thompson Ramo Woolridge, Inc. v. United States*, (1966) 175 Ct. Cl. 527, 361 F.2d 222.
[5]Cushman, Ficken, and Sneed, eds., *Construction Litigation*, p. 209.

ing and soils information did in fact exist for the interior portions of the construction but that it was intentionally left out of the contract. The exclusion of such relevant information would border on fraud. Likewise, where the owner's superior knowledge of a factor, such as the unavailability or inadequacy of a specified material, would lead to a reduced cost, improved efficiency, or simply the earlier exposure of a problem, the owner has the obligation to so advise the contractor.

(9) Offering prompt action on clarifications and changes

Change orders interfere with and disrupt the orderly sequence of the work. They accordingly must be resolved as quickly as possible to minimize their total direct and consequential impact on construction. [9]

In the case of a large public agency, this recognition may manifest itself as a prescribed set of unique procedures for handling change orders. The following passage, taken from a state Bureau of Public Works "Working Procedure" section of the General Conditions, is an example of this exception procedure:

> The Engineer will promptly investigate conditions which appear to be changed conditions. If the Engineer determines that the conditions are changed conditions and will materially affect costs, a Change Order will be issued adjusting the compensation for such portion of the work in accordance with 3-2.1.[6]

In addition to the formal recognition of the need to move the relevant information and paperwork throughout the system as quickly as possible, the informal communication network can be activated to accelerate the approval process as well. Documents can be hand delivered, special meetings with all approving authorities present can be arranged, and certain portions of the work may be authorized to proceed while the rest is negotiated.

(10) Providing "final" interpretation of the documents

Although the architect normally researches and prepares recommendations for technical matters relating to design, when involving cost or time, these corrections, changes, and interpretations are communicated to the owner who issues the final position on the matter. Actual authority for some portion of the process may rest with the architect. [11] Because no direct contract tie exists between the architect and the contractor, however, ultimate responsibility for the architect's performance in these situations rests with the owner.

Except for contract structures which incorporate unmodified versions of standard documents such as AIA Document A201, prepared by the American Institute of Architects, the owner will normally interpret all questions relating to the work, and its decision will be "final."

Understand that "final" means different things at different stages. The procedures outlined in the specifications may provide that disputes and

---

[6]Public Works Standards, Inc., *"Green Book" Standard Specifications for Public Works Construction* (2000 ed.), Change Conditions, § 3-4, p. 18.

questions be submitted to low-level owner representatives who may have many clearly defined limits of formal authority. That decision may most often be appealed to a higher level if the contractor is not satisfied with it. Procedures for a submission to a board of appeals or some other hearing authority may be specified as the final dispute resolution option, and it is this latter authority's decision that is really "final" as far as the owner is concerned. If at that point the contractor is still unsatisfied, the true final options are arbitration or litigation. Part Four offers specific recommendations in this area.

(11) Cooperating

The owner has a duty to cooperate with the contractor to the best of its ability and not to impede, hinder, obstruct, or interfere with the work. [9] Similarly, the contractor has a duty to cooperate with the owner and to not impede, hinder, obstruct or interfere with the owner's performance of its obligations under the contract. The owner is responsible for ensuring that procedures and response times do not exceed the limits expressed in the contract or implied by industry standards.

The most significant area where the effects of cooperation, or the lack of, will be felt is in the field. The construction supervisor, architect, or other inspector may decide to take an overly ambitious role. They may adhere to an overly zealous interpretation of the contract requirements.

(12) Assuming ultimate responsibility for the design professionals

The architect and its engineers have many duties and responsibilities as outlined in the following pages. Express and implied warranties exist, and professional liability is assumed. The continuing contact and dialogue between the contractor and the architect that is typical on nearly every project often confuses direct line responsibilities. A review of any of Figures 1-3, 1-5, and 1-6, however, will clarify that no contract relationship exists between the contractor and the architect. The contractor, therefore, is technically to look to the owner for ultimate satisfaction. The architect is an agent of the owner. The owner therefore ultimately assumes all liability for the architect's performance, or lack thereof. The architect is responsible to the owner. The contract relationship between those two is in theory transparent to the contractor.

In some arrangements, an owner retains an architect and/or a construction manager as its representative, and a system of multiple prime contractors is used. In this case, the owner effectively becomes the general contractor, incorporating all the responsibilities associated with the position. The many exculpatory phrases and indemnity clauses that are typically included in an owner-contractor or owner-construction manager agreement may not apply if the owner has actively interfered with or constructively altered the form of the agreement. The same is true if some owner negligence caused or contributed to a loss.

2.3.2 The Architect: 10 Categories of Responsibility

As the owner's agent, the architect is responsible to the owner for the technical design. The responsibilities outlined in this section can, therefore, accurately be thought of as an extension of the owner's responsibilities.

The details of many of the architect's responsibilities, such as review and approval of shop drawings, may be expressed clearly in the contract. In the absence of such clear definition, however, the items noted here will generally apply.

By engaging an independent design professional, the owner makes an attempt to secure the best possible design given the respective parameters. The owner should clearly and completely convey all project objectives and necessities to the architect. Owners may seek to shift complete responsibility for the design to the architect. Participation in the actual design work should then be avoided as much as possible, if the owner wishes to be sure that the design liability remains with the design professionals.

As far as the contractor is concerned, this is all strictly between the owner and designer. The contractor may consider them both as a single entity and think of itself as dealing simply with the owner.

The architect is responsible to the owner for:

1. Production and coordination of the plans and specifications.
2. Technical accuracy of all documents.
3. Specific design (not design criteria).
4. Workability of the design.
5. Code compliance.
6. Interpretation of the documents.
7. Submittal review and approval.
8. Prompt, timely response.
9. Evaluation of the work.
10. Diligence, skill, and good judgment.

(1) Production and coordination of the plans and specifications

It is the architect's responsibility to describe work clearly and completely in the plans and accurately in the specifications. In the optimum situation, this shall be accomplished comprehensively but without overlap. The designer's description should include the technical specifications for the work and clearly allocate responsibility for performing the work. [27]

Some contracts, however, attempt to shift this burden to the contractor through exculpatory language:

> The plans and specifications are complimentary. The contractor is responsible to provide all work shown on the plans, whether or not adequately described in the specifications, and all work described in the specifications whether or not specifically indicated on the plans as if called for by both.[7]

[7]Ibid.

In theory, the contractor should be able to create an individual subcontract for each specification section, "per plans and specs" with confidence that the entire project will be covered.

Additionally, the general contractor may be held responsible for obvious or glaring inconsistencies.

(2) Technical accuracy of all documents

The architect must ensure that the specifications are technically accurate. [12] For example, if a specific roof insulation R-factor thickness is shown on the plans, the contractor should not be responsible for paying for the increase in roof blocking because the roof insulation thickness shown did not meet the requirements for thermal performance. If the specifications stipulate blue, the contractor should be able to obtain blue, and the specified boiler should physically fit between the walls of the boiler room. It is not the contractor's responsibility to conform this before the specified boiler is ordered. [13]

(3) Specific design (not design criteria)

Contractors often get into design traps because their practical construction experience allows them quite competently to fill in many gaps left by incomplete designs. Despite the best of intentions, the minute a contractor initiates design changes, he or she assumes the architect's liability as it relates to that particular area. [13]

Unless there is a glaring deficiency in the design, it is the contractor's job to do, not to criticize. When a contractor begins to recommend design changes in the interest of speed or even with the noblest intentions of improvement in product quality, there may be a definite assumption of responsibility for the new detail's workability. Take the time to understand this division. Contractors have enough responsibilities and liabilities without unnecessarily assuming those of the designer.

This is not to say that in a clean general contractor arrangement the contractor should *never* recommend such changes, particularly if they improve the construction sequence. The contractor, however, should be aware of the risks of assuming this additional liability and weigh those risks against potential benefits resulting from the respective design change. As difficult as it may be to bite your tongue in those instances where you discern a design change which will benefit the project, if the potential real benefits to the contractor of the change are marginal, do the work as designed. Of course, if through a prior course of dealing a good business relationship has developed between contractor and owner, the contractor may choose to disclose the design flaw.

Contractors also become involved with design when the plans and specifications are incomplete. This situation often leads to change order requests. The components of construction assembly should be unambiguously identified. Architects or engineers sometimes cover this potential design gap by referring to applicable provisions of some accepted industry standard specification. The designer seeks to shift responsibility for omissions in the contract documents onto the contractor by referring to the technical requirements contained in those standards. In many cases, this type of reference is commonly accepted procedure and is quite legitimate. Referring to Department of Transportation standard specifications for a certain type of road construction, for example, will most often yield very specific and stringent requirements to complete the work. These requirements will incorporate exact material specifications and installation parameters to allow *nonsubjective evaluation* of performance (for example, "bank-run gravel [with the exact composition noted] compacted to a density of 95%" or "3/4-in. crushed trap rock placed to a depth of 8 in.")

The referenced standard informs the contractor of (1) exactly what material is required; (2) exactly what to do with the material; and (3) how to measure performance and compliance. Thus, with a referenced standard, the contractor will rightfully be held responsible for correct and conforming work.

Some standard specification references are, however, vague. For example, an attempt was made on one project to force a contractor to provide elaborate galvanized steel draft barriers in the large open-web roof joists of a jet hangar as part of the designed fire protection system. The basis for the owner's action was the statement in the fire protection specification that the contractor complete the system "in accordance with NFPA (National Fire Protection Association) Bulletin No. 409." NFPA Bulletin No. 409 contains relative design criteria—not the design itself. The closest thing to a material specification was statements like "made of a noncombustible material." Dimension and sizing properties were shown as a complex system of formulas which were to be applied to many different types of structures. In other words, no material specification, no assembly design, and no performance criteria were offered. The owner ultimately paid for a change order in excess of $73,000.

(4) Workability of the design

There is an implied warranty established that the professional designer's plans and specifications will create components that will in fact fit together in the same way that they have been placed on paper. [14] If a nonapparent impossibility is encountered, the contractor cannot be held responsible for it. Unless the error or omission is obvious, it probably will not relieve a designer's fault if the contractor "should have known better."

In addition, the designer is ultimately responsible for the project's design and its systems' ability to function and perform in the manner and to the extent intended. For example, it is clearly up to the designer to exactly size air handling units and to indicate specifically the precise horsepower of every pump. The size of every pipe diameter must be detailed to ensure that all systems' outputs are adequate.

The contractor's duty is to provide the materials and equipment described in the plans and specifications and to locate and install them correctly. If the installation is correctly accomplished, it is not (or should not be) the contractor's problem when the room doesn't get cool enough fast enough, or when corners in the room are dark. These kinds of problems are the results of design deficiencies.

Some contract clauses seek to impose upon the contractor responsibility for the workability of the design. This may happen through the inclusion of language such as:

> The contractor is responsible to furnish whatever is necessary to ensure a complete and properly functioning system, regardless of whether or not shown in the contract documents.

or:

> It is the intention of this section to have a complete system functioning adequately for its intended purpose. It is the contractor's responsibility to correct materials and equipment if they have not been sized properly, at no additional cost, in order to achieve this purpose.

Many courts have rejected enforcement of such clauses. However, the contractor should be aware of them and, if possible, seek to have them deleted from the contract.

(5) Code compliance

The architect is responsible for ensuring that the design complies with fire, safety, and all other applicable building codes. [15] If a door between two spaces needs to bear a fire rating, it is up to the architect to indicate in the contract the precise rating that the code actually dictates. The design professional should not leave it to the contractor to "provide all doors in accordance with the fire code."

Additionally, unless specific engineering activity is incorporated into the subcontract agreements, the designer must specify technical requirements in compliance with code restrictions. The designer is thus completely within rights to require installation and workmanship in accordance with applicable codes and standards.

(6) Interpretation of the documents

Depending upon the exact contractual relationships, the duty of an architect to interpret contract documents varies widely. [11] One tension in interpreting construction contracts is that it is not the "intent" that is priced in the original bid, but the specification as written. The "intent" in this instance may become a compensable change order.

In any event, the contract between the owner and architect may assign none or absolute authority to the architect for interpreting and making final decisions on matters relating to contract interpretation. Most often, the architect's role is to review conflicts, proposed changes, etc., and submit specific recommendations to the owner for the owner's ultimate action. Nonetheless, the contractor should read the contract documents thoroughly.

The Construction Specifications Institute asserts that AIA Document A201, *The General Conditions of the Contract for Construction*, is not a self-serving document prepared by a group of architects. According to the Institute, AIA Document A201 is a uniform, fair, and completely objective set of conditions assembled by an association of architects, owners, contractors, and lawyers, among others.[8] Yet, it does include language favorable to architects:

> The Architect will be the interpreter of the requirements of the Contract Documents and the judge of the performance thereunder by both the Owner and the Contractor.

This language makes the architect the authority in *all* matters, even where the *owner* is concerned.

Contractors should review the contract thoroughly and know what the contract says with respect to ultimate authority for interpretation of the contract; know what rights you have.

(7) Submittal review and approval

Although this is a specific activity and one that also applies to point 8, review and approval of all project submittals amounts to what is most likely the most time-consuming of all the architect's activities during the construction phase of the project. [21] It is for this reason, combined with the fact that there exists such great potential for abuse, that the subject justifies special attention.

Assuming proper and timely submissions by the contractor, the architect is to receive and act upon each submission in a manner that should be precisely described. Lacking that precise description, architect action should be within reason and within the parameters customary in the trade. That action is comprise of the following components:

(A) Conform to requirements

This is a precise, detailed review of every significant component to confirm that the item proposed meets all design and performance criteria originally specified. [44] This process is there only to confirm that the product meets the stated requirements, not necessarily as someone would now like it to be. "Standard colors" means just that, and 3/4" insulated glass does not mean 1" insulated glass. The obvious eccentricities are easy to spot. Exercise the discipline to check all the architect's modifications to your submittals and correct the most subtle changes if they have a potential for significant cost impact.

(B) Provide missing design information

In many instances, complete preparation of a submittal is not possible because design information is lacking. Sufficient dimensions and detail to allow precise location calculations may not be available. However, the precision of detailed, large-scale shop drawings may expose conflicts that may need significant redesign, or simply the designer's decision as to which alternative is preferred. [45]

When enough information had been originally included in the contract, the contractor is normally held "responsible for dimensions." If original

---

[8]Hans Meir, *Construction Specifications Handbook,* 3rd ed. (Englewood Cliffs, N.J.: Prentice Hall, 1983).

INDUSTRY AND CONTRACT ENVIRONMENTS

dimension information is lacking, the architect may become "responsible for dimensions."

This whole subject of shop drawings and missing information is so important that it is treated exhaustively in Chapter 11.

(C) Respond in a timely fashion

This requirement is included only for comprehensive treatment of the subject and is expanded upon in point 8. Suffice it to say here that all actions, favorable or unfavorable, that a designer takes with respect to submissions must be taken in a reasonable amount of time, so as not to delay the contractor. [43] The contractor must be allowed to take the next steps in time to preserve the construction schedule and other objectives.

(8) Prompt, timely response

The architect has an express obligation to take all actions "with reasonable promptness so as not to cause a delay in the work."[9] Even if "reasonable promptness" is not specifically defined in terms of the number of days considered acceptable, the durations may be established by customary practice, by confirmation or clarification at early job meetings, or by calling attention to specific requirements in the appropriate correspondence as individual situations may dictate.

Common activities that the architect will normally be required to perform at such "reasonable" speeds include the following:

- Review and approve of shop drawings;
- Review and recommend/approve change orders;
- Prepare change order designs;
- Approve requisitions for payment;
- Issue documentation (meeting minutes, transmittals, etc.);
- Make site inspection/do testing; and
- Respond to operating questions.

The architect must respond with reasonable promptness to the usual and unique situations. If not, he will risk bearing responsibility for any resulting damages incurred by the contractor.

(9) Evaluation of the work

The architect has a responsibility to satisfy himself or herself that the work is being performed in accordance with the design, as well as within applicable workmanship standards. [26] The architect is *not* responsible to be intimately familiar with every nut and bolt of construction as the work is progressing. It is the respective trade contractors who are responsible to install the work correctly in the first place. Abandonment of the site and neglect, until a problem finally comes up, are unacceptable extremes.

---

[9]American Institute of Architects, AIA Document A201, *General Conditions of the Contract for Construction*, Article 4.2.11 (1997).

The activities that make up this ongoing evaluation of the work as determined specifically in most contracts include the following:

(A) Inspection

The architect is normally responsible for making *regular* visits to the jobsite to familiarize himself or herself generally with the progress and quality of the work. [16] It is not enough to review the progress photos and try and get the picture of job progress through the correspondence. He or she must be on the site to confirm that the work is progressing according to schedule, as well as in conformance with the plans and specifications. It is not reasonable, for example, to wait until all the brick is up before the architect determines that the color of the mortar is not close enough to the sample to be considered acceptable.

Finally, the architect should be available on a frequent, periodic basis and at any other time when needed to answer questions or resolve problems.

(B) Testing

This function may actually be split between the architect and the owner. The idea is to have the testing of the work as it is being performed as an activity performed by an agent of the owner, as opposed to the contractor. This is to avoid conflicts of interest. [42]

(C) Evaluation

The architect is responsible to either determine entirely or to confirm the owner's evaluation of the fitness of the work, along with the associated dollar value. Specific activities included in the process may be the following:

- Determine, confirm, verify the amount of work in place and corresponding payment to the contractor;
- Confirm acceptable material quality and workmanship standards;
- Reject work that does not conform to the contract;
- Determine the dates of substantial and final completion; and
- Issue stop-work orders.

(10) Diligence, skill, and good judgment

It is unusual to find many express warranties of architects (and their engineers) in most construction agreements. These designers do, however, implicitly warrant that they have exercised diligence, competence, skill, and good judgment throughout the design process and contract preparation. [12] Moreover, they are required to have performed their work in accordance with the professional standards of the community where the work is to be constructed.[10] Except for the most obvious errors, the contractor has the right to assume that the information provided in the contract documents is complete and sufficient for preparation of an accurate estimate. More subtle deficiencies which become apparent as the work progresses are then the responsibility of the party who caused or contributed to the deficiency.

---

[10]Cushman, Ficken, and Sneed, eds., *Construction Litigation*, p. 212.

2.3.3 The General Contractor: 15 Categories of Responsibility

It is the general, or prime, contractor who will likely have the multitude of "express" responsibilities in a construction contract. The drafters of construction contracts combine layer upon layer of phrases, clauses, references, standards, boilerplates, and exculpatory language to help shield themselves from liability to the contractor.

These complicated dissertations on contractor's responsibilities may actually increase the likelihood of conflict—and therefore change orders. This whole idea, however, is too simple in concept to be take seriously by the document drafters.

The collection of confusing exculpatory language should not be lightly considered. Every word must be read and understood, and many inclusions may in fact be very serious and require specific attention. Clauses that will fall into this category will likely be of the indemnity or hold-harmless variety. If there is a doubt regarding the applicability or legitimacy of any inclusion, consult with a competent attorney. Confirm exactly where you stand before proceeding.

Beside the strict technical requirements of the work itself, then, responsibilities of the contractor will generally fall within these categories:

1. Duty to inquire
2. Reasonable review
3. Duty to inspect (not discussed)
4. Provide bonding (not discussed)
5. Plan and schedule the work
6. Lay out the work
7. Supervise, direct, and install the work
8. Adequate workmanship
9. Correction of patent errors
10. Coordination of all parts of the work
11. Review, submit, and coordinate shop drawings
12. Contract payments
13. Adequate insurance
14. Adherence to safety standards
15. Warranty of clear title

(1) Duty to inquire

The bid documents usually require bidders to bring any questions regarding inconsistencies, conflicts, or ambiguities found in said documents to the attention of the owner prior to the bid opening. [17] These requirements can sometimes shift responsibility for perceived contractual ambiguities to the contractor. However, it is more likely the obviousness or significance of an ambiguity that places a duty on the contractor to inquire about or request clarification regarding an ambiguity. This is true whether or not the specific responsibility to do so is clearly stated.

> It is more difficult to saddle the contractor with responsibility for ambiguities when a discrepancy was subtle in nature or hidden from a reasonable review.[11]

(2) Reasonable review

The contractor is responsible for reviewing the contract [18] and for being aware of its obligations pursuant to the contract.

Because the owner warrants the adequacy of the plans and specifications, the contractor has the right to depend upon the documents. The contractor is entitled to assume the documents to be correct except for glaring, patent, and obvious errors. The drafters of the contract had the opportunity to be sure that it was correct before asking a businessperson to sign off on it. It is this concept that underscores the rule of contract interpretation that ambiguities are to be construed against the drafter.

(5) Plan and schedule the work

There are usually any number of ways to complete a given project. The original bid will likely anticipate one reasonable sequence of construction that is achievable, given the technical and time requirements. Assuming such a workable design exists, the contractor is responsible for anticipating the various components that will eventually make up the total project. The contractor must be aware of or determine the constraints and interrelationships between the various components and, thereafter, finalize the project plan [19]. Timetables are added to transform the plan into a schedule.

If properly prepared, the resulting catalog of component dates, times, and interdependencies will represent a complete method to build the project.

If any item is misrepresented or confused in the documents, the resulting effects on the schedule may manifest themselves in the form of interruptions, resequences, accelerations, or delays.

Every schedule *will* change. Courts have begun more consistently to recognize this fact of construction life. Contemporary legal thought regarding construction schedules is that for a schedule to remain valid, it must be periodically updated. All significant effects on the schedule must be incorporated to achieve an accurate updated schedule.[12]

(6) Lay out the work

The contractor is responsible for physically laying out the work and is responsible for errors in laying out the work which should have been discovered by a competent contractor complying with common professional standard practice. [28]

Some errors may be beyond the contractor's control. For example, suppose the contract documents incorrectly indicate the position of an existing structure and a portion of the new facility is laid out in relation to the loca-

---

[11]*Mountain Home Contractors v. United States* (1970) 425 F.2d, 1260, 1264 C. Ct. Cl.
[12]Andrew M. Civitello and Anthony L. Iannone, *Construction Scheduling Simplified* (Englewood Cliffs, N.J.: Prentice Hall, 1985).

tion of the existing facility. If the remainder of the project is then laid out relative to other correct data, the contractor may be held responsible when the two pieces of the new structure don't exactly meet because, if the survey had been properly "closed" in the layout, as may be deemed a customary standard of survey practice, the error might have been discovered long before resources had been committed to building upon it.

(7) Supervise, direct, and install the work

The contractor is responsible for enforcing the construction schedule. The project must progress according to the commitments and arrangements established throughout the planning and scheduling phase of the project. This involves three phases: start-up, progression of the work, and close-out. It specifically involves orchestrating the overall game plan and job sequencing, placement of field offices, placement of material staging areas, placement of equipment controls, policing temporary protection, policing facilities, and physically installing the specific job components. [24]

It is the contractor's job to ensure that the various trades progress through the work as required by the schedule network. The target milestones must be met while avoiding unnecessary interferences on other parts of the work. In the event of schedule slippages and sequence interruptions, actions to regain lost time must be taken promptly and decisively.

As the work is being installed, it is the methods, techniques, sequences, and procedures that are the responsibility of the contractor. These are specifically different from the ownership of the design details' *workability*. In other words, the contractor begins with the legitimate assumption that the job components will in fact fit together (the workability). It becomes the contractor's assignment from that point to assemble those components in a logical, efficient sequence.

(8) Adequate workmanship
(A) Technical specification

The contractor is responsible for performing its work in compliance with the technical specifications on the project. Some specifications will incorporate precise tolerances within which the respective work is to be installed if it is to be considered acceptable, which eliminates any necessity for subjective evaluations. [20] The quality control specification for the placement of a concrete floor slab, for example, may require the finished floor surface to "be level with 1/8" in 10' in any direction." In this case, there is no question regarding the acceptability of the floor as it relates to its consistency of level; simply measure it. It either meets the 1/8" requirement or does not. The question of acceptability, in other words, is reduced to arithmetic. It's quantifiable. Some specifications do not provide precision and may provide, for example, that the same floor slab merely be "level." With this latter type of specification there may be differing opinions of what is level. Nonetheless, in either case, the contractor is obligated to comply with the specification, or seek clarification from the owner when required.

(B) Referenced standards

In the absence of a precise technical definition, the next areas in which to look for a description of an acceptable level of quality are industry standard specifications that are incorporated by reference. Some standards are more applicable than others. Precise technical references, such as requiring a particular weld to "conform to the requirements of AWS (American Welding Society) Code for Welding in Building Construction AWS D1.1-82," place absolute responsibility upon the contractor for complying with the specifications, just as if they had been directly included into the specification language.

(C) Codes and standard practice

In the event that no technical specification or referenced standard is included, it is highly unlikely that there would not be some boilerplate expressly requiring the contractor to install the work in accordance with applicable building codes as they relate to that portion of the work. If the specification was beyond this so grossly incomplete as to leave out even these most basic requirements, the law imposes other restrictions. The contractor implicitly warrants that the work will be performed in accordance with standard levels of workmanship and within the descriptions of accepted trade and practice, as defined in the community in which the work was performed.[13] [25] This idea of conformance to community standards cannot, however, be used to excuse a contractor from performing a more stringent requirement if one is unambiguously described in the specifications.

(D) Manufacturer's instructions

Specific procedures outlined in manufacturer's instructions will in nearly all cases need to be followed precisely if the warranties for those items are to remain intact. If the detailed procedures conflict significantly from the stated contract requirements, the owner and architect should be the ones to make the decision regarding the direction to take. A decision to move away from procedures recommended by a particular product manufacturer will carry with it very definite liabilities. These will best be left with the owner, if possible. Barring this type of conflict, the manufacturer's directions will generally take precedence over other noted instructions.

(9) Correction of patent errors

In most cases, contractors do not bear responsibility for the designer's or owner's errors in the contract documents. This is true unless the errors are so obvious or glaring (patent) that a competent contracting professional should undeniably have discovered them through reasonable review. [17] Therefore, where it can be demonstrated that an error can be considered patent, the contractor has a duty to bring it to the attention of the owner prior to bidding.

---

[13]*Mann v. Clowser* (1950) 190 Va. 887, 59 S.E.2d 78.

INDUSTRY AND CONTRACT ENVIRONMENTS

For example, a mason contractor's experience will likely be that most cavity walls he or she ever constructed had some kind of through-wall flashing integral to the design. If the design detail on *this project*, however, shows no through-wall flashing of any kind in any cavity wall, the defect may be considered so obvious to an experienced mason contractor that it would be incumbent upon that contractor to bring it to the attention of the owner. It remains, of course, that the definition of what qualifies as "obvious," "glaring," or "patent" is the fertile gray area that may defy easy resolution. In this situation, the discussion of trade custom and use may be relevant.

(10) Coordination of all parts of the work

Dictionaries define the verb *coordinate* as "to arrange in order" and "to harmonize in a common action or effort." [24]

The coordination activities that a general contractor is responsible for performing as they relate to the construction process amount to:

1. Assembling pertinent information from all sources that may possibly affect a certain portion of the work.
2. Correlating that information with the respective specific job requirements as indicated in the contract.
3. Distributing relevant information to those requiring it in time to incorporate it in their work.

All this must be accomplished while meeting the constraints imposed by the progress schedule. The general contractor is in a real sense nothing more than a conduit.

(11) Process shop drawings

Technically, this activity of shop drawing review, submission for approval, and distribution is included in "coordinate," but it is given special treatment because it is one of the most common and time-consuming contractor activities. [21] It is also one that affects every component of every project.

Shop drawings submitted for approval must clearly and completely describe the specific product that they represent without ambiguity. They must include exact information regarding how those respective products interface with other parts of the work. Where differences exist between the item being submitted for approval and that which is specified, they are to be highlighted as such to allow for specific consideration without the risk of oversight. Failure by a contractor to note such differences will not relieve the responsibility for them, even if the submission bears the architect's formal approval.

The contractor is responsible to coordinate all dimensions insofar as necessary to install the work properly relative to the surrounding construction. The contractor is *not* required to invent dimensions, providing dimensions originally missing from the design. That is the architect's responsibility. If a structural steel shop drawing detailer, for example, takes it upon himself or herself to decide upon a column location as might seem logically correct but not necessarily as a result of calculation or extrapolation of other data on the plans, it *must* be called to the attention of the designer for the architect's verification and approval. Otherwise, the detailer, and therefore the contractor, will bear the full responsibility for how well (or how poorly) that "missing" dimension fits within architectural, plumbing, ductwork, or other considerations.

Finally, the confirmed and approved submittal information must be forwarded to all who might require it in time to incorporate it in stride. If an approved pump is mounted on a beam and requires 208 volt three-phase power, the information must be distributed to the steel subcontractor in time to complete its shop drawings of the affected beam and to the electrical subcontractor prior to conduit installation in the area. The rule to remember is, "When in doubt, send it."

(12) Contract payments

The contractor is normally responsible for paying for all costs directly related to the physical completion of the work. [23] These include the cost of labor, materials, equipment, tools, machinery, transportation, and sales and consumer use taxes. In addition (in a straightforward general contract arrangement), the responsibility for all costs to provide temporary heat, light, and utilities will likely be the responsibility of the contractor. Know what the requirements are on *your* project and manage them accordingly.

(13) Adequate insurance

As a standard article in construction contracts, the contractor will be required to obtain predetermined amounts of insurance coverage prior to any work beginning on the site. The categories normally included will be:

- Workers' compensation insurance.
- Fire and extended coverage insurance against fire and other risks normally included in standard coverage endorsements.
- Public liability and property damage insurance with express limits.
- Contractor's protective public liability insurance covering the operations of subcontractors.
- Automobile insurance for owned or hired vehicles.

In addition, the contractor may be required to provide a performance bond or a payment bond. A performance bond ensures that the project will be completed, even if for some reason the contractor is unable to complete the project. The payment bond protects subcontractors', material suppliers' and laborers' right to payment for services, material or labor supplied to the project.

(14) Adherence to safety standards

The contractor is responsible for regularly, at all times, and without the need for any specific notice taking all necessary precautions for the safety of all people on a construction site as well as the public. [22] Safeguards for the protection of both the workers and the public as may be required by job conditions, progress of the work, and applicable authorities such as OSHA may include warning signs, barricades, scaffolding, lights, fire extinguishers, proper ladders and walkways. Workers are to wear hard hats and proper clothing and are to use correct safety equipment in appropriate situations.

(15) Warranty of clear title

This contractor responsibility is usually in the form of an express warranty located in the general conditions of the contract. The guaranty of clear title is the owner's assurance that the project is free of liens for all materials, labor, and equipment incorporated in the work. [29] A typical clause may read as follows:

> Contractor represents and warrants that the title to all work, materials, and equipment for which payment shall have been made to Contractor shall vest in Owner upon said payment, free and clear of all liens, encumbrances and adverse interests of any kind whatever; however, Contractor shall remain liable for damage and loss to said work, materials, and equipment, until such time as the Project is fully completed and accepted by Owner, and final payment is made pursuant to the Contract.

In addition, there is usually an agreement to assign to the owner all warranties on equipment and materials that have been received by the contractor as part of the project, a general guarantee of material and workmanship for a specific period (most often one year), and any special warranties for specific material or equipment.

2.3.4 A Final Note

In the interest of clarity and simplicity, the discussions in Sections 2.2 and 2.3 were for the most part confined to the owner, architect, and contractor, without much specific reference to engineers or subcontractors. This was done by design and within the context of an idea referred to as the "pass-through" principle (that will be elaborated upon in Chapter 3). Simply stated, the pass-through idea means that the engineers are responsible to the architect in the same way as the architect is responsible to the owner (and vice versa). Correspondingly, the subcontractors are responsible to the general contractor in the same way that the general contractor is responsible to the owner (and vice versa).

Last, it is very difficult to read and write about many construction topics without encountering on a repeated basis words such as "reasonable" or "acceptable" or other such words that appear on the surface to clarify expected levels of quality or performance. In practical application, however, whenever an adjective is used to describe some measure of acceptability which is not specifically definable (e.g., "flat" or "hard," as opposed to "level within 1/8" in 10' in any direction" or "95% compaction"), a flag should pop up in your mind that a gray area has been encountered. The word "reasonable" has been and will continue to be the epitaph on many contractors' profits and change order requests.

3
Proven Strategies for Applying Construction Contracts

3.1	Strategic Interpretation: Applying Contracts to Secure Power Positions	37
	3.1.1 Contract Law Concepts	37
	3.1.2 Construction Law Considerations	38
3.2	The Contract Documents: Simplified Descriptions to Prevent Oversight	38
	3.2.1 Contract Components	38
	3.2.2 The Four C's of Contracts	40
3.3	Rules of Contract Interpretation: The Cards up Your Sleeve	41
	3.3.1 Introduction	41
	3.3.2 Standard of Interpretation: Reasonable Expectations	41
	3.3.3 Ambiguities Resolved Against the Drafter	42
	3.3.4 Right to Choose the Interpretation	43
	3.3.5 Specific vs. General	43
	3.3.6 Usage of Trade Custom	43
3.4	Applying Construction Contracts Without Resistance	44
	3.4.1 Introduction	44
	3.4.2 Change Clauses	44
	3.4.3 The Pass-Through Clause	46
	3.4.4 The Dispute Clause	48
	3.4.5 Authority (Formal/Constructive)	49
	3.4.6 "General Scope" of Work	50
	3.4.7 "Reasonable Review"	51
	3.4.8 "Intent" vs. "Indication"	51
	3.4.9 "Performance" and "Procedure" Specifications	52
	3.4.10 Equitable Adjustment	53
References		54

3.1 STRATEGIC INTERPRETATION: APPLYING CONTRACTS TO SECURE POWER POSITIONS

In the preparation of any complicated end item, there is a certain risk that any of the basic components that make it up will contain flaws. With construction contracts, each of the separate contract sections has probably been (1) prepared by a different person that the other sections and often by entirely different companies; (2) assembled by unqualified people; (3) lifted out of boilerplate that does not apply; or (4) rushed in an eleventh-hour effort to complete the contract.

Conflicts, gaps, or inconsistencies are most likely to occur where the various sections of a construction contract intersect, thereby resulting in change orders and disputes. It is an extremely difficult job to make all of the individual contract components comprehensive and correct in themselves, without layers of redundancies. That difficulty is multiplied almost exponentially when the task is to coordinate all these pieces into a single, unified set.

Construction contracts have become packed so tight with terms and conditions that those who assembled them do not even clearly understand their intent. This chapter will begin to expose several effects resulting from the influx of "creative" clauses and how many are applied, or misapplied. We'll also see that as the number of the complex, and even confusing, clauses increases, the probability of errors, conflicts, ambiguities, and change orders goes up as well.

3.1.1 Contract Law Concepts

The concepts of contract law interpretation are based upon the idea of protecting and fulfilling the reasonable expectations of the parties who have relied upon the promises of other parties to the contract. [30] When these reasonable expectations are violated by one of the parties, the other party is entitled to compensation for damages resulting from the other party's failure to fulfill its obligations under the contract. [32]

The key to ease or difficulty in applying these objectives revolves around the definition of "reasonable expectations." In interpreting construction contracts it is important to develop the ability to determine the division line of "reasonable" and "extra" and to improve our ability to persuade others of the legitimacy of *our* definition of "reasonable." Developing these skills should help to improve a contractor's approval rate on submitted change orders.

Contract requirements must be clearly represented and completely understood if such reasonable expectations can be defined. Some matters must be quantified to the point where some real basis for evaluating performance can be established. If the responsibilities, rights, and liabilities of the various parties to a construction contract are not so described, it is likely that a major dispute will lead to litigation or arbitration. It is helpful to have an explicit mechanism built into the contract to clarify the steps to be taken toward resolution of the inevitable conflicts.

Good contracts meet this fact of construction life head-on. They at least attempt to make an effort to remove gray areas by determining who is expected to bear the risks of the various uncertainties.

Contractors should have a good working knowledge of the general rules of contract interpretation. A second-nature ability to apply these rules within the context of specific construction agreements will help a contractor to perceive potential and actual problems as well as the details of action options.

Approving authorities, courts, and arbitrators will look first to the contract in attempting to resolve changes and disputes. It is therefore imperative that all related contract provisions are complied with. Variance from contractual administrative procedures has been and will continue to be one of the most common grounds for the denial of change orders. If these technical kinds of difficulties can be avoided, at least that much risk can be removed from the process of getting changes formally approved and paid for.

3.1.2 Construction Law Considerations

A contract might not expressly provide procedures for resolving changes on a project, or the stated procedures may be ambiguous. If this is the case, the principles and procedures that are specific to the construction industry should guide the parties. By resolving disagreements in the contract, the parties ensure that disputes will be resolved more quickly. When contractual uncertainties are not resolved, contract disputes will likely be referred to a third party such as a judge.

3.2 THE CONTRACT DOCUMENTS: SIMPLIFIED DESCRIPTIONS TO PREVENT OVERSIGHT

To speak to a single construction contract may confuse the definition of the entire agreement. A construction contract is actually comprised of a large number of items integrated into what is collectively referred to as the *contract documents*. These individual components are consolidated to make up the entire agreement between the prime contractor and the owner. The typical components in nearly every construction contract are listed in Figure 3-1.

3.2.1 Contract Components

The Agreement is often a standard form document, such as the AIA Document A101—*The Standard Form of Agreement Between the Owner and Contractor*. It includes descriptions of the parties, the work, time of commencements, time of completion, the contract sum, and (most important) references to the other documents. It is the mechanism that consolidates the various individual provisions into a single contract.

The General Conditions section defines the legal relationships among the parties. It catalogs the components that make up the entire list of ultimate responsibilities.

INDUSTRY AND CONTRACT ENVIRONMENTS

The General Conditions and Instructions to Bidders have been described as "intricately designed, highly machined and honed boilerplates."[14] They contain terms which reflect principles and risk allocations that are generally accepted in the construction industry.

BID DOCUMENTS
Instructions to Bidders
Prime Contract Bid
Subcontract Bids

THE AGREEMENT
Prime to Owner
Prime to Subcontractors
Amendments Issued Prior to Execution of the Agreement
Changes Executed After the Original Agreement

PLANS
Architectural
Site
Structural
Plumbing
Fire Protection
Heating, Ventilating, A/C
Electrical
Specialty
Shop Drawings
Clarifications
Changes

SPECIFICATIONS
General Conditions
Supplementary General Conditions
Special Conditions
Working Procedure
Technical Specifications
Referenced Technical Standards

STATE OR FEDERAL
Regulations
Restrictions
Labor Standards
Environmental Standards, as applicable

FIGURE 3-1 THE CONTRACT DOCUMENTS

The General Conditions are then tailored to a specific project with the addition of any or all of the following:

[14]Hart, *Practical Problems and Legal Trouble Spots in Construction Agreements,* Construction Contracts 52 (Practicing Law Institute, 1976).

- Supplementary General Conditions.
- Special Conditions.
- Working Procedure.
- Many other direct requirements that are specific to the project and the owner.

The Plans and Specifications *together* describe the technical requirements and relationships of the project. They incorporate physical descriptions, locations, dimensions, and quality criteria, and, it is hoped, the ultimate result intended in the grand scheme.

The contract documents will grow throughout the project to include "all modifications issued after the execution of the contract."[15] Modifications are then clarified to be *written* amendments made after the execution of the contract. They may manifest themselves as:

- Formal change orders.
- Interpretations of specific sections of the contract.
- Clarifications.
- Directives of any kind.

By definition, contract agreements purport to be sophisticated and finely tuned works of legal art. The truth is, however, attorneys ordinarily have little or no input in the preparation of the plans, specifications, and even the general portions of the contract. These are prepared by the designer. This is the paradox of the contract documents.

The plans and specifications are the primary source of scope-of-work disagreements. The lawyer's input prior to contract execution is often as narrow as simply drafting or reviewing the Agreement portion of the contract. In some instances, the General Conditions portions may be reviewed. Probably more often, however, substantial construction contracts may be produced completely without significant legal review.

3.2.2 The Four C's of Contracts

Any evaluation of construction agreements will be kept in perspective if you apply the four C's of contracts as a test of the caliber of the document. If a contract is to be applied with a minimum number of disputes and disagreements, its terms and conditions should fulfill the four C's:

Clear

The terms of a contract must be unambiguous. The meaning and description of the terms, components, and references should be without confusion.

Complete

The contract should include adequate details. For example, material specifications should have physical parameters described with sufficient partic-

---

[15] American Institute of Architects, AIA Document A101, *Standard Form of Agreement Between Owner and Contractor Where the Basis of Payment Is a Stipulated Sum* (1997).

ularity. Contract clauses should contemplate pertinent considerations. All this should be accomplished without unnecessary duplication or gaps.

Comprehensive

The descriptions and references for contract provision and construction detail should adhere to the following:

1. Be broad enough to encompass the affected areas in a manner consistent with the design intent;
2. Be effectively correlated with other provisions, where necessary; and
3. Combine to give adequate attention to the entire project.

Comprehensible

The terms of a contract and its components must be comprehensible. The requirements of a specification and of the detail drawings must be described in a manner consistent with the reasonable level of competence that can be expected from those who will be applying those requirements. They must be understandable and intelligible. They should avoid language that is designed to mislead, confuse, or be more "sophisticated" than the situation requires.

3.3 RULES OF CONTRACT INTERPRETATION: THE CARDS UP YOUR SLEEVE

3.3.1 Introduction

There exist fundamental rules of interpretation that apply to all contracts, regardless of the industry or subject. These principles form the initial basis for assessing how conflicts in the contract language and format will be decided. Application of these rules will also facilitate accurate assessments of the suitability of various interpretation options. Finally, a working awareness of these basic considerations will make the application of Section 3.4, "Applying Construction Contracts Without Resistance," complete.

The ideas presented here are those that are most common and, therefore, most useful in determining the limits of the existing contract and where change orders begin.

3.3.2 Standard of Interpretation: Reasonable Expectations

The first objective of contract law is to protect the reasonable expectations of the parties. [30] The first step then becomes somehow to define clearly what those reasonable expectations are.

In an effort to reduce the subjectivity in this determination, the courts have sought to develop some objective standard that could be applied as the first step in contract interpretation. To reduce uncertainty, the court, rather than the jury, interprets the contract language. The jury determines the facts, but the court gives the jury the law.

The standard of "reasonable expectation" has been described to be the meaning that would be attached by a normally intelligent person competent in his or her profession, with complete knowledge of the facts and circumstances. [33]

Application rules

The following rules will aid the contractor in applying the "reasonable expectation" standard of contract interpretation:

1. The ordinary meaning of language is given to words unless circumstances show that a different meaning is applicable. [31]
2. Technical terms and works of art are given their technical meaning unless the context indicates a different meaning. [33]
3. A writing is interpreted as a whole, and all writings forming a part of it are interpreted together. [34]
4. If the conduct of the parties defined a particular interpretation, that meaning is adopted. [35]
5. Specific terms are given greater weight than general language. [36]
6. Separately negotiated or added terms are given greater weight than standardized terms or terms not specifically negotiated. [37]

3.3.3 Ambiguities Resolved Against the Drafter

The person or parties who draft contracts should have a clear idea of their intent. In addition, they have every opportunity to be sure that those intentions are defined clearly enough to allow reasonably competent people to understand and accommodate such instructions.

The law assumes that those responsible for drafting the contract will provide for their own interests and will have reason to be conscious of uncertainties. The owner and architect, for example, typically draft the contract and have full control over the terms and are, thus, in the position to ensure that the terms are clear and complete. As such, in the event an ambiguity arises, the contract will be interpreted against the owner [38].

It is important to be aware, however, that this rule only applies in the case of adhesion contracts, and *not* where the terms were the product of the joint efforts of both parties to the contract. Adhesion contracts are those that are produced by one party and are presented to the other party on a "take it or leave it" basis, without any reasonable opportunity to negotiate. Most contractors can therefore be comfortable with the knowledge that this rule will apply to lump-sum bids submitted to the owner on complicated forms that refer to the plans and specifications as is.

Nonetheless, for a contractor to recover costs and damages resulting from an ambiguity, it must demonstrate that the interpretation was in fact relied upon.

3.3.4 Right to Choose the Interpretation

If an adhesion contract or its terms are subject to more than one reasonable interpretation, the contractor has the right to choose between the reasonable interpretations. [38] The reasons for this policy are generally the same as those for ambiguity resolution. It also provides that if a reasonable interpretation is available, there is no duty to continue to search for other "reasonable interpretations" that may or may not exist.

3.3.5 Specific vs. General

Specific terms, conditions, and requirements are given greater weight than general language designed to cover multiple circumstances. [36] The thought is that specific language was prepared to accommodate a unique circumstance. It is, therefore, reasonable to assume that it is more likely to consider the precise conditions.

3.3.6 Usage of Trade Custom

The Uniform Commercial Code defines "usage of trade" as follows:

> A usage of trade is any practice or method of dealing having such regularity of observance in a place, vocation, or trade as to justify an expectation that it will be observed with respect to the transaction in question. (U.C.C. Sec. 1-205(2))

As anyone who has ever worked with lumber knows, a 2 × 4 stud is not 2″ × 4″, but $1^1/_2″ \times 3^1/_2″$. An 8″ concrete block is really $7^5/_8″$, and so on. Customs always seem unusual at first, but then through familiarity eventually seem to make sense. There is, therefore, a necessity for learning how things are usually discussed and what people really mean before a contract is interpreted in the proper context.

In applying "trade custom" to a situation to establish a trade custom as the definition of a meaning, contractors must demonstrate that the custom is followed with *absolute regularity*. It is not enough that it is "usually" done in a certain manner. It must be a usage that is observed in virtually *all* cases in the area where the contract is to be performed, except where it has been expressly altered. Just because the mason doesn't usually provide the portion of a masonry tie that is welded to a steel column, for example, doesn't mean that it is not a requirement of the contract being considered.

Finally, it is important to realize that trade custom cannot be relied upon to excuse a contractor from the performance of a contract requirement if that requirement is clear and subject to one reasonable interpretation. [39] Trade custom cannot be used to torture an interpretation into one that is illogical or inconsistent.

3.4 APPLYING CONSTRUCTION CONTRACTS WITHOUT RESISTANCE

3.4.1 Introduction

Beyond the rules of general contract interpretation, the construction industry has developed a wide arrangement of distinct contract clauses, each requiring special consideration. Unique to the industry, these contract rights and requirements are very consistent throughout the majority of contemporary construction contracts. It is this fairly universal application throughout the industry that has allowed the courts to test the applicability of the language and develop a set of guidelines to assist us all in interpreting construction contracts.

3.4.2 Change Clauses

Description

The Change Clause authorizes the owner to alter the work if the contemplated change falls within the general scope of the original agreement. Without this express provision, there is no inherent "right" of the owner to make changes that the contractor will be obligated to perform. The Change Clause gives the owner that right.

The Change Clause provides both the means by which the owner may adjust the plans and specifications and the mechanism by which the contractor may incorporate suggestions. It will outline the procedure to organize and present claims for additional compensation for extra work.

The Change Clause should be coordinated closely with an efficient Dispute Clause that details the procedure for settling serious problems.

Basic elements

Although the specific language varies, most change clauses incorporate the same or similar provisions. Typically, the requirements will include the following terms:

1. Adjustments to the contract may only be made by a change order;
2. Change order must be in writing, signed by both the owner and the architect;
3. Change order must specify both the adjustment in contract price and net effect on the *project* time;
4. Change orders will be for work contemplated by the original scope of the contract; and
5. No changed work is to be performed without a properly executed change order. (This provision is most often expanded to exclude those instances where a contractor must act in an emergency to prevent injury or damage to property.)

Beyond these provisions, the Change Clause may go on to describe administrative procedures associated with change orders, or refer to appropriate instruction for actual processing.

Problems of change clauses

Common difficulties that arise in the application of change clauses often fall within the following categories:

1. *Disagreement on whether the contemplated change does in fact fall "within the scope of the contract."* Is the type of work sufficiently similar? Is the dollar amount of the change significant to those who are to perform it?
2. *Must the change be in writing?* Even if there is no question in this regard, at what point must it be committed to paper? Can the authorized entity direct the work to proceed and then follow up with the paperwork, or must everything be properly executed before any work is performed?
3. *Who has the real express or implied authority to authorize the change?* The clause usually empowers only the owner to approve changes affecting time or money. Does the architect, clerk of the works, or inspector have any power? If for convenience the owner has allowed any of these agents to approve minor changes without the owner's direct involvement, at what point, if ever, does this act as a waiver of a contract clause which provides that only the owner can approve changes?
4. *Is the consent of a surety needed if a significant change alters the bonding capacity of a contractor or entitles the surety to additional premiums?*
5. *Is there a complete mechanism to resolve the problem with a minimum amount of difficulty, if a price cannot be agreed upon?* Does the work stop? Does the contractor proceed "under protest" and file a claim?
6. *Is the additional time requirement as stated in the change order request accurate?* Does it only incorporate the additional work into the original work sequence, or does it accommodate alternative logic arrangements.
7. *Do markups apply to net price changes (adds + deducts) with the effect of lowering the total allowed markup?*
8. *If the percentage markup reduces as changes increase in price, will the owner lump several distinct changes together?* If so, although the changes must be managed individually, the lower aggregate markup might create an inequity in the form of disproportionate overhead.
9. *The incorporation of a large change order will have the net effect of reducing the allowed percentage of the items to be billed on contract work.* This may occur when the "general conditions" items in the original schedule of values and progress billings are paid as a percentage of total work in place. This is an issue to consider when preparing change order requests and should be clarified before the final change order price is submitted.

Sample change clauses

Actual Change Clauses may require up to several pages. In that case, they may go into extreme detail, outlining the procedure for many conceivable circumstances. They may even develop into Dispute Clauses.

The clause listed here represents common language fulfilling the primary functions of the clause:

a. A construction change is a written order prepared by the architect and signed by the owner and architect directing a change in the work prior to agreement on adjustment, if any, in the Contract Sum or Contract Time, or both. The Owner may by Constructive Change Directive, without invalidating the contract, order changes in the Work within the general scope of the Contract consisting of additions, deletions, or other revisions, the Contract Sum and the Contract Time being adjusted accordingly.[16]

3.4.3 The Pass-Through Clause

Description

The Pass-Through Clause (sometimes referred to as the Conduit or Flow-Down Clause) is to the prime contractor or the construction manager one of the most important provisions in any subcontract agreement. It incorporates into a subcontract by reference all or part of the rights and responsibilities of the prime contract to the owner as they relate to the respective subcontract. [40]

For example, a common clause in the General Conditions makes "the contractor" responsible to furnish and maintain all tools, scaffolding, hoisting equipment, and so on as necessary to perform the work. The Pass-Through Clause narrows this to mean that the plumbing contractor provides all these items as necessary to perform the plumbing work, the plaster subcontractor provides these items as necessary to perform the work of that section, and so on. If a clause in the general contract requires that construction debris be cleaned up daily and removed from the site (barring a separate agreement otherwise), the Pass-Through specifies that the mason pick up and remove the masonry debris, the roofer do the same for roofing waste, and the kitchen equipment subcontractor remove those packing materials from the site.

The standard subcontract and trade contract forms of both the Associated General Contractors (AGC) and the American Institute of Architects (AIA) contain a form of the Pass-Through. It binds a subcontractor or trade contractor to the same rights and responsibilities that the prime contractor or construction manager has concerning the owner that apply to the work of the respective subcontractor.[17]

Application

There is a simple and accurate method to apply this idea cleanly to your construction contracts. As you read through a particular general specification requirement, simply insert the name of the subtrade being considered wherever the word "contractor" appears. Likewise, simultaneously insert the words "prime contractor" or "general contractor" wherever the corresponding word "owner" appears and accordingly read the clause in that context.

For example, if the General Conditions Clause requires that:

---

[16]American Institute of Architects, AIA Document A201, *General Conditions of the Contract for Construction*, Article 7.3.1 (1997).

[17]Flow-down clauses do not incorporate the general contractor's administrative obligations to the owner pursuant to the prime contract, and subcontractors are thus not required to fulfill the general contractor's responsibilities with respect to change orders, notice requirements and claims. Flow-down clauses make subcontractors responsible for prime contract terms which relate to performance of their work.

INDUSTRY AND CONTRACT ENVIRONMENTS

> "The contractor will remove all rubbish ..." and "in the case of dispute the owner may remove the rubbish and charge the cost of such removal to the contractor,"

the Pass-Through will read this same clause as it relates to the millwork subcontractor as:

> "The millwork subcontractor will remove all (millwork) rubbish" and "in the case of dispute the general contractor may remove the rubbish and charge the cost of such removal to the millwork subcontractor."

The technical specifications themselves reinforce the firm application of the Pass-Through. The various specification sections usually begin with a distinct incorporation of the General Conditions and Division 1 into the requirements of the respective section. Language such as:

> The General Conditions and Division 1 shall form a part of this and all other sections of the specifications.

or:

> Applicable provisions of the General Conditions and Division 1, General Requirements shall govern the work under this section.

or something similar binds the respective trade contractor to *all* requirements of the front-end documents, even without a clearly expressed Pass-Through provision in the subcontract.

As clear as this idea may seem to some, a surprising number of prime and subcontractors still do not clearly understand the pass-through principle. Learn to make the Pass-Through Clause work for you, and your contract administration will run much more smoothly.

Exceptions

Be careful when the words "general contractor" or "prime contractor" (as opposed to simply "contractor") are encountered in the original General Conditions clauses. They may represent an intentional modification of the Pass-Through in the interest of job efficiency or for some other objective. The most common of this type of exception is when the *general* contractor is called upon to furnish, erect, maintain in a safe condition, and remove staging to be used by *all* trades. Specific attention will be called to this type of *general* contractor requirement if it would not otherwise make cost or efficiency sense to make each trade responsible for its own provision.

For instance, if the mason were to erect staging around the building for its work and dismantle it, only to be followed by the plaster subcontractor erecting staging in the same location to allow the construction of the exterior soffits, it would not only add to the cost of the project but to the time of completion as well. Additional examples of this kind of exception may also apply to the categories of temporary heat, light, and power; a common dumpster; or the erection and removal of safety constructions.

Practical requirements may also dictate limits to the Pass-Through. General Conditions clauses are written to apply to a broad spectrum of project sizes and types. On some types of projects, then, a blind application of the Pass-Through to every instance where it may technically apply may not be the best use for practical reasons. A strict application of the Pass-Through, for example, may required each subtrade to provide its own sanitary facilities. For practical considerations on a small project, this just wouldn't make sense. In this case, the prime contractor is likely to bend the rule to provide sanitary facilities for use by all trades.

Finally, if you are a sub- or trade contractor, it is wise to be cautious. Often, subcontract forms designed by general contractors, prime contractors, and construction managers contain the Pass-Through as described, binding the sub- or trade contractor to the responsibilities. They may, however, noticeably omit similar provisions giving the sub- or trade contractor the corresponding rights. General contract law will most likely come to your aid in this circumstance because this type of form will be considered unreasonable. However, if your review of the subcontract exposes this type of dealing, you can save much aggravation by having the language corrected before the document is executed.

Sample clauses

The Contractor and Subcontractor shall be mutually bound by the terms of this Agreement and ... the Subcontractor shall assume toward the Contractor all the obligations and responsibilities which the Contractor, under such Documents, assumes toward the Owner and Architect.[18]

The trade contractor agrees to be bound to and assume toward the construction manager all of the obligations and responsibilities that the construction manager by those documents assumes toward the owner.

3.4.4 The Dispute Clause

Description

The Dispute Clause is the contract mechanism that provides the specific procedure for resolution of serious differences. Ordinarily its purpose is to provide a method to settle differences efficiently and expeditiously for the benefit of both parties. It may detail a progressive approach through a series of steps that must be followed, or it may simply outline the ultimate resolution option. An example of such a series of steps may be some platform to allow negotiation, with an appeal process through boards or commissions, at the end of which is arbitration or litigation. The simplest of clauses may provide simply that arbitration will be the final determinant.

The clause may stand by itself and be easily identified. An example of this type of treatment is found in the AIA Document A201, Article 2.2.12:

Any claim, dispute, or other matter in question ... shall be subject to arbitration upon written demand of either party.

The clause then goes on to describe the conditions and procedures that must apply to invoke the article.

---

[18]American Institute of Architects, AIA Document A401 (1997), *Standard Form of Agreement Between Contractor and Subcontractor*, Article 2.1.

The other common treatment is to combine the Dispute Clause with the Change Clause (see Section 3.4.2). It then becomes the final step in the progressive change order negotiation procedure to ultimate resolution of deadlocks.

3.4.5 Authority (Formal/Constructive)

Description

It is critical to determine who or what entity possesses authority to approve various activities on a project. Contractual descriptions of authority are typically found throughout the documents. Frequently, authority is confused based on expectations arising from previous projects or by the constructive actions of the parties. However, extreme care must be exercised on this issue and a contractor must confirm precisely where *formal* authority lies with respect to the specific issue under consideration.

Catch-22

The General Conditions section is likely to contain some language that (1) the architect will be the interpreter of the requirements of the plans and specifications and (2) the architect shall interpret the requirements of change orders and shall decide all other questions in connection with the work.

Another paragraph of the same article may include something to the effect that the architect shall have no authority to approve or order changes in the work which call for an extension of time or change in the contract price.

On public contracts, the language may place all interpretation responsibilities and authority with the owner or its agents, with the architect being reduced to an advisory capacity.

AIA Document A201 authorizes the designer to approve changes that do not affect contract time or price. Change orders approved in this manner have been interpreted as being "constructively approved" by the owner despite the owner's lack of knowledge of the change at the time.[19]

Application

> If the contractor has received payment for past work ordered by the designer at a variance from the change order mechanism, a court is likely to find that the contractor is justified in relying upon the designer's authority to act in this manner.[20]

Even so, there is always a risk that recovery will be denied for failure to follow the contractually defined procedure.

Read and become thoroughly familiar with the exact wording of the specific authority as outlined in the contract. Know precisely which avenues are to be cultivated. Avoid any assumption regarding anyone's real power to act—see it in writing first.

Once a confident knowledge of the specifications is secured, application of strategies and tactics to move change order requests through the informal

---

[19]Ibid., p. 82.
[20]Jeremiah D. Lambert and Lawrence White, *Handbook of Modern Construction Law* (Englewood Cliffs, N.J.: Prentice Hall, 1982), p. 82.

authority structure can be considered. This idea is treated in Chapter 12, Winning in Change Order Negotiation.

3.4.6 "General Scope" of Work

It is critical that the scope of work that a contractor must complete to fulfill its contract obligations is clearly and completely defined. In addition to work which is explicitly required, the scope will be interpreted to include work that is plainly and obviously necessary for a complete installation.

Change Clauses give the owner the right to include modifications to the contract without invalidating it if such changes fall "within the general scope of the contract." Strictly speaking, this narrows the potential work of the change to the same or similar type of work originally incorporated:

1. The relative dollar size should not cause a hardship or inequity for any contractor, whether the change is an addition or deletion.
2. The type of work added should be within the expertise of the contractor if that company is to be reasonably held responsible for it (and if the owner can reasonably expect to be satisfied with the end result).

Special considerations

Construction contracts do not define what qualifies as being "within the general scope" of any portion of the work. All proposed changes are therefore usually treated as if they comply with the requirement, at least initially.

(1) Additions

As a practical matter, a change that doubles the work of a contractor is seldom argued with. This is true as long as the contractor is compensated equitably for the corresponding overhead (which may become disproportionate) and the proportional increase in necessary management attention.

Most often, however, the contractor pricing the additional work is not in the intense competitive situation as during the preparation of the original quote. As such, a contractor will typically submit a less economical estimate than it would when competing with other bidders.

(2) Surety involvement

An example of an increased cost that is treated differently by owners is the treatment of performance and payment bonding requirements. Some contracts may consider bonding as a direct compensable cost. Others may consider it overhead included in the allowed percentage markup. If the scope of a contractor's work is changed from $1,000,000 to $2,000,000 (clearly significant), the contractor's bond premiums may increase disproportionately (if the company can secure the additional bonding at all). In addition, because large change orders normally have smaller markup percentages than do smaller ones, the inequity of this situation may be compounded.

(3) Deletions

Changes that significantly reduce the work of a contractor may do more damage than might first appear. Even if the owner will openly entertain a "lost profit" item to compensate for at least a portion of the profits that would otherwise have theoretically belonged to the contractor, it does nothing to accommodate the previously committed bonding capacity and opportunity costs. [41] These and other considerations will be detailed in Chapter 8, Designing and Constructing Effective Change Order Proposals.

3.4.7 "Reasonable Review"

Description

Contractors are responsible for reviewing all bid and contract documents as they relate to their work and for submitting a proposal that is responsive to the requirements. [18] What can be read, seen, or *reasonably* inferred gets priced. What has been intentionally or unintentionally left to the imagination does not find its way onto an estimate sheet in a competitive situation.

It was clarified in Chapter 2 that contractors are responsible for disclosing obvious "patent" errors. In addition, the introductory provisions of the General Conditions will normally require the contractor to visit the site prior to bid and accommodate all considerations that are readily apparent. The contractor is *not* responsible for performing a complete search of documents, seeking out hidden flaws in the contract, making subsurface explorations, or undertaking any other extreme, costly, or time-consuming investigations. This kind of detailed attention should have been given to the contract by the owner and designer.

3.4.8 "Intent" vs. "Indication"

Description

In a scope of work dispute, the "intent" of the contract is likely to come up at some point and may be relied upon in interpreting the contract. Some typical contract clauses provide as follows:

1. The intent of the Contract Documents is to include all items necessary for proper execution and completion of the work. (AIA Document A201—1976, Article 1.2.3); and
2. All interpretations ... shall be consistent with the intent of and reasonably inferable from the Contract Documents. (AIA Document A201—1976, Article 2.2.10)

Such clauses may be used as a basis for filling contractual gaps and to induce the contractor to perform additional work.

Defense

Design "intent" should apply only if there is a true ambiguity or gap in the contract documents and the gap being bridged is so obvious that a professional contractor would not normally overlook it. Refer back to Section 3.3.6, "Usage of Trade Custom," and Section 3.4.7, "Reasonable Review," for more detailed treatment.

Applicability test

There are methods to test whether design "intent" is an appropriate method for determining the contractual scope of work. One may start the analysis by asking the following questions:

1. Can the remaining work shown be completed without the extra work in question?
2. Is there more than one way to complete the extra work?
3. Is the extra work not usually encountered by the type of trade now being considered to construct it?

An affirmative answer to any of the above questions will cast serious doubt on the applicability of "intent" as a basis for determining the defined scope of work. For "intent" to apply, the work in question, relative to the surrounding construction, must be considered necessary with regularity and consistency in the construction industry.

3.4.9 "Performance" and "Procedure" Specifications

Description

A performance specification is one that describes the ultimate *function* to be achieved, leaving the means and method completely to the contractor. It describes performance characteristics on completion of a component or system, without indicating the precise materials to be used in their achievement. If a requirement, for example, directed that the foundation perimeter insulation "have a $k = .20$ @ 75 degrees F and minimum compressive strength of 20 lb/ft," it has been left to the contractor to provide and install whatever material is necessary to meet the stated design criteria. With regard to performance specifications, each of the following applies:

1. The ultimate performance is the design objective;
2. The particular material used is of no consequence, given the proper level of quality, and will be the contractor's option; and
3. It is the contractor who bears the risk of selecting the materials that will meet the stated objective of the performance specification.

In contrast, procedure specification explicitly describes the *material* to be used and its physical relationship to the remaining construction. It details the properties, qualities, composition, and assembly of materials and components. A procedure specification for the same foundation insulation might read "extruded, closed-cell polystyrene board as manufactured by XYZ Corp., 2" thick × 24" wide." There is no question as to the material desired.

Application

There is nothing inherently "wrong" with either type of specification. On the one hand, performance specifications may shift design risk onto the contractor, but they decrease the designer's control. On the other hand, a typical contract will contain a much larger percentage of procedure specifications.

Practicalities of product descriptions may, however, weave performance requirements in with specific product descriptions. Because this in effect duplicates the product specification, it becomes another interface of possible inconsistency—and change orders. This is treated specifically in Part Three.

Performance specifications are most commonly used in the procurement of mechanical and electrical equipment. In such a case, the performance characteristics can be readily measured and tested. Another common area is concrete compressive strength (3,000 p.s.i. at 28 days).

Procedure specifications must be correct in every respect to avoid problems. The sufficiency of the architect's design must be relied upon by the contractor to achieve the ultimate intended result. If the procedure specification is properly followed, the risk of design success lies completely with the architect and/or engineer.

Cautions

When a product is specifically described *and* performance requirements are indicated, the contractor is performing up to his or her obligations by procuring and installing the exact material detailed. There is no duty (aside from the duty to note "obvious errors") of the contractor to be sure that the exactly described material also meets the performance requirements listed. It was up to the designer to confirm this before the specifications were finalized. When a difficulty between the two arises, the designer or owner may want to reference the "performance" portion of the specification, claiming that it governs over the "procedure" portion. Contractors should dispute this position.

3.4.10 Equitable Adjustment

Description

Equitable adjustment is the concept that a contractor can be made "whole" in the face of a changed condition that affects its work. The contractor should be left in the condition that it would have been had the owner-caused change not occurred.

Contract changes do more than alter the scope of work. They also do each of the following:

- Disrupt orderly sequences;
- Interfere with planned deliveries;
- Disrupt prior coordinations;
- Change schedule logic;
- Change methods for work not otherwise directly addressed by the change;
- Cause a contractor to remain mobilized on the site longer than originally anticipated; and
- Contribute to disproportionate administrative costs resulting from backtracking and rework.

In theory, the contractor is entitled to recover any direct or indirect costs that occur as a result of the respective owner-caused change.

Basic elements—the three costs

To identify clearly the elements included within a contractor's compensation for changed or extra work, visualize the two scenarios being constructed, i.e., the work as originally planned without the subject interruption, and the work with all the impacts of additional scope, resequences, payment disruptions, etc. Then catalog all extra cost items as they apply to the three cost elements of every change:

1. *Direct costs* (labor, materials, supervision, etc.—the hard costs).
2. *Indirect costs* (home office overhead, delays, opportunity costs, lost profit—the soft costs).
3. *Consequential costs (damages)* (interference, disruption, resequence).

These three cost components of every change order are detailed specifically in Chapter 8. Their applicability to a particular change is a matter of degree. Their appropriate inclusion in the respective change proposal is a matter of business judgment, applied within the context of the owner–contractor relationship.

REFERENCES

1. *United States v. Spearin* (1918) 248 US 132; Michie, *Construction & Design Law*, Warranties §20.2, pp. 11-14; and §20.2c.2, pp. 14-17 (1998).
2. *J.L. Simons Co., Inc. v. United States* (1969) 412 F.2d 1360; *J.J. Welcome Construction Co.* (1986) 86-3 BCA ¶19,502.
3. *Premier Electrical Construction Co. v. United States* (1973) 473 F.2d 1372; *Darvo Corp.* (1978) 79-1 BCA ¶13,575; Michie, *Construction & Design Law*, Warranties §20.6b, pp. 69-70 (1998); *Kent v. United States* (1965) 343 F.2d 349.
4. *P.J. Maffei Building Wrecking v. United States* (1984) 732 F.2d 913; *Foster Construction & William Bros. Co. v. United States* (1970) 435 F.2d 875, 875; *E.H. Morrill Co. v. State of California* (1967) 65 Cal. 2d. 787; *Wunderlich v. State ex rel Department of Works* (1967) 65 Cal. 2d. 777; *Frederickson & Watson Construction Co. v. Department of Public Works* 28 Cal.App. 3d 514; Michie, *Construction & Design Law*, Warranties §20.e and e.1, pp. 25-26 (1998).
5. *See* Carter, Cushman, Palmer, *Construction Litigation Representing the Contractor*, Bid Protest and Bid Mistakes, p. 65 (1992).
6. American Institute of Architects, AIA Document A201, Article 2.2.3 (1997); Michie, *Construction & Design Law*, Owners, §6.2 a.2, pp. 8-9 (1996).

7. American Institute of Architects, AIA Document A201, Articles 2.2.2 and 2.2.3 (1997); Michie, *Construction & Design Law,* Owners §6.2b.2, pp. 11-16 (1996).

8. *Panamint, Inc.* (1987) 87-2 BCA ¶19,927; Michie, *Construction & Design Law,* Owners, §6.2c, pp. 16-17 (1996).

9. *Restatement (Second) of Contracts* §205 (1981); Michie, *Construction & Design Law,* Owners, §6.2f.2, p. 31 (1996).

10. American Institute of Architects, AIA Document A201 Articles 2.2.1 and 14.1.1.4 (1997); Michie, *Construction & Design Law,* Owners §6.2e.1, pp. 26-27 (1996).

11. American Institute of Architects, AIA Document A201 Article 4.2 (1997); Michie, *Construction & Design Law,* Contract Documents, §11.7, pp. 82-83 (1994).

12. *Hotel Utica, Inc. v. Ronald G. Armstrong Engineering Co.* (1978) 404 N.Y.S. 2d 455; Michie, *Construction & Design Law,* Design Professionals, §4.6b.4, pp. 84-85 (1995).

13. *Wunerlich Contracting Co. v. United States* (1978) 351 F.2d 956; *Stuyvesant Dredging Co. v. United States* (1987) 11 Claims Court 853; Michie, *Construction & Design Law,* Warranties, §20.2d.1, pp. 18-19 (1998).

14. *Hill v. Polar Pantries* (1951) 219 S.C. 263; *but see 530 East 89 Corp. v. Unger* (1977) 431 N.Y. 2d 776; Michie, *Construction & Design Law,* Design Professionals §4.6.a.5, pp. 51-53 (1995).

15. *Huang v. Garner* (1984) 157 Cal. App. 3d 404; Acret, *Architects & Engineers,* Liability Arising from Statute, §3.03, Failure to Comply with Building Codes, pp. 92-93 (3d ed. 1993); Cushman & Hedemann, *Architect & Engineer Liability:* Claims Against Design Professionals (3d ed. 1993), Liability for Negligence, §7.2, p. 133 (2d ed. 1995).

16. American Institute of Architects, AIA Document A201 Articles 4.2.2, 4.2.9, 9.9.1 and 13.5 (1997). *City of Columbus v. Clark Dietz & Associates-Engineers, Inc.* (1982) 550 FS 2d. 610; *United States ex rel Los Angeles Testing Lab v. Rogers & Rogers* (1958) 161 FS 132.

17. *REDM Corp. v. United States* (1970) 428 F.2d 1304; *J.D. Hedin Construction Co. v. United States* (1965) 347 Fd. 2d. 235; *Wiechmann Engineers v. State of California ex rel Department of Public Works* (1973) 31 Cal. App. 3d. 741.

18. American Institute of Architects, AIA Document A201 Article 3.2 (1997); Michie, *Construction & Design Law,* Contractors, §7.2a, pp. 47-49 (1998).

19. American Institute of Architects, AIA Document A201 Article 3.10; Michie, *Construction & Design Law,* Contractors, §7.2j, pp. 58-59 (1998).

20. American Institute of Architects, AIA Document A201 Article 3.5 (1997); Michie, *Construction & Design Law,* Warranties, §20.3a, p. 27 (1998); *Pacific Coast Builders v. Antioch Live Oak Unified School District* (1956) 143 Cal. App. 2d. 125.

21. American Institute of Architects, AIA Document A201 Article 3.12 (1997); Michie, *Construction & Design Law,* Contractors, §7.21.4 (1998); *Waggoner v. W&W Steel Co.* (1982) 657 P.2d 147.

22. *Teitge v. Remy Construction Co.* (1998) 526 N.E. 2d 1008; Michie, *Construction & Design Law,* Contractors, §7.1b.2, pp. 19-23 (1998).

23. American Institute of Architects, AIA Document A201 Article 3.4.1 (1997); *Southern Dredging Co.* (1992) 92-2 BCA ¶24,886.

24. American Institute of Architects, AIA Document A201 Articles 3.3.1 and 3.10.1 (1997); *Christ Lutheran Church v. Equitable Church Builders, Inc.* (1995) 909 S.W. 2d 451; Michie, *Construction & Design Law,* Contractors, §7.2B, pp. 49-50; §7.2a, p. 55; and §7.2i, pp. 56-58 (1998).

25. American Institute of Architects, AIA Document A201 Article 3.5; Cushman & Hedemann, *Architect & Engineer Liability:* Claims Against Design Professionals, Liability for Negligence, §7.2, pp. 133-134 (2d ed. 1995); Michie, *Construction & Design Law,* Design Professionals, §4.2b.1, p. 6 (1995).

26. American Institute of Architects, AIA Document A201 Articles 4.2.5, 4.2.6, 4.2.7 (1997); *Huber, Hunt & Nichols, Inc. v. Moore* (1977) 67 Cal. App 3d. 278; Michie, *Construction & Design Law,* Contractors, §7.2t (1998).

27. J. Sweet, *Sweet on Construction,* Major AIA Documents §12.7 (1996 Ed.); Michie, *Construction & Design Law,* Contractors, §7.2a, p. 48 (1998).

28. American Institute of Architects Document A201 Articles 3.2.1 and 3.3; Michie, *Construction & Design Law,* Contractors, §7.2j, pp. 58-59; and §7.2n, pp. 66-67 (1998).

29. American Institute of Architects Article 9.3.3 (1997); Acret, *California Construction Law Manual,* Mechanic's Liens and Stop Notices, §6.09, p. 374 (5th ed. 1997).

30. *3 Corbin on Contracts* §538 (1960); *4 S. Williston* §§607, 618 Contract, (3d. Ed. 1961); *J.W. Bateson Co. v. United States* (1971) 450 Fd. 2d. 896; *Royal Indemnity Co. v. Kenny Construction Co.* (1975) 528 Fd. 2d. 184.

31. *3 Corbin on Contracts* §535 (1960); *Wholesale Tire & Supply Co. Ltd.* (1992) 92-2 BCA ¶24,960.

32. *3 Corbin on Contracts* §538 (1960).

33. *17A CJS Contracts* §302(1); *Kenneth Reed Construction Corp. v. United States* (1973) 475 F.2d 583.

34. *3 Corbin on Contracts* §549 (1960).

35. *Restatement (Second) of Contracts* §203.

36. *Restatement (Second) of Contracts* §203(c); 3 Corbin on Contracts §547 (1960).

37. *3 Corbin on Contracts* §548 (1960).

38. *3 Corbin on Contracts* §559 (1960).

39. *3 Corbin on Contracts* §556 (1960).

40. Acret, *California Construction Law Manual*, Subcontractors' Work, §2.90 (5th ed. 1997).
41. J. Sweet, *Legal Aspects of Architecture, Engineering, and the Construction Process*, §25.04(h) (4th ed 1989); *General Contracting & Construction Co. v. United States* (1937) 84 Ct. of Cl. 520; *ACS Construction Co.* (1987) 87-1 BCA ¶19,660.
42. American Institute of Architects, AIA Document A201 Article 13.5.
43. American Institute of Architects, AIA Document A201 Articles 4.2.7 and 4.2.11.
44. American Institute of Architects, AIA Document A201 Articles 4.2.3, 4.2.6, 4.2.7.
45. American Institute of Architects, AIA Document A201 Article 3.2.1.

Part Two
Change Orders Exposed

4
Change Order Diagnosis

4.1	A Normal Part of the Construction Process	62
4.2	Clarification or *Change?*	62
4.3	Reasons for Change Orders (Additions and Deductions)	63
4.4	Change Order Categories	63
	4.4.1 Owner-Acknowledged Changes	64
	4.4.2 Constructive Changes	64
	4.4.3 Consequential Changes	66

4.1 A NORMAL PART OF THE CONSTRUCTION PROCESS

The right to make unilateral changes to a contract is unique to the construction industry and contrary to the principles governing most business contracts. However, the right to make changes is a necessary part of all construction projects. There are many reasons for changes and there are those that are more obscure. On a project of any size, it is important to understand that changes will *definitely* occur on every type of project, large or small. Owners, designers, and contractors will be way ahead if they accept the fact that changes are a normal part of every project so that time and energy can be diverted away from improper arguments and paper shuffling and be devoted to prompt settlement of the three critical change order components: Scope, Dollars and Time. (Interestingly, most owners, and lenders on private projects, acknowledge the certainty of changes by providing a contingency fund to accommodate potential changes. This contingency fund is typically 10% of the total project cost.)

One major difficulty in the acceptance of this fact is that a large percentage of changes are the result of some kind of error or deficiency. Additionally, a large portion of the changes will, by their nature, assign a mistake to someone or some company.

In these situations, change orders bruise reputations and egos. They tend to deflate the "professional" self-image of the designer, and they invariably cost the owner money and time. For these reasons, change orders create friction.

4.2 CLARIFICATION OR *CHANGE?*

By the time the contract has been executed, the owner has already taken full advantage of the available competition to get a favorable price for the work that is depicted on the plans. The contractor is normally responsible for meeting "job conditions," but only up to a point. That point is when a "clarification" becomes ambitious and includes additional work which is beyond the work shown or reasonably anticipated. In preparing its bid, the contractor had to assume the most competitive method to construct any unclear details to get the job. The contractor has to bid on what is shown and not price what is not shown. If the contractor does not take this approach, the contractor will not be competitive and the job will be lost to the competition.

As mentioned previously, every contractor must know the scope of work contemplated by the bid. Clarifications that include additional or different work typically involve extra time or additional cost or both. Clarifications can be issued in many different forms, including supplemental drawings or details, responses to Requests for Information, or review comments on shop drawings. All clarifications should be scrutinized in minute detail. Determine if the scope of work has been modified and, if so, get paid for it.

4.3 REASONS FOR CHANGE ORDERS (ADDITIONS AND DEDUCTIONS)

The causes that form the basis for nearly all changes are relatively few. They include:

1. Design errors.
   a. Contradictions, discrepancies, impossibilities, inconsistencies.
2. Changes in market conditions.
   a. Specified product becomes unavailable.
   b. New products become available, offering price advantages or other benefits.
   c. New information becomes available, affecting the choice of specified materials.
3. Change in the owner's requirements.
   a. Scope of work.
4. The uncovering of undisclosed existing conditions.
5. The uncovering of unknown latent conditions.
   a. Unexpected soil variations.
   b. Unknown or unanticipated conditions in existing structures.
6. Suggestions to initiate better, faster, or more economical construction.
7. Change in designer preference.
8. Change in external requirements.
   a. Building codes.
9. Final coordination with owner-furnished equipment.
   a. Space.
   b. Mechanical and electrical provisions.

Chapters 5, 6, and 7 will deal with specific examples of each of these situations.

4.4 CHANGE ORDER CATEGORIES

Most change orders will apply to one of three different categories. These categories are:

1. Owner-acknowledged (change initiated or approved by the owner);
2. Constructive; and
3. Consequential.

The applicable category will determine the best process and approach for presenting the change and the probability for resolution and payment without conflict.

4.4.1 Owner-Acknowledged Changes

Description

This is the category that most commonly comes to mind when the subject of changes is considered. An owner-acknowledged change may be the result of a decision by the owner to make a change for its own convenience or an owner's recognition that the plans or specifications were incomplete or unclear. It is the only change type that will become incorporated into the owner's budget. In most cases, this category of change will present the least conflict, even if the change is the result of an error or omission on the part of the design team.

The primary objective of proper change order management is to convert the change orders of the other classifications into this type. This is necessary because this is the only category in which the owner accepts liability for the change. Since the owner has accepted liability, the owner will be more likely to resolve the monetary aspects of the change, and write the check, without being ordered to do so by the courts.

Difficulties

Total change order costs often encompass:

1. Cost of the work actually performed;
2. Costs related to any change in method sequence or schedule;
3. Costs related to the effect of changed work on other contract work; and
4. Other "impact" or "interference" costs (delay, acceleration, etc.)

In most cases, the owner will acknowledge and authorize payment for the cost of actual work performed, and possibly *some* related expenses. The other costs are likely to be left to be "absorbed" or to be recovered, if possible, through litigation.

The real talent in change order resolution lies in the ability of the contractor to move the owner to accept and authorize the payment of the consequential cost items that have traditionally been rejected.

4.4.2 Constructive Changes

Description

A constructive change order is one that occurs when the owner or its authorized representative acts in such a way that causes a contractor to perform additional work. These kinds of actions or inactions may take the form of verbal and written directives (outside any specific change order procedure), but may also include any act or omission (failure to act) that has the ultimate effect of changing the work from the work which was specified.

Some of the most common types of constructive changes include changes which arise from:

1. Defective specifications;
2. Alteration of the "as planned" method of performance;
3. Misinterpretation of specifications;
4. Overinspection;
5. Rejection of conforming work;
6. Rejection of "or equal" submissions;
7. Problems with owner-furnished equipment; and
8. Acceleration, by virtue of the owner's failure to properly adjust the contract time.

Cautions

In most constructive change situations, there is not any formal recognition by the owner that a contract change does in fact exist. Since a change order is not initiated, the contractor must be quick to recognize that the action or inaction on the part of the owner or his representative has resulted in a change to the work. Thereafter, the nature of the change must be identified and documented as early as possible to preserve the right to additional compensation and, if appropriate, extra time. Once the constructive change has been identified and the contractor's rights secured, the ultimate objective becomes the conversion of the constructive change into an owner-acknowledged change as described in Section 4.4.1.

Recommended Action

As soon as a constructive change is recognized to have occurred, thoroughly research the contract documents to confirm the existence of the change. Pay particular attention to any notice requirements so that the right to request additional compensation and time is preserved.

Upon such confirmation, immediately prepare and submit a complete change order proposal, in compliance with all contract requirements. Treat the submission with as much dignity as an owner-requested change so that the constructive change is viewed as a legitimate proposal. Since the burden of proof with respect to the constructive change is on the contractor, get the owner's conduct down in writing and included into the official project record. Try to make the additional work seem obvious and be sure to point out what aspect of the owner's conduct necessitated the additional work. If the owner rejects the constructive change proposal, the proposal and the project record will be the cornerstone in any future dispute resolution procedure involving the change. If the change order is completed properly, the contractor will have removed or minimized any risk arising from the notice and record keeping provisions in the contract.

4.4.3 Consequential Changes

Description

A consequential change is similar to a constructive change in that the change may not be readily apparent or recognized by the owner. A consequential change differs from a constructive change in that the consequential change involves additional work that becomes necessary or additional costs incurred as a direct result of some usually more obvious change. The change is the result of the domino relationship of cause and effect.

The cost components of a consequential change may include any or all of the components of any of the other change types:

1. Direct cost of additional or changed work.
2. Interference costs (disruption, rework, etc.).
3. Impact costs (delay, acceleration, etc.).

Difficulties

As with constructive changes, the difficulties facing the contractor are twofold. First, the contractor must recognize that a change has occurred early enough to comply with the notice requirements, and second, the contractor must convince the owner to convert the consequential change to an owner-acknowledged change. Once again, the contractor must try to build as much legitimacy into the proposal as possible.

5

Understanding How Change Orders Arise

5.1	Introduction	68
5.2	Defective Specifications	68
	5.2.1 Cut-and-Paste	68
	5.2.2 Silly Specifications	69
	5.2.3 Old and Outdated Specifications	70
	5.2.4 Inconsistencies	70
	5.2.5 Impossibilities	71
5.3	Nondisclosure	72
5.4	Lack of Coordination Among Design Disciplines	72
5.5	Incomplete Design	73
5.6	Latent Conditions (Defects)	73
5.7	Owner Changes	74
5.8	Improved Information	75
5.9	Improvements in Workmanship, Time, or Cost	75
5.10	Illegal Restrictions	76
5.11	Nonapplicable Boilerplate	77
5.12	"Intent" vs. "Included"	77

5.1 INTRODUCTION

Chapters 2 and 3 detailed what is *supposed* to happen throughout the performance of a construction contract. They were included as important background against which to compare what actually happens on the project. Chapter 4 identified the general categories of changes as an introduction to the concept.

In this chapter we will begin to deal with "how to" recognize changes by elaborating on the categories introduced in Chapter 4. The process will be completed in Part Three, where we discuss several specific problem areas.

5.2 DEFECTIVE SPECIFICATIONS

The term "defective specifications" seems at first to be almost too general to be of real use. It is, however, one that describes an area that is probably the most fertile ground for changes. In most instances, changes which result from defective specifications will present the least difficulty in securing approval and payment. The term "defective specifications" simply refers to documents containing flaws.

The most common reasons for defective specifications are:

1. Cut-and-paste preparation methods;
2. Silly specifications;
3. Old and outdated specifications;
4. Inconsistencies in specifications; and
5. Impossibilities.

5.2.1 Cut-and-Paste

Cut-and-paste describes the process by which most specifications are actually "written." Whether manually or electronically, most specifications begin with some general standard format. It might be either some existing specification that the architect used in the past on a similar project, or it may be a commercially prepared standard general format (such as the one produced by the Construction Specifications Institute). In either case, most specifications did not start out with a particular project in mind.

Cut-and-paste begins with specifications prepared for one or more other projects, which may have been the product of a prior cut-and-paste project, and continues until the specifications are deemed complete and suitable for use on the proposed project. If done manually, the process is literally to cut out those paragraphs and phrases that do not apply to the current contract, and paste in those additions that will supposedly add the new requirements that do apply. If done electronically, the process is the same, although neater. The problems that each technique will breed, however, are identical.

During this time-saving process, there remains a high risk that:

1. Provisions that do not apply will remain, and
2. Important terms that are necessary for a comprehensive description will be overlooked.

Two additional factors that increase the probability that the cut-and-paste method will cause errors in the documents are:

1. *The technical specifications are usually an architect's **secondary priority.*** Throughout the process of design development, specific thought is given to the precise materials desired and their configuration. The design process, bringing the initiated concept through detailed design, gives the architect the greatest satisfaction. After the design is complete, the technical specifications and the contracts themselves are assembled. Although important, many architects view this task as an unfortunate, tedious necessity because it does not involve their creative skills. The cut-and-paste technique is viewed as a cost-effective method for completing a tiresome task quickly. Unfortunately, these specifications, by incorporation into the construction contract, become part of an important legal document.

2. *The technical specifications and the front-end documents are **assembled at the eleventh hour.*** The design process itself usually continues until the last possible moment. It is only when the deadline of the bid solicitation date looms that serious attention is diverted to completing the specifications and construction contract. In many instances, the specifications are likely to be "written" at the last minute, between the "real" design work, and by junior architects or engineers who lack sufficient experience or familiarity with the proposed project.

It is bad enough that the documents may have originally been written for another project and made to "fit" this particular project. Add to this situation the possibility that the specifications were assembled in the final stages of the design process by someone other than the principal architect, and the likelihood of errors and omissions increases.

5.2.2 Silly Specifications

Silly specifications are related to cut-and-paste in that silly specifications originate in other areas and are dumped into the requirements for a particular project without proper forethought. Silly specifications can come from many sources, but the most common examples include:

1. *Product requirements lifted from manufacturers' brochures.* The detailed specifications for a preferred item will be meticulously listed word for word, exactly as they appear in that manufacturer's brochure. These brochures often contain sample specification sections designed to make it easy for an architect to lift and include the product description and requirements in the project specifications. The difficulty arises when several other manufacturers are listed as acceptable, and the words "or equal" are included. The "equal" manufacturers will have different specifications. If the differences are slight, they are ignored. If the differences are significant, an "or equal" product may not be suitable and the contractor may be entitled to a change order if forced to provide the "specified" item at an increase in cost.

2. *Specifications written by manufacturers' representatives.* Because many architects view the preparation of technical specifications as unpleasant, architects are usually receptive to a representative's offer to prepare the product specifications. The product representative may do an excellent job of specification preparation but for one consideration: The specification section is prepared in a vacuum and without considering the project as a whole. Similarly, the manufacturer's representative usually lacks familiarity with the design development process experienced for *that* job.

3. *Specifications written without a specific product in mind.* On occasion, a specification will be written without a specific product in mind. Desirable qualities or requirements will be listed without ever confirming if the requested qualities are available in a single product. Although this problem may be discovered during the bidding phase, in many instances the problem is not discovered until the submittal or procurement phase. Frequently, some qualities will be found in one product and some qualities will be found in others. If this ambiguity results in an increase in cost, the contractor must identify the change and be prepared to submit a change order proposal.

5.2.3 Old and Outdated Specifications

Old and outdated specifications are usually found on public and private projects which have been "shelved" for two, three or more years due to an unfavorable political or economic climate or a lack of funding. In the meantime, products change. Specified items become unavailable and superior and/or more economical products become available.

Changes accommodating the new products are easily justified and quickly approved. If pushed through too quickly, however, important coordination issues involving other parts of the work (to be constructed at a much later date) might be overlooked. If *all* these effects are not completely anticipated by the contractor at the time of original change order approval, the contractor will have an uphill battle trying to reopen the change order at a later date. In Chapter 8, we address the issue of "reservations" and "exclusions" in change order requests.

Finally, a project's time on the shelf increases the possibility that priorities will change. The agency controlling the project may have a different chief and the people who designed the job may be working for somebody else. The market conditions which formed the basis for the project concept may also have changed. The net effect of the foregoing is that the original project concept will become watered down. Specifically, the reasons for certain design considerations will be lost, accountability for design success will become diffused, and the motivation for decisive action which is necessary to keep a project on schedule may be nonexistent.

5.2.4 Inconsistencies

As mentioned previously, if a specification is subject to more than one reasonable interpretation, the contractor has the right to choose the interpretation. Such inconsistencies may take the form of any of the following:

1. Discrepancies between requirements of the plans and the specifications;
2. Differences between small and large details;
3. Differences between planned and finished schedule requirements;
4. Differences in plan dimensions and equipment locations among design disciplines; and
5. Differences between the actual equipment costs and those details originally shown in the contract documents.

Many construction contracts will have some mechanism (a set of rules) to determine precedence among conflicting details. If such a mechanism does exist, it most often is nothing more than a restatement of the rules of common sense. A sample precedence clause is set forth below:

> **Precedence of Contract Documents.** If there is a conflict between Contract Documents, the document highest in precedence shall control. The precedence shall be:
> 1. Permits from other agencies as may be required by law;
> 2. Special Provisions;
> 3. Plans;
> 4. Standard Plans;
> 5. Standard Specifications;
> 6. Reference Specifications.
>
> Change Orders, Supplemental Agreements, and approved revisions to Plans and Specifications will take precedence over Items 2) through 6) above. Detailed plans shall have precedence over general plans.[1]

5.2.5 Impossibilities

Every contractor, at one time or another, has restated the phrase "It's easy to draw lines on a sheet of paper." A designer may be blindly following engineering rules, fitting together components without consideration for following construction activities or coordination with adjacent work.

Some examples of difficulties which result from poor planning or coordination include:

1. Foundation wall configurations that don't allow removal of forms.
2. Anchors that are to be installed in inaccessible areas.
3. Available access that will not accommodate specified equipment (it won't fit through the door...).
4. Illogical construction sequences.
5. Insufficient physical space.
6. Equipment that cannot be accessed (after installation) for maintenance.

---

[1] *Green Book*, Section 2-5.2.

5.3 NONDISCLOSURE

The failure to inform a contractor of design or construction information that is significant to the completion of the project is know as nondisclosure. If intentional, nondisclosure is an unethical tactic. The tactic is employed by unscrupulous owners out of fear that full disclosure would limit the number of bidders or result in an increase in price. If unintentional, nondisclosure can be the result of a deficiency in the design effort or simply the failure of an owner to understand the significance of the information to the construction process. In either case, the contractor had been denied the opportunity to consider the information at the time of bid and, as such, the contractor should be entitled to additional compensation as a result of the effects of the owner's nondisclosure. Both intentional nondisclosure (concealment) and unintentional nondisclosure (negligent) may be used as the basis for a separate legal cause of action against the owner if the effects arising from the owner's acts or omissions are not resolved through the change order process.

Some examples of the kind of information that would cause an unanticipated hardship on the contractor, if withheld from the contractor or not disclosed during bidding, include:

1. A seasonal watercourse that is not apparent in a prebid site investigation;
2. Changes in the funding climate that will impact the timing or amount of contract payments;
3. The presence of rock in the way of excavation;
4. The presence of material with unsuitable bearing capacity (such as silt or an old disposal area) at the site; and
5. An unusual or restrictive procedure for site access.

5.4 LACK OF COORDINATION AMONG DESIGN DISCIPLINES

The architect orchestrates the assembly of the individual designs necessary for a complete facility. In the final composite, the walls envelop the steel columns, the beams fit on the masonry, and the roof pitches toward the roof drains. The ductwork, heat piping, water piping, insulation, drain and vent piping, light fixtures, and electrical conduit all fit neatly above the ceiling, without having to cut holes in structural beams. The underground piping layout shows the sink drains in the same location that the architectural plans show the sink. When the toilet rooms are physically laid out to the dimensions given on the plans, the water closet does not fall exactly where the column is, and the wall does not wind up in the middle of the window.

If everything went along like this, there would be no problems. Unfortunately, projects are becoming increasingly complex, design budgets are often tight and review times are short. Difficulties also arise because the separate plans and details are developed independently by individual consultants. For these reasons, among others, designs may be completed without full consideration for the other disciplines or they are completed with too many assumptions.

Some examples of improper or inadequate coordination include:

1. The structural engineer assumes that the architect will provide the dimensions;
2. The architect leaves an apparent conflict in the plans with the hope that it will get "worked out" in the field, or through shop drawings;
3. The duct drawings are done without regard for steel beam locations; or
4. The structural engineer is not advised that the air conditioning units are being mounted on the roof.

Problems of this type arise because the job of coordinating the respective designs just was not done properly. Either complete information was not given to the consultants, or the individual designs were simply thrown together as a package without the benefit of some kind of conflict identification or design coordination effort.

5.5 INCOMPLETE DESIGN

Incomplete design is a term that describes the failure of a designer to specify adequately all project components to the level necessary to allow the contractor to proceed with the work. This type of deficiency can appear anywhere in the contract documents and can include simply neglecting to adequately specify certain equipment or failing to verify that supplementary information is in fact being provided by someone else on the design team. The net effect is that information is left out entirely.

Notations like "See Specs" instead of "See Section 05500," or "See Structural Plans" instead of "See 4/S3" are clear indicators that an *assumption* was made that somebody provided the relevant information. They indicate that somebody did not take the time do his or her homework and verify that the supplementary information was included at a particular location in the plans or specifications. The assumed information is just as likely to be missing altogether.

5.6 LATENT CONDITIONS (DEFECTS)

Latent conditions are those conditions that are unknown to everyone (the owner, the designer and the contractor) at bid time. If the proper relevant information is unobtainable by reasonable means, the presence of latent conditions may be nobody's fault. Inasmuch as latent conditions are unknown, their impacts are not included in the contractor's bid and the owner's approval of the resulting change order should be straightforward.

Please be aware that most contracts, and some statutes, have specific procedures if a latent condition is uncovered. For example, differing site conditions on public works projects in California are now governed by California Public Contract Code Section 7104. This section provides in pertinent part that:

> Any public works contract of a local public entity which involves digging trenches or other excavations that extend deeper than four feet below the surface shall contain a clause which provides ...
> (a) That the contractor shall promptly ... notify the public entity, in writing, of any:
> (1) Material that the contractor believes may be material that is hazardous waste ...
> (2) Subsurface or latent physical conditions at the site differing from those indicated.
> (3) Unknown physical conditions at the site of any unusual nature ...
> (b) That the public entity shall promptly investigate the conditions, and if it finds that the conditions do materially so differ, or do involve hazardous waste, and cause a decrease or increase in the contractor's cost of, or the time required for, performance of any part of the work, shall issue a change order ...

Latent conditions are commonly classified into the following two broad categories:

1. Type I: Subsurface or latent conditions at the site differing materially from those indicated in the contract documents; or
2. Type II: Unknown physical conditions at the site, of an unusual nature, differing materially from those ordinarily encountered and generally recognized as inherent in the work described in the contract documents.

Some common examples of latent conditions include:

1. *Subsurface site conditions* that were undetected by testing or research, such as unanticipated or unusual soil composition, high water table or natural springs, undisclosed rock, abandoned disposal area or mislocated or unrecorded utility.
2. *Hidden conditions in an existing facility,* which is the subject of an alteration project and which could not have been disclosed prior to demolition. Contractors engaged to modernize an existing facility often encounter structurally unsound materials, deficiencies in existing equipment or building configurations that are different from the conditions shown on the existing plan or as-built drawings.

5.7 OWNER CHANGES

It is the owner's project and, assuming the proper contract language (refer to Section 3.4.2), the owner is within its rights to order changes in the work. However, the owner's right to order changes in the work is limited by the following two principles:

1. Changes must be within the general scope of the contract; and
2. The owner must be prepared to adjust the contract price and/or time on account of the change.

Subject to these two broad limitations, the owner may:

1. Add more space to a project;
2. Make a project more luxurious;
3. Require the installation of better equipment.

Owner-initiated changes may also have a negative impact on the project. For instance, changes may alter the owner's ability to pay for the work and may lead to a reduction in the scope of work.

5.8 IMPROVED INFORMATION

During the course of some projects, new, improved or additional information becomes available to the owner, designer and/or contractor. From the contractor's standpoint, the contractor may want to give the owner and designer the benefit of the doubt when they say the information was not available when the bid documents were prepared. The availability of this new and improved information may be the result of a variety of different factors, including:

1. Previously unknown conditions may become evident after the commencement of construction or demolition;
2. New or better methods to achieve a desired function may be developed; or
3. Specific information is developed which corrects previous assumptions and allowances.

When initiated by the owner or the designer, each of these examples should result in a contract change which should be approved easily. In the event the contractor proposes a new method to attain a desired function, the contractor will be required to explain the benefits to the owner (i.e., longer life expectancy, reduced maintenance or operational costs) before the contract change will be approved.

The contractor should be aware that the guise of "improved information" may be used by a designer to cover up what is really an incomplete design or defective specification. This characterization by the designer may not be necessarily bad for the contractor in that it allows the designer to maintain his or her self-respect and to justify the change to the owner with a minimum of criticism. A shrewd contractor will spot such an opportunity to remove the friction from the change order process. This will enable the contractor to get paid for the extra work while keeping the peace.

5.9 IMPROVEMENTS IN WORKMANSHIP, TIME, OR COST

Changes of this type will normally be initiated by the contractor. As such, if they result in additional costs, it will be extremely difficult to get the owner to accept even a portion of the responsibility.

One way to overcome this difficulty may be to tie the proposed change to one or more of the other categories described in this section. If this transition can be achieved, it will become much easier to convince the other party of his or her responsibility for payment.

Those cases that will be accepted and paid for by the owner include:

1. Cost reductions achieving the same or similar function and/or appearance;
2. Improvements in time (acceleration) where the current delays are the result of the owner's actions or failure to act;
3. Acceleration where the owner now desires an earlier completion date (the competition is opening across the street); or
4. Significant improvements in appearance, workmanship, and quality that will add value.

5.10 ILLEGAL RESTRICTIONS

This category involves the inclusion of requirements or the deletion of some option contrary to a procedure outlined by law. This inappropriate measure requires the contractor to complete a portion of the work more expensively than otherwise allowed.

The most common example of this type of interference is the use of a proprietary specification on a public project. Such a specification is one that is restricted to unreasonably narrow choices (usually a single source of supply). On many public projects, the use of proprietary specifications is prohibited except in very limited circumstances (i.e., match existing fixtures). Further, certain statutes provide that several sources of supply be permitted and that the contractor be afforded the flexibility to propose additional items as equal to those listed. If a single product is listed, it may be an indication that the specification has been unlawfully restricted.

Additional examples may be failure to comply with:

1. Zoning regulations;
2. Property line restrictions;
3. Building code requirements;
4. City ordinances;
5. Military regulations; or
6. Owner's company policy.

If these kinds of restrictions are present, they must be exposed and resolved as early as possible. Prompt action is required not only for the potential change order impact, but because if left outstanding, they may ultimately interfere with and delay project close-out. That would lead to a different kind of problem.

5.11 NONAPPLICABLE BOILERPLATE

Nonapplicable boilerplate generally is a result of a fast cut-and-paste (see Section 5.2.1). It is contract language that has been included in the contract documents, but that just does not apply to the work at hand. Material references may be inappropriate, inconsistencies may be included, and conflicts may be developed. The boilerplate is excess baggage, causing inaccuracies and misunderstandings.

In many cases, an architect or owner may attempt to rely on such inappropriate language in a scramble to justify a position. Thus, it is important to expose these errors when they are discovered to try to get the inappropriate language stricken *before* problems develop. This will take the wind out of the sails of those who might otherwise try to use the boilerplate term at a later date to strain some uncomfortable interpretation.

5.12 "INTENT" VS. "INCLUDED"

Supplementary General Conditions often include a project description. Contained in this description is likely to be a remark to the effect that it is the "intention" to have a "complete" project. In addition, the separate technical specification sections may have their own clauses stating a similar objective. Statements like, "It is the intent to have a complete and operating system in every respect..." permeate the documents.

There are certain statements of intent that are specific and completely legitimate. The notation in an electrical specification, for example, that the

> ... electrical drawings are symbolic and diagrammatic in nature and are intended to show only the general scheme, equipment involved, and ... the approximate locations of outlets and equipment.

is consistent with trade custom and usage. Most important, however, it is clear as to the *manner* in which the work is represented.

Statements that purport to place a design responsibility on the contractor are another story. It may be the "intent to have a complete and operating system," and all persons involved with the project can appreciate the desire and need for a complete system. However, the contractor cannot afford to allow the owner or designer to stretch this kind of statement to cover up design flaws. Specifically, the "intent" provision should not be interpreted as "The contractor is required to supply all the components of a complete system, whether or not they have been properly called for, sized, or coordinated."

Do not let this become the interpretation of the "intent" term. If an item is now shown or specified as expected, there may be a basis for extra review. Research the documents thoroughly to confirm that it has not been specifically covered elsewhere; then follow the recommendations for change order submission detailed in Part Four.

6

Using the Change Order Process to Your Maximum Advantage

6.1	The Six P's of Change Orders	79
6.2	Prospecting for Change Orders (Discovery)	79
6.3	Preparing the Change Order	80
	6.3.1 Establishing the Change Order File	80
	6.3.2 Change Order Research	81
	6.3.3 Change Order Research CHECKLIST	82
	6.3.4 Notification	86
	6.3.5 Sample Notification LETTER TO THE OWNER on Changes	87
	6.3.6 Notice Components	89
6.4	Pricing the Change Order	89
	6.4.1 Now or Later	89
	6.4.2 Pricing Methodology	92
	6.4.3 Selecting the Proper Tone	92
6.5	Presenting the Change Order	92
	6.5.1 Proposal Submission	92
6.6	Performing the Work	93
	6.6.1 Tracking Project Effects	93
6.7	Change Order Payment	95
	6.7.1 Billing and Payment	95
	6.7.2 Claims and Disputes	96

6.1 THE SIX P'S OF CHANGE ORDERS

Despite appearances, all change orders lend themselves to nearly the same analysis, irrespective of the type or the size of the project, and the change order life cycle will progress through the same steps in virtually every case. Depending upon the specific change, the individual divisions in the process may be more or less clearly defined and the relative importance of each part of the process will vary depending upon the type or value of the change (refer back to Section 4.4).

For these reasons, the contractor's objectives with respect to the change order process may be boiled down to the following:

1. Identify the steps in the process;
2. Assess relative values to their importance;
3. Proceed methodically through each step; and
4. Monitor performance in time to avoid unacceptable results.

The Six P's describe this process. They are:

1. Prospecting (discovery)
2. Preparing
3. Pricing
4. Presenting
5. Performing
6. Payment (getting paid for)

6.2 PROSPECTING FOR CHANGE ORDERS (DISCOVERY)

The first step in the change order process is the clear identification of items of work that are beyond the scope of the original contract. The origins and the timing of their discovery will ultimately determine the net impact on the contract work, as well as the ease with which the requested change will be approved and paid.

It is easy to "discover" additional work when suddenly faced with it and interruption is imminent (or has already occurred). Immediate resolution becomes critical if the most dramatic impact on both the contractor and the owner is to be avoided. Last-minute efforts create tension that minimizes options and clouds thinking. Interestingly, even though this scenario does not seem to make much sense when we sit back and coolly analyze the process, it is precisely this pattern that is allowed to happen over again by most contractors.

The real talent in prospecting for changes lies in the ability to expose extra cost items in sufficient time to allow their incorporation into the work in stride. If the contractor can learn to take the initiative to actively search and discover potential extra cost items before they affect construction activities, there will be time enough to work the process through in a less hurried manner. The contractor will have the luxury of being able to contemplate the best strategies and to consider those tactics that will get the change order approved with a minimum of friction. It is the old business of making it happen instead of letting it happen.

There may be instances in which a last-minute presentation to the owner will be *calculated* to force a quick decision (see "Quick Deals" and "Acceptance Time" in Chapter 12 for explanations of this tactic). However, this kind of action should not be employed unless the contractor has had enough time to analyze the requested change and evaluate its impact on the project. The time constraints imposed by the project, coupled with this type of approval request, may result in a quick and presumably less thoughtful response to the request. This approach is often employed in legislature, for example, where literally hundreds of bills are introduced in the last days of a session. Quick action becomes forced without the opportunity for extensive research or debate. Tactically, it may be an advantageous approach if the contractor has had the time to prepare and analyze the change and the owner and designer have not.

Chapter 5 outlined, among other things, changes originated by the owner and the architect. Although these types of easily identifiable, easily paid for changes are relatively few, the first step is still up to the contractor. The documents must be systematically researched to discover where the changes lie.

Prospecting for changes is such an important part of effective change order management that an entire part (Part Three) of the book is devoted exclusively to defining the specific areas in which to search. The search involves:

1. Breaking the project into its component parts;
2. Reviewing the design and approval processes to determine if they were properly or improperly completed;
3. Dissecting the contract itself to expose potential conflicts; and
4. Researching the documents to uncover the errors and omissions.

6.3 PREPARING THE CHANGE ORDER

6.3.1 Establishing the Change Order File

Immediately upon the discovery of a potential change, it is important to record the event. The full-size forms and their specific uses are included in Part Five, Change Order and File Administration. Here it is important to understand that a change order file can always be voided if it is later determined that the item is already part of the contract. Documentation of each potential change must be started as early as possible. The cost associated with some extra paperwork is a small price to pay to eliminate the possibility of overlooking important facts in a fourth-quarter backward analysis. It is far worse to allow a compensable change to get shuffled back into the project and lost until the project is about to be closed out. An easy, systematic documentation procedure will prevent these kinds of costly oversights. Such a procedure is detailed in Chapter 9.

A separate general file is established to track all *potential* project changes through their eventual resolution. (Again, "potential" is stressed to be sure that even questionable issues continue to receive attention either until they are confirmed to be contractual or until they get paid.) Within the general file are individual files for each potential change. These individual files will include all important documentation, including:

1. Date of discovery;
2. All circumstances of discovery;
3. Company/person responsible for first identification;
4. All written documentation (reports, invoices, letters, meeting minutes, schedules, memos, notes, etc.);
5. Photographs; and
6. Plans, sketches, surveys, specifications, diagrams, and so on.

Finally, the general file will include a separate section for some kind of change order log, such as the Change Order Summary Sheet in Chapter 11. This is an important summary record to maintain a clear overall visibility of the current status of all changes. A good log form will also facilitate fast analysis of apparent trends in the timing and relative success of the change order administration.

6.3.2 Change Order Research

All change order research begins with a clear understanding of the issue. After the issue is identified, the contractor should begin a general fact finding effort concerning the potential change, which effort should proceed methodically through the discovery of the specific facts which support or refute the change.

The potential for friction exists in every change. This is because implicit in the identification of a potential change is an observation that some error or oversight exists. When identifying or preparing a change, it is wise to double-check the circumstances surrounding the change and the research supporting the requested change. Read and reread the documents to confirm or correct original impressions. This is particularly true with the contractor's initial change order request on a project. The extra precaution will first establish and then maintain credibility in the introduction of future change orders. Remember the Six P's of project administration: "Proper project planning prevents poor performance."

6.3.3 Change Order Research Checklist

Good habits can be developed over time that will help pattern thoughts into an almost routine research procedure. Even so, it is likely that even basic considerations can still be overlooked. This is because:

1. Changes of a given type usually do not occur very close to each other;
2. Heightened activity at a construction site may steal time from those responsible for change order management; and
3. The way that one change order was handled successfully in the past may no longer apply because of differences in people, contract provisions, value, timing or the disposition of the owner or the architect.

The use of a tailored checklist will provide a mechanism by which all possible considerations can be evaluated without the risk of an oversight. The applicability of each item can be assessed with clear thinking, and the process can be completed in a minimum of time.

Having completed the Change Order Research Checklist, all important considerations will have been given direct attention. What does not apply will have been discarded by design, and not simply overlooked. What does apply will have been given the recognition that it deserves.

More important than all this, however, is that the Change Order Research Checklist becomes an exercise in a thought *process*. Rather than pretending to lend itself to some algebraic conclusion that will neatly add up to an approved change, the checklist forces a thorough contemplation of all aspects of a change. Having considered and answered each inquiry on the Checklist will consistently place the contractor in a position of superior knowledge, and as Sir Francis Bacon said, knowledge is power.

Finally, once the process has been completed, the real issues surrounding the change will become clear. The transition to Chapter 12, Winning in Change Order Negotiation, will then become smooth.

CHANGE ORDER RESEARCH CHECKLIST

	YES	NO

A. CONTRACT
1. Form of contract: (a) Adhesion (see Sections 3.3.3 and 3.3.4). _____ _____
    (b) Negotiated _____ _____
2. Is the subcontract scope of work:
    (a) Owner-defined? _____ _____
    (b) Contractor-defined? _____ _____
3. Does contract language exist defining the situation? _____ _____
4. If A.3. is yes, can it be opposed with contract law (see Chapter 3)? _____ _____
5. Can trade practice be used by you or the owner to stretch an interpretation (see Section 3.3.6)? _____ _____
6. Are rules of precedence clearly stated in the General Conditions (see Section 3.3)? _____ _____

B. SPECIFICATIONS
1. Does a specification section exist? _____ _____
2. If B.1 is yes:
    (a) Is it complete? _____ _____
    (b) Is it subject to more than one reasonable interpretation? _____ _____
    (c) Can it be deemed to fall into any "per plans and specs" trade or subcontract? _____ _____
    (d) Can the work be done precisely as specified? _____ _____
3. If B.1. is no, does General Conditions boilerplate exist that:
    (a) Clearly describes the work? _____ _____
    (b) Might be used by your opponent to strain an interpretation? _____ _____
4. Does more than one specification exist? _____ _____
5. If B.4 is yes,
    (a) Are they included in different sections? _____ _____
    (b) Do conflicts exist between the requirements of each? _____ _____
    (c) Does one make any more sense that the other? _____ _____
6. Are there references to industry standard specs? _____ _____
7. If B.6 is yes:
    (a) Do they contain precise material descriptions (as opposed simply to design criteria)? _____ _____
    (b) Do they conflict in any way with other stated (specified) requirements? _____ _____

C. PLANS
1. Do any specific notes exist related to work? _____ _____

2. Do general notes exist that might be used to strain an interpretation (boilerplate)?
3. If C.2 or C.3 is yes, do they conflict with the requirements included in the respective specification section?
4. Are specific details included?
5. If C.4. is yes:
   (a) Have you checked *all* cuts and references to details, elevations, plans, specs, etc.?
   (b) Is all relevant information included?
   (c) Do conflicts exist between any items of C.5.a?
   (d) Are they subject to more than one reasonable interpretation?
6. Have shop drawings been approved that differ from specified requirements?
7. If C.6. is yes:
   (a) Were the differences clearly highlighted and understood at the time of approval?
   (b) Are the differences subject to more than one reasonable interpretation?
   (c) Are errors or omissions evident?
   (d) Have they been properly coordinated with all other parts of the work?
   (e) Have all "by others" and "not by _____" notes been addressed?
8. Has the job been fast-tracked?
9. If C.8 is yes, do the dates on the current documents match those originally included in the contract?
10. If C.8 is no, are there significant differences?

D. SITE
1. Should the work have been apparent in a prebid site investigation?
2. Are the changed conditions the result of owner or architect nondisclosure?
3. Are the conditions the result of some previously undetectable latent (hidden) condition(s)?
4. Is the site information given in the plans:
   (a) Accurate?
   (b) Complete?
5. Are any site conditions different now from those at the time of bid?

E. SUPPORTING DOCUMENTS
1. Is the changed work included in the Schedule of Values?
2. If E.1 is yes, did whoever prepared the schedule have a justification for it?

3. If E.2 is yes, can it be credibly dismissed in any way? _____ _____
4. Are there any prior discussions, meeting minutes, letters, quotes, etc., that confirmed that extra cost will or will not be applicable? _____ _____

F. ADMINISTRATION

1. Date of discovery ___/___/___
2. Company and person responsible for first identification _____
3. Persons notified: _____ Date: ___/___/___
    _____ Date: ___/___/___
    _____ Date: ___/___/___
4. Have all circumstances of discovery been recorded? _____ _____
5. Has all relevant documentation been assembled?
    (a) Field reports _____ _____
    (b) Letters and transmittals _____ _____
    (c) Telephone logs and miscellaneous notes _____ _____
    (d) Material invoices and payroll records _____ _____
    (e) Meeting minutes _____ _____
6. Are before-during-after photographs necessary? If so, have they been arranged? _____ _____
7. Have all relevant plans, sketches, surveys, diagrams, etc., been assembled? _____ _____
8. Has the schedule impact been analyzed? _____ _____
    (a) Has interference occurred? Date: ___/___/___
    (b) If not, what is the date of anticipated interference? Date: ___/___/___
    (c) How many extra days can be assigned to the change? _____
9. What is the date that change order approval is required before the schedule is affected? Date: ___/___/___
10. Have all trade contractors and suppliers even remotely affected been advised of the change? _____ _____
11. If E.10 is yes, have all cost changes (adds and deducts) been assembled? _____ _____
12. If E.11 is yes:
    (a) Does each positively indicate that the schedule is or is not affected (and by how much)? _____ _____
    (b) Is each properly broken down to allow meaningful evaluation? _____ _____
    (c) Does each have all substantiating documents attached? _____ _____

6.3.4 Notification

Notice Requirements

Although change order proposals are not disputes until they are refused by the owner, it is wise to make every effort to meet all notification requirements contained in any Dispute Clause (see Section 3.4.4). In this way, the contractor's position will be maintained and the owner will not gain a technical advantage over the contractor as a result of the contractor's failure to comply with the notice requirements in the contract.

Dispute Clauses normally indicate a specific time requirement for notification after the discovery of a change giving rise to increased time and/or money. It is common for some owners to argue that the contractor has not satisfied the notice requirement and that the contractor's inaction has prejudiced the owner. Invariably, the owner will begin the time calculation from the earliest date that the change may have become evident, in an attempt to limit the contractor's rights. The contractor's calculation, however, will begin from the date that it became known that the change will affect time and/or money. It is at *that* point that the potential change will no longer be considered a job condition.

Hardball/Softball

Notification can be a funny thing. If you strictly follow some attorneys' advice, the language in the notification letter will convey that the contractor's primary objective is to cut the owner's legs off at the knees. The inclusion of all blunt notifications and reservations of rights to every conceivable claim in the first owner communication will not make any friends and may create an adversarial relationship with the owner for the duration of the project. Such a one-sided communication may eliminate any type of informal communication and prevent an important level of trust.

On the other extreme, a contractor paralyzed by the fear of offending the owner may fail to issue the proper notification and may maintain good relations right into the red column.

The talent is in the ability to formulate effective communications that are courteous, but that also clearly protect the contractor's issues. If done properly, the contractor will be respected for his or her professionalism. If the courteous and professional communications do not produce the desired result (or elicit a proper response), the contractor will be able legitimately to take blunt, decisive action. It will be only at that point that the contractor might still manage to be admired as a fighter, and not be criticized as an opportunist.

If this kind of hardball tactic is begun too early in the game, and without any prior justification (without having been abused at least once by the owner), making dramatic statements may cause the contractor to appear more frivolous than serious. This will be a difficult impression to change if it is the first impression made by a contractor on a project.

6.3.5 Sample Notification Letter to the Owner on Changes

Technically, there can never be too much notification. Depending upon the existing contractor–owner relationship, as well as the predispositions of each, you may consider an early tactical maneuver of one-upmanship, the Sample Notification Letter to the Owner on Changes.

The purpose of this letter is to set the tone and pace of the job's change order process. It can be one of the first communications to the owner, and as discussed, its timing is a matter of business judgment. It advises the owner (and the architect who will be on the letter's distribution list) that:

1. The job's time and cost were figured based upon the information in the contract documents;
2. Changes will disrupt the orderly "as planned" sequence of the project;
3. The owner is expected to make every effort to resolve problems quickly; and
4. The contractor intends to be compensated for all effects of any change.

The letter is designed to help the owner recognize that changes are a normal part of the construction process and that fair treatment and equitable adjustment will be expected in all cases. As a tactical matter, it is probably best to wait until there have been at least a few minor errors or discrepancies discovered in the documents before the letter is sent. If done in this way, it is more likely that the contractor will appear to have been prompted to action by having to experience the effects of a deficiency in the contract documents. This will make it more difficult to fault the contractor.

SAMPLE NOTIFICATION LETTER TO THE OWNER ON CHANGES

LETTERHEAD

(Date)

To: (Approving authority)　　　　　RE: (Project no.)
　　　　　　　　　　　　　　　　　　　(Project title)

　　　　　　　　　　　　　　　　　SUBJ: Procedure to minimize change order disruptions

Mr./Ms. (　　　　　　　):

　As you know, multiple changes always disrupt orderly work performance, causing loss of productivity and increased costs.

　Inasmuch as there have been (insert appropriate number) changes to date, we urge you and the architect to review the documents for additional errors, omissions, impossibilities, and previously unforeseen needs now. We ask this so that the required changes may be resolved and executed in time for the work to be incorporated with a minimum of interference and associated costs.

　If a fair and reasonable price and time cannot be negotiated prior to performance in some situations, your authorization will be expected to allow us to proceed on a time and material basis. This will be in accordance with the provisions of (insert the appropriate description of the Change Clause included in the specific contract).

　Thank you for your prompt response.

Very truly yours,

Project Manager

cc: Architect
　　Owner's Field Representatives

6.3.6 Notice Components

The general notice of Section 6.3.5 may have set the stage, but in all likelihood will not cause anyone to dig frantically into the contract documents to comply immediately with your request. The primary objective of the letter is to document the fact that the plans and specifications have their flaws and that the owner and/or the designer will be expected to resolve these matters promptly. The letter represents the first of many nails in the coffin. It will become part of the project record and may be used in a later negotiation or dispute to support the statement: "I warned you…."

After the general notification of Section 6.3.5, each respective change proposal will incorporate its own individual notice components. Part Four includes the actual language recommended for various situations. To complete the treatment of the subject here, it is important to understand that each individual change order proposal will include some language that notifies the owner that:

1. A change has occurred or will occur that is above and beyond the original scope of work.
2. The change will cost extra money and require additional time.
3. The cost will be:
    a. Submitted now, with all accommodations for the unknown.
    b. Submited now, but indirect and consequential charges will be submitted when all costs are known (after performance).
    c. Computed on a time and material basis.
4. All rights are reserved to claim additional costs resulting from unanticipated work, unforeseen effects, and job delays. These costs will be submitted when known (presumably near the end of the project).

6.4 PRICING THE CHANGE ORDER

Change order pricing begins with the determination of the category (see Section 4.4). As with any negotiation, the overriding strategy will be initially to present the maximum position that can be justified.

6.4.1 Now or Later

In keeping with the maximum position objective, the contractor must first determine whether the change must be approved and signed before any work is performed, or if it is more advantageous to start the work pending the change finalization. This can be an easy or difficult decision, depending upon a variety of factors. Some of the factors which must be considered include:

1. Total cost of the change;
2. Level of trust between the parties;
3. Limits of authority of the parties who do trust each other;
4. Available time before schedule impact;
5. Cost and complexity of schedule impact;

6. Ease or difficulty in demonstrating responsibility for schedule impact;
7. History of owner action on past change proposals; and
8. History of owner action relative to past promises and commitments.

If it is decided that it is important to finalize the change approval before the start of any work, it is nearly certain that the decision will be technically supported by the owner's Change Clause in the contract (see Section 3.4.2). There it is likely to specifically state that:

> a. A construction change is a written order prepared by the architect and signed by the owner and architect directing a change in the work prior to agreement on adjustment, if any, in the Contract Sum or Contract Time, or both. The Owner may by Constructive Change Directive, without invalidating the contract, order changes in the Work within the general scope of the Contract consisting of additions, deletions, or other revisions, the Contract Sum and the Contract Time being adjusted accordingly.

These clauses support the contractor's contention that it is the owner's *contract* that is preventing the commencement of the changed or extra work without a signed change order. The contractor can go on to explain that the clause was placed in the contract for the owner's benefit, to guarantee that the owner will receive exactly what is contemplated by the finalized change, without any surprises.

There are generally three circumstances in which performing the work before the change is finalized may be advantageous. The first case involves relatively small change orders that carry with them the possibility of having a disproportionate impact on construction sequence. Such small changes have a way of:

1. Stopping previously confirmed deliveries;
2. Stopping or otherwise interfering with orderly (productive) work sequences; and
3. Delaying planned production until new material is delivered and installed.

If left alone to run out through the contractual procedure, the interference and delay portions of the change proposal will dwarf the actual cost portion. The total proposal will then appear so aggressive that cost components are nearly certain to wind up in a court battle.

Under this circumstance, it may be smart to make minimizing the actual project impact the greater priority, rather than sitting tight through bureaucratic delays.

The second case is one in which the contractor makes the judgment that doing the work under the circumstances and *then* negotiating is the best and/or fastest way to secure approval and payment. However, this strategy can be a risky proposition even if the contractor is entirely correct and justified. In this situation, there will be little or no motivation for the owner to expedite the approval and payment of the change if the work is in and the potential job impact is gone. The contractor has lost any leverage arising from the critical nature of the work.

This kind of action is often considered when the owner either is not available or persistently does not respond in time to preserve some job objective. If it is small and reasonable, performing the work should not present much of a problem. If it is large and/or complex, the decision may only be justified if:

1. Not starting the work will cause a marginally effective owner to procrastinate until it becomes serious; or
2. Waiting for approval and payment is not a problem; or
3. The contractor is confident in its legal position, and is willing to spend the time, effort, and money in arbitration or litigation to get it back.

The third case involves work that is unusual, broken up, or for whatever reason cannot be priced easily or quickly. In this situation the choices are:

1. Assume excessive risk and submit what is believed to be a "reasonable" proposal.
2. Provide "contingencies" for the unknown in the proposal (and hope they were large enough).
3. Proceed with the work on a time and material basis.

The first option may allow the contractor to submit a proposal that might "look" right. However, presenting a proposal that simply "looks" right should not be the contractor's foremost consideration. If this option is selected, the change may required an unusually large amount of management attention to be sure that costs do not go out of sight.

The second option might work if the owner will accept a price with an allowance for the "contingencies." Unfortunately, it is unlikely that the owner could ever see the change and all its effects from the same perspective as the contractor, and all the contingencies included to accommodate the absolute worst-case scenario will run the total direct cost portion of the proposal into low orbit. This can lead to long and exhausting negotiations or to flat rejection of the proposal, leaving the contractor back at the starting block. Being placed in this position, however, may be the blessing in disguise. After having demonstrated through the negotiations the concerns giving rise to the need for contingencies, the contractor will appear very reasonable to suggest the third (and least risky) option, "time and material" (or T & M).

A time and material (or cost-plus) option is usually provided for very clearly in the contract's change clause. AIA Document A201—1976, Article 12.1.4, provides for the cost-plus alternative, "if none of the methods set forth...is agreed upon." Most other Change Clauses will contain something similar, outlining T & M or cost-plus as the final method to determine costs—only after all more definite cost-establishing methods fail. For the contractor, T & M minimizes the risk. The owner, however, perceives the method (and sometimes rightfully so) as a blank check.

6.4.2 Pricing Methodology

As mentioned above, many contracts include a methodology for pricing change orders. For instance, the *Green Book* provides that changes will be priced as follows:

(i) if the change involves an item of work covered by a unit price, the adjustment will be based on the change in quantity and the unit price;
(ii) if the changed work is not covered by a unit price, then the adjustment shall be determined by the owner and the contractor; and
(iii) if the owner and contractor cannot agree on the adjustment to the contract, the owner will pay the contractor based on the cost of the changed work. In this regard the contractor is required to maintain meticulous records and submit certain records to the owner on a daily basis.

In the event the contractor cannot price the work easily, the best approach is to proceed on a time and materials basis and to maintain the requisite records. Be aware, the failure to maintain the proper records may preclude the recovery of all amounts due on account of the changed work.

6.4.3 Selecting the Proper Tone

The tone of your change proposal gets back to the hardball/softball of Section 6.3.4. Part Four will develop specific programs and include sample language for each proposal type, once all the components of the change order proposal have been analyzed. Each change proposal must include important qualifications, notifications, and reservations of rights. Moreover, these inclusions must become a part of every proposal, no matter how redundant they may seem. These inclusions are discussed in Chapter 8, Designing and Constructing Effective Change Order Proposals. The judgment to be exercised by the contractor in the preparation of the proposal concerns the tone rather than the content of each proposal. If the owner must be inflamed to achieve the desired objective, then so be it.

The contractor should consider the discussion of professional communications in Section 6.3.4 before sending a communication which will create unnecessary animosity. The contractor should also double-check his or her motivation for the inflammatory tone before proceeding. As mentioned previously, it may be possible to accomplish all proposal objectives without creating any more friction than is absolutely necessary.

6.5 PRESENTING THE CHANGE ORDER

6.5.1 Proposal Submission

Once the timing, methodology, and tone of the change order price have been decided upon, attention can be focused on the proposal itself. Section

3.4.10 described the three basic elements of every change order. To repeat here for convenience, the elements are:

1. Direct costs (labor, materials, supervision—the hard costs)
2. Indirect costs (home office overhead, delay costs, opportunity costs, lost profit—the soft costs)
3. Consequential costs (damages) (interference, disruption, resequence)

These cost categories, as well as one additional element, must be addressed in the preparation of *each* proposal. That last element is the change to contract time. The final portion of a complete change order proposal is the substantiating backup for each cost or statement included in the change. Chapter 8 outlines the procedure to assemble powerful change proposals. Word-for-word letters are then given for each possible situation. Chapters 9 and 10 contain specific recommendations to back up each proposal component. When the process is complete, the change proposal form will have been selected for its appropriateness, and its parts will have been assembled in the most effective and convincing manner.

6.6 PERFORMING THE WORK

6.6.1 Tracking Project Effects

By the time the changed work is begun at the site, accurate documentation becomes absolutely necessary. This is true whether or not the change order has been finalized. There are many reasons for this consideration, some of which are more clear than others.

a. The change order has been finalized.

In cases when the change has been finalized, the first consideration is to maintain a comparison of actual versus estimated costs. If everything is proceeding in an acceptable manner, and the actual costs are in line with the estimated costs, the need for further action may end with the completion of the change. If costs are beginning to get out of control, however, it becomes critical to be aware of the situation at the earliest possible moment. The contractor must determine the cost of the overrun and take whatever actions are required to protect his or her position under the contract. Just because the change order has been executed, it does not necessarily follow that the contractor will be bound to the existing bottom-line price. Any number of effects may have been introduced into the actual work that can justify the reopening of the issue. These effects might include:

1. Cost of overruns due to conditions unforeseen at the time of the original change proposal;
2. Interferences beyond those anticipated;

3. The owner or architect not having complied with some condition(s) of the proposal (such as shop drawing approval time- or owner-furnished equipment); or
4. "Allowance" items in the proposal have become inadequate.

It is for these kinds of effects that the proper qualifications and reservations of rights become imperative in every change proposal. They are included to leave open doors through which any item of each "finalized" change can be reaccessed. As soon as an effect altering the cost of the change is evident, research the approved proposal to confirm the existence of reserved rights and any qualifications of the proposal that may have been breached. Refer then to Chapter 12, Winning in Change Order Negotiation, for suggestions on effectively reopening the change.

b. The change order has not been finalized.

Whether the work is proceeding on a time and material basis or the change has not been finalized for any other reason (refer back to Section 6.4.1), precise documentation is *the* rule that must not be compromised. *All* direct cost items must be tracked, starting with the effort that went into the proposal preparation. These historical records may become important if it becomes necessary to demonstrate just how much time the change has been taking away from other important project activities.

"Equitable adjustment" (see Section 3.4.10) is premised on the ideal that a contractor is entitled to receive compensation for *all* costs that would not have been experienced had the change not occurred. Chapter 8 will go into these descriptions in precise detail. The obvious costs to be tracked include items such as:

1. Material invoices;
2. Payroll records; and
3. Schedule impact (including logic changes and milestone start delays).

The more subtle kinds of compensable costs may include:

1. Long-distance phone calls;
2. Excessive management time devoted to research and administration;
3. Loss of production efficiency due to outside effects (adverse weather conditions);
4. Increased surety involvement and costs;
5. Delay to contract payments that cause excessive carrying costs.

If these are not completely documented along with their respective justifications, they will *not* be paid. That is the rule. Do not forget it.

c. In either case...

Whether the change has or has not been finalized, it is important to realize that the foregoing discussion revolved around direct costs. Even more important than the documentation of direct costs is the documentation of the indirect and consequential costs. Since these components are subject to the greatest criticism (and scrutiny), they must be cataloged in explicit detail. Their justification must be absolute, and their calculations must be unchallengable.

To accomplish these objectives, the following two management systems must be in place:

1. A good cost accounting system; and
2. Good schedule documentation techniques.

In the design and application of these systems, do not confuse "sophistication" or level of complexity with *effectiveness*. Whether or not they are computerized, generate impressive-looking reports and charts, or command extensive company resources, the keys to their effectiveness lie in their ability to:

1. Permit easy research, identification, and consolidation of all *relevant* information, while allowing the deletion of information which does not apply;
2. Demonstrate clearly cause-and-effect relationships; and
3. Maintain a clear trail of personal accountability.

One such inexpensive and easy-to-use scheduling and cost-tracking system that clearly pinpoints responsibilities is detailed in the book, *Construction Scheduling Simplified,* Prentice Hall, 1985.

6.7 CHANGE ORDER PAYMENT

6.7.1 Billing and Payment

Change order billing and payment will normally follow the same procedure and timing as the procedure for contract work, unless specific conditions are negotiated as part of the final proposal. The attention given to any special billing circumstances will depend upon a variety of factors, including:

1. Size of the change order (and corresponding retainage amounts).
2. Any money required "up front" to finance the change (such as a payment to a utility before their work is even scheduled).
3. Delays in the ability to invoice for contract work.
4. Delay in the date of substantial completion or other significant date that determines when the retainage on the entire contract amount can be either reduced or eliminated.

There can be any number of other considerations, depending upon the contract itself or even the current billing format. For example, it is common for the Schedule of Values to include an item for General Conditions or something similar. If so, it may be billed each month in the same percentage as the total percentage complete for the project. If this is the case, a very large change order that has just been added to the books but that has not yet been completed will have the arithmetic effect of lowering the total percentage complete for the entire project, given the new increase in total project cost. For the same amount of physical work, the General Conditions item will actually be billed at a *reduced* rate.

The point is that it is important to be able to forecast these kinds of effects and accommodate them in the most reasonable way in the initial change proposal.

6.7.2 Claims and Disputes

Part Six deals with the specifics of reversing change order rejections, and ultimately winning in arbitration and litigation. It is important to understand that legitimacy and credibility must be maintained above all else. This will apply to whatever strategy, tactic, or countermeasure that may be employed during the course of the negotiations and eventually in the dispute resolution process.

The first consideration in this regard is to be sure to research the contract and specifications thoroughly. Before *any* action is taken, determine if there is any specific procedure outlining the precise steps to be followed in the event of a serious problem. Start with the Change Clauses (see Section 3.4.2), and then move to the Dispute Clause (see Section 3.4.4). If nothing becomes apparent, continue the search through the remainder of the contract. Remember, most contracts have some form of dispute resolution procedure. The dispute resolution procedure may be located in a clause entitled "claims" or in an arbitration clause (such as an appeal to a higher level or a "board of contract review," and so on).

After the applicable procedure has been located, then it should be followed in a literal sense. In the unlikely event that a dispute resolution procedure is not included in the contract, then it may be wise to advise the owner in the first correspondence that the action is predicated on necessity, given the fact that the contract is deficient in this regard.

The net effect of all this effort is that the contractor will be clearly communicating to anyone who is interested that:

1. The contractor has taken the time to read the specifications;
2. The contractor is making every effort to proceed in accordance with the terms of the contract, and the owner is expected to do the same; and
3. The contractor knows the owner's procedure and will take the actions necessary to protect his or her contractual rights.

Providing professional notices in a timely manner and in accordance with the procedures outlined in the contract will cause the other parties on the project to take the notices seriously.

On the other hand, the quickest way to lose a good amount of respect is to jump the gun. Taking dramatic steps without completing the proper research will only result in that exact procedure being handed down in a "you should have known better" treatment. An important edge will have been lost, along with the momentum necessary to move the change proposal to approval. Additional effort will become immediately necessary to salvage credibility.

As set forth above, the contractor must be sure that the owner understands that:

1. The contractor is prepared and knows what he or she is doing;
2. The contractor knows the owner's procedure; and
3. The contractor is willing to do whatever is necessary to preserve his or her rights.

People who are willing to work will command more respect than those who are not. Their opinions will carry greater weight, and their actions will be taken seriously. In the dispute resolution process, threats to proceed with arbitration or litigation will be considered seriously. Finally, the well-prepared, professional contractor will be perceived in the proceedings themselves as one who has exhausted all contractual possibilities for the resolution of the dispute before winding up in arbitration or trial. Be prepared, credible, and confident, and acceptable results should follow.

Part Three

Prospecting for Change Orders and Their Components

7

Where and How to Find Potential Change Orders

7.1	Introduction	103
7.2	Predesign	103
	7.2.1 Adjacent Properties	103
	7.2.2 Boring (Subsurface) Data	104
	7.2.3 Building Code Compliance	105
	7.2.4 Easements/Rights of Way	106
	7.2.5 Special Agency Approvals	107
	7.2.6 Interference from Utilities Not Properly Shown	108
	7.2.7 Plan Approvals (Building Permit)	109
	7.2.8 Temporary Utilities—Availability Within the Contract Limit Lines	110
7.3	The Contract and Bid Documents	111
	7.3.1 Award Date	111
	7.3.2 Named Subcontracts	113
	7.3.3 Sample LETTER TO THE OWNER Regarding Obligation to Determine Responsibility for Questionable Work	115
	7.3.4 Sample LETTER TO THE SUBCONTRACTOR Regarding Owner's Decision Directing Work	117
	7.3.5 Price/Bid Allowances	119
	7.3.6 Contract Time	119
7.4	Plans and Specifications	121
	7.4.1 "As Indicated"	121
	7.4.2 Ceiling Spaces (Conflicts)	123
	7.4.3 Sample LETTER TO SUBCONTRACTORS Regarding Coordination of Work in Ceiling Spaces	125

	7.4.4	Changed Existing Conditions	127
	7.4.5	Column and Beam Locations	128
	7.4.6	Design Change Telltales	130
	7.4.7	Design Discipline Interfaces	131
	7.4.8	Duplication of Design	132
	7.4.9	Sample LETTERS TO THE OWNER Regarding Design Duplications	135
		LETTER 1	136
		LETTER 2	137
		LETTER 3	138
	7.4.10	"Fat" Specifications	139
	7.4.11	Finish Schedule vs. Specification Index	140
	7.4.12	Inadequate Level of Detail	140
	7.4.13	Light Fixture Locations	142
	7.4.14	Match Lines and Plan Orientations	144
	7.4.15	Mechanical, Electrical, and N.I.C. Equipment	145
	7.4.16	Sample LETTER TO SUBCONTRACTORS Regarding Material and Equipment Coordination	148
	7.4.17	Sample LETTER TO THE ARCHITECT Regarding Contract Equipment Coordination	150
	7.4.18	Sample LETTER TO THE OWNER Regarding N.I.C. Equipment Coordination	152
	7.4.19	Numerous Details and Dimension Strings	154
	7.4.20	Performance and Procedure Specifications	155
	7.4.21	Proprietary Restrictions (Public)	156
	7.4.22	Sample LETTER TO THE OWNER Regarding Equal to Proprietary Item	158
	7.4.23	Sample LETTER TO THE OWNER Regarding Rejection of Equal for Proprietary Item	160
	7.4.24	Specification Section "Scopes"	162
7.5	Site		162
	7.5.1	Introduction	162
	7.5.2	Grades, Elevations, and Contours	163
	7.5.3	Sample LETTER TO THE OWNER Regarding Changed Site Conditions	165
		LETTER 1	166
		LETTER 2	168
7.6	Change Order Discovery CHECKLIST		169

7.1 INTRODUCTION

Chapter 5 outlined the general sources of conflicts giving rise to change orders. This chapter will develop the process by segregating the major areas of the contract. Each important component will then be broken down into specific items of possible conflict. The potential origin of each possible change order will then be exposed in detail. Precise action steps will be included wherever appropriate for the most effective treatment of each circumstance. Finally, a Change Order Discovery CHECKLIST is provided at the end of the chapter to make the actual process of completing the detailed review systematic and efficient.

7.2 PREDESIGN

7.2.1 Adjacent Properties

Description

The significant characteristics of properties directly adjacent to a construction site may not be apparent in a reasonable prebid site investigation. For them to become visible may be strictly a case of being in the right place at the right time. If the owner is aware of possible effects, there may be reasons why their inclusion in the bid documents may have been considered unnecessary or inadvisable. On the other hand, even the owner and designer may have not been aware of the adjacent property's characteristics or their impact on the proposed construction.

Examples of the kinds of conditions that have the potential to seriously affect a project are:

1. A seasonal watercourse that drains precipitation from several acres of adjacent land into the site;
2. Heavy traffic that restricts perimeter site mobility (in urban areas);
3. An adjacent, independent construction excavation that presents unanticipated problems for your own excavation.

Action

These kinds of effects are among the most difficult to spot before they become painfully apparent. Accordingly, once the hint of such a condition appears, the contractor must be quick to identify it for what it is—an unanticipated interference. Fast action becomes critical to:

1. Confirm the condition;
2. Engineer possible corrective action (or, preferably, alternative corrective actions that can be presented to the owner for selection).
3. Immediately prepare a detailed change order proposal (incorporating the suggestions in Part Four), with direct attention given to specific acceptance time requirements. The proposal should also note the schedule

impact that has already occurred, as well as the extent of damages and delays that will result from procrastination in approval.

7.2.2 Boring (Subsurface) Data

Description

Boring data is provided to give important information regarding relevant subsurface characteristics of the site. Experienced estimators know to research the precise definitions of technical terms to gain a clear understanding of the exact conditions described in the data. From those characteristics, the ease or difficulty in working on the site can be determined.

A large percentage of "fine" particles and the presence of a small amount of water may, for example, lead an estimator to conclude that construction vehicle traffic in the area may cause a "pumping" condition. The water and fine particles would quickly ooze to the surface, making the area impassable. If this is anticipated, it may be necessary to construct a temporary road of gravel or crushed stone and maintain it throughout the construction period.

Whatever the conclusions are, the important considerations as far as changes are concerned are:

1. Relevance of the boring locations and data to the construction area, and
2. Consistency of the data throughout the site.

If the boring data is taken from borings around the perimeter of the site, the soil condition in the middle of the excavation is still an unknown condition.

If the boring locations within the building lines are not consistent (i.e., many in one corner, with few or none in other areas), is someone trying to hide something (rock or water)?

At the time that the estimate was prepared, it is only reasonable to require that the contractor interpolate the information between boring points to infer the material characteristics in those areas. No one can expect the contractor to invent information that is unavailable. When rock, a dump, dinosaur footprints, or whatever are encountered where there was no definite indication of such, the contractor should request and probably receive compensation for the interference.

Action

Review all boring data and soil characteristic information provided as soon as it becomes available. Look specifically for:

1. Inconsistencies in the depths of the borings;
2. Erratic boring locations; and
3. Relevance of the boring locations to the planned construction.

In addition, be critical of all water table information and be sure to consider the dates on which the tests were taken. For instance, borings taken in late August may not be representative of the water table which will be encountered in the spring.

If the information looks the least bit suspect, be skeptical. This is not to criticize the owner's intent; they may be any number of innocent reasons for a peculiarity. (An example may be that the building location was shifted after the original borings were done, and no additional borings were provided.) Whatever the reason, the effects can be severe. The justification for extra costs will be predicated on the lack of or the inaccuracy of the information provided by the owner or the engineer.

7.2.3 Building Code Compliance

Description

It is the architect's responsibility to be sure that the project meets the code requirements of all governmental entities having jurisdiction over the project. Contractors are normally responsible for *performing* the work in a code-approved manner. This criterion usually applies, however, to workmanship-type considerations. The designer remains responsible for the *design's* compliance to the applicable codes. Whether it is headroom, provisions for the handicapped, the number of exits, or handrail configurations, the design is up to the designer.

Mechanical and electrical contractors have slightly more responsibility in this regard, but for the most part, the workmanship consideration is still the same (unless some design responsibility is specifically included in the contract, such as sprinkler head layout).

In any event (if not specifically indicated), no one should be required to tell the properly prepared contractor that "you should have known better" in that:

- The elevator shaft needs a vent at the top.
- There should be a two-hour fire-rated wall over there.
- There are not enough exit lights.
- There are not enough parking spaces.
- You need ADA-compliant fixtures.
- That's the wrong height.

Refer to Section 7.2.7, "Plan Approvals," for related discussion and action steps.

Action

Advance warnings of a designer's failure to meeting building code requirements will often come from a contractor's *construction* experience. Something just may not look right. The detail may be completely contrary to the way that the contractor has always seen it done in the past. Once an apparent code violation has been observed, the next steps may or may not proceed in a straightforward manner, depending upon the owner's and architect's dispositions.

The procedure will be:

1. Advise (or "question") the architect of the perceived violation (the design may have accommodated the code in some other, less obvious, way);
2. Confirm that the design in question does indeed violate the code;
3. Upon such confirmation, advise the owner;
4. Take the initiative in leading the architect through alternative solutions as soon as practical;
5. Be sure the finalized corrective action is completely detailed *by the architect* and that the level and detail of the documentation are appropriate for the correction; and
6. Submit a complete change proposal, in accordance with the recommendations in Part Four, as soon as possible.

7.2.4 Easements/Rights of Way

Description

Access to the site should be clearly incorporated into the project's work description and the bid documents. Any restrictions on site access should also be mentioned. If adequate site access is apparent, the contractor is justified in acknowledging the case and moving on to the next bid consideration. There is no further duty to inquire (see Section 2.3.3). This is so even when easements or rights of way are specifically identified. If the access information has been given in the contract, the owner warrants its correctness (see Section 2.3.1).

The potential change in this situation involves a partial or total restriction to the use of such access. If it can be demonstrated that the restrictions are contrary to the information provided in the bid documents or could have been reasonably anticipated at the time of bid, compensation for the detrimental effects of the restrictions is warranted.

Action

When confronted with an adequate site access issue, the contractor should take the following actions:

1. Review the documents to determine if the potential for restricted access existed at the time of the bid. Look for items such as:
   a. Designated easements;
   b. Constraints imposed by adjacent properties (see Section 7.2.1);
   c. Traffic patterns and traffic flow in and around the site; and
   d. Parking, traffic patterns, businesses, and so on at the immediate perimeter of the site.
2. If an easement was indicated, request a complete description of all conditions of the easement from the owner prior to submitting the bid.

3. Review the details of the perceived restriction with the estimator(s) to confirm whether or not it has already been considered in the estimate.
4. If it was not considered or accommodated in the estimate, confirm that it was reasonable that it was "missed" under the circumstances.
5. If (4) is confirmed, proceed with an immediate change proposal per the requirements and recommendations in Part Four.

In most cases, however, access interference actually occurs with little or no warning. Production and efficiency are dramatically reduced because of sudden, unexpected complications in moving about the site. In this event, it cannot be stressed enough that steps (3), (4), and (5) be completed as expeditiously as possible.

7.2.5 Special Agency Approvals

Description

By the time the owner authorizes the commencement of performance on a contract, all approvals by any agency having jurisdiction over the design and the site should have been properly obtained. Among these special governmental agencies are the Inland Wetlands Commission and the California Coastal Commission. The rules themselves vary in complexity and severity, depending upon circumstances, design infringements on protected areas, and the community in which the project is being built. The point is, if the project's design encroaches upon (or even comes close to) a protected area, there is a good chance that formal approval of at least those portions of the facility had to have been obtained from the applicable agency. The contractor has every right to expect that necessary approvals had been obtained, without the need to confirm the case before the work actually begins.

Action

If the site is located in or adjacent to a protected area, or will affect the area during and/or after construction, it can present a real dilemma for the contractor. If inquiries are conscientiously made to the authorities, the contractor may wind up stopping his or her own job if the approvals are not in order. On the other hand, if the contractor does not take any action to confirm the existence of the necessary approvals or that the project will not be stopped by the governing agency, the contractor's work on the project may be interrupted, delayed or stopped altogether. Additionally, some owners may try to shift responsibility for this shutdown to the contractor, based on the boilerplate provision which states that the contractor shall be responsible for procuring all approvals and permits related to the work.

One way to approach this situation is to send a written correspondence to the owner that:

1. The effect of the project on the protected area is apparent;
2. All governing agencies may require a permit or approval before the commencement of the work;

3. The work is proceeding in accordance with the executed agreement and the owner's instructions; and
4. The owner should confirm, in writing, that all required approvals have been received and that the work will not be interfered with.

This will at least offer a certain amount of protection in the event that the work does become affected due to the failure of the owner to secure the proper approvals.

7.2.6 Interference from Utilities Not Properly Shown

Description

The locations of the utilities (storm, sanitary, telephone, power, water) existing on a site are normally indicated on the plans. This information is provided to the designer by the respective utility companies. Errors in the indicated information can therefore come from two sources:

1. Errors in the information as it exists in the file maintained by the utility or in its transmittal to the designer; and
2. Errors in transposition or in the representation of the information on the plans by the designer.

Based on the information depicted on the plans, a contractor is free to plan activities, locate temporary facilities, stockpile materials, and sequence the work around or between the utilities. Before any digging is actually performed, however, there is usually some entity to contact to reconfirm the exact locations of the utilities. It is an important precautionary measure to:

1. Confirm that the information on the plans is consistent with the information on file with the respective utility; and
2. Confirm that the information has not changed (that the utility has not been modified in some way) since the time that the information was provided to the designer.

If the work begins in the vicinity of a utility without verifying its most current characteristics with the utility that owns or maintains the structure, the contractor will be assuming the risk associated with damage to the utility structure. Moreover, many contracts require the contractor to confirm the location and depth of existing underground utilities by "potholing." If the contractor proceeds with the work without confirming the location of the existing utilities, the responsibility for any damage to the utility may rest with the contractor.

On the other hand, the verification procedure may expose different information regarding the utility. If your work is adversely affected by the new or different information, the justification for the resulting change order proposal will be clear.

Action

In all cases where excavations are to be performed in or near the vicinity of an existing utility, the contractor should:

1. Verify all characteristics of the existing structure with that utility company.
2. Have a company representative review the details of the existing utility *at the site.*
3. If nothing is substantially different from the information shown on the plans, proceed and hope for the best.
4. If, however, the verification process exposes something different from the condition depicted on the plans and the new condition will cause an increase in the contract time and/or cost, additional compensation will be justified.
5. If additional time and/or money is required to address the condition uncovered during the verification process, notify the owner immediately of the changed condition and proceed with the submission of a change proposal in accordance with the recommendations in Part Four.

7.2.7 Plan Approvals (Building Permit)

Description

Plan approval is similar to building code compliance (see Section 7.2.3), in that it is the designer's responsibility to incorporate all building code requirements into the documents. The contractor's application and securing of the building permit should then be nothing more than the clerical formality of delivering the plans, specifications, and fee to the building department. If delays to commencement are experienced because the building department discovered a serious design flaw, the delay is the responsibility of the owner. More common, however, is having the permit granted, on the condition that the owner or designer make some modification to the design. In that event, there will not be any holdup in the job start, "just" additional work.

Action

The contractor should properly plan for the commencement of the work. The contractor should not wait until he or she is ready to move. If the contractor is responsible for obtaining the building permit, the contractor should follow these steps:

1. File the documents and permit application immediately upon contract award and availability of the required plans and specifications.
2. If a problem is encountered and a delay in the permit issuance is evident, clarify with the responsible building department representative that mobilization and temporary office setup can proceed, pending corrections to the design.

3. Make every effort to demonstrate to the building officials that the building department will be kept advised every step of the way.

4. In the case of a design violation, it is important that all building inspectors immediately perceive that every effort will be made to do things right.

5. Before leaving the building department, establish with the building inspector the fastest way to satisfy all interested parties that the changes will be incorporated. (Will a visit from the architect be necessary? Do the plans need to be altered and resubmitted? Is a confirming letter sufficient? If so, from whom?)

6. Involve the architect immediately (preferably with a phone call from the building inspector's office) so that the architect can explain if there was some other design consideration given to the situation that may not have been apparent from the initial review of the plans.

7. If a change is indeed necessary before the permit will be granted, have the architect confirm the nature of the change and the procedure for the correction of the applicable documents.

8. Advise the owner of everything that is going on as soon as possible, and confirm that a change order proposal will be submitted as soon as the design change is finalized. (Remember, the notification should refer back to the initial notice to the architect during the plan check review and the change proposal should follow the steps contained in Part Four).

9. The size, complexity, and circumstances of the change, will determine whether the work will or will not proceed without the expected change in hand. (Refer to Section 6.3, "Preparing the Change Order," and Section 6.5, "Presenting the Change Order.")

10. Be certain that any decision is clearly represented in the change notification/proposal to the owner.

7.2.8 Temporary Utilities—Availability Within the Contract Limit Lines

Description

Even in the face of unusual circumstances, the lack of available sources for temporary utilities at the site will be very difficult to get accepted as a change unless some qualification was clearly included and accepted in the bid proposal or the original agreement. This is because a professional estimator really should have known enough to include an allowance for all utilities necessary for construction and field offices (water, power, light, heat, sanitary) in the estimate. Plain assumptions without a reasonable amount of investigation are not justified.

Extra costs for temporary utilities may be justified in the case of changed conditions (refer to Section 5.6, "Latent Conditions," and Section 7.5.2, "Grades, Elevations, and Contours," for related discussions and action steps). If it can be demonstrated that the conditions observed at the time of bid with respect to the temporary utility situation are different from those

PROSPECTING FOR CHANGE ORDERS AND THEIR COMPONENTS   111

existing now, the contractor will be justified in requesting compensation for additional costs experienced.

Action

In the event the contractor wishes to pursue additional contract time or compensation on account of changed conditions which impact the cost or availability of temporary utilities, the contractor should:

1. Establish what the conditions were at the time of bid. (Check with those who prepared the estimate to see exactly what was anticipated, along with the specific reasons.)
2. Confirm exactly what is required now:
    a. Do you need additional telephone poles?
    b. Is the utility company inventing charges that they hope will finance their new power plant?
    c. Is power available at all (without generating equipment)?
    d. Is the previously anticipated use of existing facilities suddenly prevented?
    e. Are temporary lighting requirements more elaborate because other design changes require more stringent quality control?
    f. Are temporary heat and protection now required because of owner-caused delays (or acceleration)?
    g. Is water available in sufficient amounts as required for construction?
3. Calculate the net differences in cost, along with the total effect on project sequences and time.
4. Submit a detailed proposal in accordance with the recommendations in Part Four.

7.3 THE CONTRACT AND BID DOCUMENTS

7.3.1 Award Date

Description

Most bids are submitted with some understanding as to the planned start date for the work. If an owner wants to extend the contract signing date, it usually begins with a telephone request. In that conversation, the contractor will be asked to confirm that there is no problem and that the bid price will not change. Under these circumstances, the contractor may have three paths to pursue:

1. Refuse to extend the date and possibly lose the project;
2. Accept the request, and possibly turn a profitable project into a loser; or

3. Present the owner with a change proposal on account of the requested extension.

There are many considerations which need to be evaluated as part of the contractor's response to the owner's request to extend. For instance, was the contractor selected for the project because of price or performance considerations, or both? The characteristics of the owner and the project itself will determine the relative importance of each. If it is a public project, it's all price. If it is a private job, the contractor's reputation for superior performance may be a larger consideration. In any event, consider the circumstances surrounding the project and the owner's request carefully to determine what leverage or power is available in this situation. Some additional considerations include the price of the next lowest qualified bidder and the owner's ability to give the contract to a competitor at this stage if a request for additional compensation is submitted.

Action

When requested to extend a signing/contract award date, the contractor should consider the following:

1. Try to avoid giving an immediate answer.
2. Assess your position relative to the competition and try to make the assessment as complete and as accurate as possible. This assessment should consider:
    a. How far below the next bidder is your price?
    b. Have you indicated (or can you indicate) that you can complete the facility in less time than the next bidder?
    c. Are the scopes of work and contract structures between competitors identical (as in a public bid), or is it possible that contract terms are or can be different?
    d. Have you been involved in the project development and design, and does this involvement afford you some advantage?
3. Consider the owner's motivations.
    a. Where is the pressure? Does the owner need you more than your competition for any reason?
    b. Is the owner tied to you in any way? Has your involvement in the design process increased your value to the owner?
    c. Where is the time pressure? Why is the owner requesting the extension? Are the reasons beyond the owner's control? Will the time associated with bringing another contractor on-line present more serious complications to the owner?
    d. What are the owner's options? How easy is it for the owner to change bidders at this point?
4. After answering the questions contained in steps 2 and 3, the contractor should be able to assess whether he or she possesses enough power to present a price increase to the owner.

5. If the result of the foregoing analysis is positive, prepare the cost increase in accordance with the recommendations in Part Four, and proceed with as much courteous professionalism as possible. Give the owner the price increase face-to-face and provide every opportunity to talk it around in circles as many times as necessary to remove as much animosity as possible. Allow the time necessary for acceptance (see Chapter 12, Winning in Change Order Negotiation).

6. If, after considering the answers to the inquiries set forth in steps 2 and 3, it is determined that the contractor is not in a position to request additional compensation at the time of the request, consider the possibility of qualifying the agreement to extend the date based on conditions that are apparent now. Try to include a provision in the agreement that will leave the door open for a price increase in the future if conditions change.

7.3.2 Named Subcontracts

Named subcontracts, or owner-selected subcontractors, refers to a special procedure for subcontractor selection by the owner.

Owners use this approach generally for two reasons. First, the procedure may be used to reduce or eliminate the practice of "bid shopping" by prime contractors.

The second reason comes down to an attempt by the owner to exert some (at least perceived) control over who will be performing portions of the work, as well as the contractual provisions that will govern them. Typically, the arrangement will go as far as to dictate the precise terms of the agreement between the subcontractor and the prime contractor.

This process, however, may create a serious problem for the owner in that the owner's selection of the subcontractor removes the prime contractor's responsibility (and liability) for the respective subcontractor selection. When the subcontractor's performance degenerates, the prime contractor dutifully turns to the owner and says, "What do you want to do with *your* subcontractor?" The owner will then (at least initially) refer to the old boilerplate in an attempt to convince the prime contractor that the prime contractor has responsibility for any problems caused by the subcontractor. If the owner selected the subcontractor, the subcontract form and the scope of work included in the subcontract, the prime contractor cannot be expected to take complete responsibility for the problem.

One of the simplest problems that can develop under these circumstances is when the lowest apparently qualified bidder becomes unavailable after using the bidder's price to calculate the amount of the prime bid. Because it was originally the owner's selection, and a simple mathematical relationship, the approval of the change order request for the additional cost of bringing the next lowest bidder on board should be straightforward. Straightforward, that is, for the net difference in price between the two prices *only*, and not so straightforward for consequential or interference costs.

Securing owner recognition and approval for changes involving deficiencies in owner-defined work is generally more difficult. This is particularly true when faced with the gray areas at the interfaces between two or more owner-defined subcontracts. (Is the work described in both places, or not at all?)

For example, a specification in Division 16 which provides that the underground electrical duct bank to be "encased in concrete in accordance with the requirements of Section 03300" could be interpreted to mean that the electrical contractor places the concrete, using Section 03300 as a guide, or that the Section 03300 concrete contractor is to place the concrete. If both of the subcontractors were selected by the owner, the contractor must notify the owner of the ambiguity and demand that the owner designate the responsible subcontractor. Sections 7.4.8, "Duplications of Design," and 7.4.12, "Inadequate Level of Detail," develop this idea further.

Often a disclaimer exists in the front-end documents or in the contract itself that tries to limit or remove the owner's liability for the owner's subcontractor selection. Depending on the actual language, however, it may actually help your case. For example, one specification stated that:

> The [specification] divisions and sections do not operate to make the owner or the architect an arbiter to establish the limits to the contracts between the general contractor and the subcontractor, except in the case where owner-selected subcontracts establish such limits.

If this paragraph, or something similar, is included in the General Conditions, the contractor can use this provision as a basis for requesting a determination by the owner. If just the first portion of this paragraph is included, the contractor's rights will be generally the same, but the contractor will have to be more persuasive in his or her attempt to secure a determination by the owner.

Action

In cases involving owner-selected subcontractors or owner-defined work scopes, the contractor needs to:

1. Determine if the agreement or the scope of work for a subcontractor has been owner-defined.
2. Identify and analyze any disclaimers or other qualifying language that exists in the owner–prime contractor contract.
3. Ask the *owner* to determine and confirm exactly what specification section covers the work and to *name* the subcontractor "responsible." Refer to the sample letter in Section 7.3.3 for an example of this procedure.
4. Advise the named subcontractor that the *owner* has determined the work to be within the scope of that subcontractor's responsibility and direct the subcontractor to proceed with the work. The notice should mention that the owner is issuing the directive to proceed and a copy of the notice should be sent to the owner. Refer to the sample letter in Section 7.3.4 for an example of this kind of notification.

5. When the subcontractor objects, advise the owner of the subcontractor's position. Try to avoid stepping into the line of fire. The general contractor's role in this situation should be limited to the communication and coordination of the owner's directives with the subcontractor's actions. (Refer to Section 12.17, "General Contractor As a Conduit.") Try to keep liability for the owner's actions as well as liability for the subcontractor's actions where they belong.

6. Keep up the momentum. Immediately submit a detailed change order proposal in accordance with the recommendations in Part Four. If it is possible to submit a properly substantiated proposal along with the subcontractor's objection mentioned in step 5, do it.

7.3.3 Sample Letter to the Owner Regarding Obligation to Determine Responsibility for Questionable Work

The sample Letter to the Owner Regarding Obligation to Determine Responsibility for Questionable Work is designed to accommodate Section 7.3.2, step 3. Specifically, the sample letter:

1. Notifies the owner that by all appearances, the work in question is not included in any of the owner's subcontract scopes.

2. Maintains that nobody intends to do the questionable work without further direction.

3. Reminds the owner that the contract provides that the owner is responsible for making the final, precise determination.

4. Requests a definite reference to a precise specifications section. (This may become an extremely important piece of information in the event of a subsequent appeal of the owner's decision.)

5. Asks that the responsible party be determined by *name*.

6. Notifies the owner that a proper, timely response is required by a certain date or job progress will be adversely effected.

These notifications and requests for precise information by exact dates will become an important basis for subsequent action by all parties.

If the owner's response to the letter is less than complete, it should trigger an immediate reaction on your part. The owner should be advised of the deficiencies in the owner's response, reminded of his or her contractual obligations and notified of the resulting interference.

Refer to Section 7.4.8, "Duplication of Design," action step 6, for additional related discussion.

**SAMPLE LETTER TO THE OWNER
REGARDING OBLIGATION TO DETERMINE
RESPONSIBILITY FOR QUESTIONABLE WORK**

LETTERHEAD

(Date)

To: Owner (or architect, depending on specification direction)

RE: (Job no.)
(Job name and description)

SUBJ: 001 (Brief description of work)
Responsibility determination

Mr. (Ms.)   :

    Per our conversation of (insert date), both (name subcontractor 1) and (name subcontractor 2) take exception to the responsibility for (brief description of questionable work).

    General Conditions Article (   ) directs that the owner (or architect, as described) interpret the requirements of the contract. Accordingly, please advise of the party responsible for the subject work, indicating the applicable specification section and responsible contractor by name.

    Your complete written response is required by (insert date) to maintain job progress and avoid (additional) interference.

Very truly yours,

Project Manager

cc: Architect (or Owner)
    Subcontractor 1
    Subcontractor 2
    File: 001

7.3.4 Sample Letter to Subcontractor Regarding Owner's Decision Directing Work

The sample Letter to Subcontractor Regarding Owner's Decision Directing Work is designed to accommodate Section 7.3.2, action step 4. Specifically, the letter:

1. Notifies the subcontractor that the *owner* is directing the subcontractor to perform the work.
2. By copy, notifies the owner and the architect that the subcontractor is being directed to perform the work *in the owner's name.*
3. Requires positive action by the subcontractor on precise dates.
4. Notifies the subcontractor that if the subcontractor fails to take timely action or takes some inappropriate action, the burden of delays and/or job interference will shift onto that subcontractor.

The subcontractor's response to the letter must be monitored closely. Any rejection of the owner's directive must trigger an immediate reaction. The next step in the contract-defined process for dispute resolution must be set into motion. Refer to Section 3.4.3, "The Pass-Through Clause," Section 3.4.4, "The Dispute Clause," and Part Four for related discussion. Section 7.4.8, "Duplication of Design," action step 6, will also be of use.

**SAMPLE LETTER TO SUBCONTRACTOR
REGARDING OWNER'S DECISION DIRECTING WORK**

LETTERHEAD

(Date)

To: (Subcontractor named in owner response to Section 7.3.3)

RE: (Job no.)
(Job name and description)

SUBJ: 001 (Same description used in Section 7.3.3)

Mr. (Ms.)   :

Attached is the (date) (owner's name) determination of (subcontractor name) responsibility for the subject work.

The work is required to be complete by (date) to avoid (additional) schedule interference and associated charges. Accordingly, please immediately:

1. Confirm acceptance, and
2. Advise this office by (insert date) of your proposed schedule for completing this work.

Very truly yours,

Project Manager

cc: Owner
    Architect
    Subcontractor 2
    File: 001

7.3.5 Price/Bid Allowances

Description

Allowances for bid items can be initiated by the owner or negotiated by the contractor. The justification for allowances is usually tied to the fact that either:

1. Subcontractor prices received by the owner are unacceptably high, in which event, the item may be redesigned and put out for rebid; or
2. The design for that item is not complete, or the exact scope is otherwise unclear.

In either case the owner may direct the contractor to include an allowance price for the item in the bid for budgetary reasons. Only after enough information is made available can final prices for the work be solicited. Assuming proper substantiation, the net difference (add or deduct) between the allowance amount and the final price will be accommodated with a change order.

Allowance items will remain straightforward if they are reconciled in enough time to avoid interference with the construction schedule. Otherwise, interference, delays, and disruption will complicate the issue just as with any other change.

Action

If the opportunity presents itself, many contractors and construction managers would prefer to negotiate as many "allowance" items into a contract as possible. While this strategy all but eliminates the possibility of a significant cost savings (and profit) in the buy-out, it also eliminates any over-budget risk. The pressure is not as acute as with many other hard-dollar items, and the work will not need to be managed as tightly. There's less of a necessity (from the contractor's viewpoint) to expend great amounts of time and effort to get the most competitive prices.

Whatever the reason for including the allowance item in the contract, the issue must be resolved as soon as it is feasible. Use the construction schedule as a guide to indicate the latest acceptable dates for the resolution of each allowance item and the award of the respective contract. Do not procrastinate. Get all allowance items resolved at your earliest opportunity.

7.3.6 Contract Time

Description

When the time required to complete a project is stipulated in the contract documents, the owner implicitly warrants that the project can, in fact, be completed (in a logical manner) within that duration. A contractor, therefore, has a right to expect that, barring some obvious or glaring deficiency (refer to "Duty to Inquire" in Section 2.3.3), the work can be planned in reasonable sequences and that the work will be completed at the end of the contract time.

Conditions either inherent in the design or changed between the time of bid and contract execution or the start of construction may, however, create a difficulty with the time requirement.

Assume, for example, that the contract allows 200 working days. On the basis of that information, estimates are prepared for mobilization, home office overhead, cost of retainage withheld for the period, and so on. Estimates are also developed for project manager, site superintendent and other key personnel (executives, engineers, etc.) for the same time period. Assume further that it is subsequently discovered that some exotic (specified) piece of equipment actually requires 190 working days for delivery. If the schedule logic cannot be rearranged to accommodate this delay without the employment of extreme or unusual means, it is not reasonable to hold the contractor responsible for the delays and additional costs attributable to this condition.

Many times, the issue will not be as clear as a single, long lead-time delay. Most often, it may involve schedule logic sequences and/or activity relationships with effects that were not apparent to the contractor at the time of bid. Due to the increased complexity of real-world scenarios, the success of any requests for additional time and/or money as a result of a problem with contract time, duration or sequencing will be directly dependent upon:

1. A demonstrably reasonable and accurate initial schedule logic; and
2. A mechanism that will incorporate the change information and display, *clearly* and *definitely*, a direct cause–effect relationship between the changed information and the requested time extension.

A complicated and confusing CPM or PERT schedule and its updates may not be enough to demonstrate the cause and effect relationship. The presentation must be clear enough and well enough understood to convince approving authorities that the request really is more than a scheduling technician's sophisticated con.

In this regard, the contractor must consider a particular owner's level of sophistication and disposition. (An easy-to-learn and -use scheduling and change documentation system is presented in *Construction Scheduling Simplified*, Prentice Hall, 1985. There, the use of the MOST (Management Operation System Technique) scheduling method is detailed in a step-by-step process. The technique requires no computer, and schedule updates can be completed quickly and easily, regardless of project size. It is exceptional in its ability to simplify a complicated job record and to display cause–effect relationships in a visible, *understandable* manner.)

Action

The basis for any meaningful action for the variety of circumstances that can occur boils down to securing complete documentation from the best available sources. From that point, a convincing demonstration of cause–effect is necessary. To accomplish these objectives:

1. Be sure that the project files are up to date and correct.
2. Confirm all circumstances regarding:
   a. Contract time;
   b. Contract award date;
   c. Contract start date; and
   d. Any extensions in the award or start dates.
3. Establish the item(s) interfering with the contract time and confirm all circumstances involving the item(s) in detail.
4. Confirm all in-house circumstances:
   a. The item was ordered (or scheduled to be ordered) at the earliest possible opportunity;
   b. Shop drawings were (or will be) processed in both directions as expeditiously as possible; and
   c. Precisely calculate the as-planned delivery time, the current delivery schedule and the impact of the current delivery schedule on the overall project schedule. (Be sure that the logic that affects the end date is clear and correct.)
5. Secure written certifications from the companies whose products or services are creating the interference and confirm:
   a. That better deliveries times are not possible; or
   b. The cost and circumstances associated with a better delivery schedule, if possible.

   (Get the information and certification from the highest level possible. The gravity of the situation will help to determine the organizational level from which such a confirmation should come.)
6. Notify the owner. Concisely describe the problem and the reasons for it. From that point, the contractor should consider either:
   a. Recommending some other product, design, or whatever alternative construction that may be available that *will* allow you to meet the original time requirement;
   b. Calculating the resulting increase in time and expense;
   c. Consulting with a competent attorney to determine if it is either necessary or advisable to submit the actual change order proposal now or later for all cost increases associated with the time extension. Refer to Chapter 8 for each change order component to consider.

7.4 PLANS AND SPECIFICATIONS

7.4.1 "As Indicated"

Description

The "As Indicated" or "See Specs" kinds of notes on the plans can be an indication that final design coordination checks were completed in a responsible manner. They may expose *assumptions* on the part of the designer that somebody accounted for the detail. The supporting details were assumed to be completed only because they were *supposed* to be completed.

The fact is, if a designer has not been able to exercise the discipline to coordinate methodically all such details, the plans may very well be in much worse shape than they first appear.

Other telltale phrases that similarly point to a lack of design follow-up include:

- "As Shown"
- "See Structural"
- "See Plans"

and any other vague language that does not contain a precise reference.

In contrast, the indication that the design coordination process was done more correctly is the presence of exact references such as "See 3/S5" instead of "See Structural" or "See Section 04200" instead of "See Specs."

The presence of these kinds of remarks is a sign that someone *did* cross-check the documents to be sure that the respective completing details really do exist.

Action

With each new project, the contractor should follow these steps:

1. Review each plan included in the contract set and identify every vague cross-reference that exists. Notes such as:
   - "As Indicated";
   - "As Shown";
   - "See Plans";
   - "See Architectural";
   - "See Structural";
   - "See Specs"; and
   - "Per Specs"

   should all be marked with a highlighter.

2. Rather than allowing each of these vague cross-references to affect the job one at a time as the respective details become needed, take the time to research one or more of the references to determine if the complete details actually exist.

3. If the completing details cannot be located after a thorough search, catalog each incomplete, conflicting, or missing reference and arrange each reference on a priority basis.

4. Each incomplete, conflicting or missing reference should be viewed. Consider each missing detail as a potential change order to be addressed individually or together. (Refer to Section 7.4.12, "Inadequate Level of Detail," for further direction on specific situations.)

7.4.2 Ceiling Spaces (Conflicts)

Description

There probably has not been a project designed and built that has not had some kind of problem with everything that was supposed to fit so neatly into the restricted spaces above the ceilings. Design/cost parameters create mutually exclusive objectives: minimum distance between floors (to minimize cost) and adequate ceiling heights. Satisfying these competing objectives results in all structural, plumbing, heating, fire protection, insulation, light fixtures, wiring, and so on being crammed into the same small area. The smaller the space, the greater the importance of effective design coordination to minimize the inevitable conflicts. The smaller the area and the more complex the building systems, the greater the probability of conflicts and change orders.

Action

In reviewing and analyzing the potential for change orders as a result of conflicts in the ceiling spaces, begin with the assumption that conflicts will definitely occur among:

- Ductwork.
- Water pipe.
- Sanitary, waste, and vent piping.
- Sprinkler mains and branches.
- Sprinkler head locations.
- Light fixtures.
- Structural steel beams.
- Structural concrete.
- Ceiling suspension systems.
- Architectural soffits and other special shapes.
- Electrical duct banks.

1. Research the General Conditions, the subcontracts, and any other appropriate documents to confirm the party responsible for ultimate coordination of the respective systems. (Remember the Pass-Through Clause described in Section 3.4.3 and the general contractor's objective to keep responsibilities where they belong.)

2. Based on the Pass-Through Clause (and any other contract language that you can find to support your position), notify the respective subcontractors of the potential for conflict and make it clear that it is *their* responsibility to coordinate *their* work in the ceiling space. Refer to the Letter to Subcontractors Regarding Coordination of Work in Ceiling Spaces in Section 7.4.3 for an example of this kind of notification.

3. Regardless of how formal contract responsibilities fall, review the plans and consider each area of potential conflict, including:

a. What are the high and low elevations of all drain lines, considering required pitch?

b. What are the elevations of the plumbing lines at the points they cross structural, steel and concrete beams?

c. What is the distance between beams and the top of the ceiling construction at the points where ductwork is shown to cross?

d. Where are the areas in which multiple items are occupying the same space (for example, water pipes crossing sprinkler lines—all directly above a light fixture)?

e. What is the total space necessary for all items in (d)? What is the space available?

f. What are the sizes of the heating, ventilation, and air conditioning ductwork at the points where they cross beams or other work? What is the available space?

g. What are the total heights of all light fixtures? Will they all fit?

h. Are there special architectural shapes either penetrating ceiling spaces or requiring structural support above the ceiling? Are plumbing, electrical, or other building systems shown to penetrate the ceiling? If so, what are the *actual* sizes of everything?

4. Catalog all potential conflicts discovered by subcontractors in response to step 2, and by your own investigation in step 3.

5. Consider each conflict to determine all available or potential solutions. Can things be moved, or do you have to cut a hole through something?

6. Assign recommended solutions to each conflict and be sure the project record documents every correction, even if the solution will cost no additional money. As mentioned previously and as discussed more fully in Chapter 10, maintaining proper project records is important for many reasons. In this instance it is important because:

   a. Although the contractor has fixed the owner's mistake, any liability associated with the redesign should remain with the owner; and

   b. As the conflicts begin to pile up, they will ultimately interfere with the work.

7. The number, cost, and complexity of the changes discovered through this process will determine the best way to present the change request(s) to the owner. The request should be prepared in a manner consistent with the recommendations in Part Four. The list which was prepared as part of step 4 and the records which were maintained as part of step 5 will tell a convincing story when trying to demonstrate the extra effort which was expended to coordinate the architect's design and entitlement to additional compensation. Refer to the Sample Notification Letter to the Owner on Changes included in Section 6.3.5 to consider whether its use may be appropriate at this time.

Refer to Section 7.4.13, "Light Fixture Locations," for additional related discussion of coordinating work in ceiling spaces.

7.4.3 Sample Letter to Subcontractors Regarding Coordination of Work in Ceiling Spaces

The Sample Letter to Subcontractors Regarding Coordination of Work in Ceiling Spaces is designed for use in Section 7.4.2, action step 2. Specifically, it notifies the respective subcontractors that:

1. It is *their* contractual responsibility to coordinate *their* work and this obligation applies not only to all portions of their own work, but to coordinating their work with the work of all other trades as well.

2. Probable conflicts exist in the ceiling spaces. (It is much more convincing if you've already begun Section 7.4.2, step 3. It will then be possible to say that conflicts definitely exist in the ceiling spaces.) Because of this, the subcontractor's responsibility to coordinate becomes critical. It is not to be treated in a casual manner if they want to avoid responsibility for failure to coordinate.

3. Definite action is required on the part of subcontractors by a specific date if they do not want to risk becoming responsible for any delay and/or interference caused by their lack of attention.

SAMPLE LETTER TO SUBCONTRACTORS
REGARDING COORDINATION OF WORK IN CEILING SPACES

LETTERHEAD

(Date)

To: (Separate letter to each subcontractor)

RE: (Project no.)
(Project title)
SUBJ: Coordination of work in ceiling spaces

Mr. (Ms.)   :

General Conditions Article (insert number of the Pass-Through Clause and any other direct reference) and paragraph (insert number) of your subcontract require you to coordinate your work with the work of all other trades.

As you know, the spaces above the ceilings are very restricted. Close and timely coordination of all building systems is therefore critical to avoid unnecessary interference. To date, there have been (insert number) conflicts discovered in the ceiling spaces. You are therefore cautioned to review all work in the ceiling spaces in detail now, and confirm to this office all potential conflicts that you identify.

Your response by (insert date) is required to minimize impact and delay of any conflict exposed by this process. Please be advised that failure to respond by (insert date) may leave you responsible for costs resulting from your lack of coordination.

Thank you for your cooperation.

Very truly yours,

Project Manager

7.4.4 Changed Existing Condition

Description

This situation is one in which conditions or circumstances at the site become different from those that were documented or observed at the time of the bid. They can apply to anything beyond the contractor's control at any time during the project's life cycle and can range from the plainly obvious to the obscure.

Examples of certain changed conditions that a contractor may encounter and that will result in a change order may include:

1. Another contractor has been allowed on the site and has constructed something that restricts or impedes access.
2. Utility lines have been relocated, or the conditions of their use have been altered.
3. Someone fails to complete work or arrangements that were scheduled to be completed prior to the start of the work (demolition, traffic rerouting, etc.).
4. The site has been used as a disposal or storage area.

Refer to Section 7.5.2, "Grades, Elevations, and Contours," and Section 7.5.3, "Letters to the Owner Regarding Changed Site Conditions," for related discussion, sample letters and action steps.

Action

The discovery of these kinds of changes prior to the start of the work so as to minimize real or potential interference is the primary objective, albeit a difficult one. This is due to the fact that the discovery procedure is usually not implemented until immediately prior to mobilization. It occurs at a time when dozens of other critical start-up activities are all competing for attention. As such, the contractor should take the following actions at the earliest opportunity:

1. Review every portion of the estimate that interfaces with an existing condition in any way. Pay particular attention to:
    a. All sitework considerations (refer to Section 7.5.2);
    b. Any interfaces with existing structures;
    c. Any noted conditions concerning other construction activities;
    d. Locations, extent, makeup, and conditions of existing utilities;
    e. Traffic patterns and site access;
    f. Anticipated storage and staging areas; and
    g. Parking and security arrangements.
2. Arrange to have the estimator(s) responsible for those portions of the bid listed in step 1 meet at the site to review all *current* existing conditions in detail.

3. Evaluate the site as if bidding the job all over again based on current conditions. (Differences at this point should jump out at the estimator(s).)
4. Catalog all apparent changes and make immediate arrangements to determine:
   a. The cost of each change and
   b. The impact of each change on planned sequences.

   Be aware that the heightened activity surrounding the job start may make it impractical to complete accurate job impact estimates at this time. This may be due to the fact that the detailed construction schedule may not even be completed yet. Even so, start-up sequences should have been established and to the extent that detailed activity sequences have been determined, the effects should be documented as accurately and completely as possible. It will be only from that basis that any meaningful or convincing calculation of additional time requirements can be accomplished.

5. Follow the recommendations in Part Four for proposal preparation and submission. If the net impact on job sequence and project time cannot be determined and displayed at this time with complete confidence, do not include the calculation with the proposal. Do not risk compromising the validity of the proposal by using bad or questionable information. Instead, note in the proposal that:
   a. The time and activity sequences are affected; and
   b. The detailed calculations and corresponding dollar amounts will be submitted when known.
6. Refer to Section 7.5.2, "Grades, Elevations, and Contours" for related discussion, and Section 7.5.3, "Letters to the Owner Regarding Changed Existing Conditions," for word-for-word examples of the procedure to consider.

7.4.5 Column and Beam Locations

Description

Although sized and located principally by engineering parameters, all structural accommodations (beams, columns, plates, etc.) are coordinated exactly with architectural, mechanical, and electrical designs. Once column and beam locations have been confirmed:

- Walls are laid out between them;
- Plumbing is arranged around them;
- Windows are located to miss them;
- Ducts are placed below them; and
- Equipment is located on top of them.

Errors occur as a result of inconsistencies in the size and location information indicated on the structural drawings and the balance of the contract set. The inconsistencies themselves occur from:

1. Errors in information given to or received from the structural engineer;
2. Errors in transposition into the architectural plans of design and location information;
3. Failure to clarify responsibilities for providing designs and details (see Section 7.4.1, "As Indicated," for a related discussion);
4. Wishful thinking—that the contractor will somehow "work it out in the field."

Action

When evaluating the potential for changes as the result of possible inaccuracies in the location and/or size of columns and beams, the contractor needs to:

1. Begin with a detailed review of the structural plans:
   a. Are columns laid out in a consistent matrix, or are they out of parallel or otherwise erratic?
   b. What are the column line dimensions? Are any unusually long spans apparent?
   c. Are there unusual shapes, slopes, angles, or connections?
   d. Are elevation changes smooth and consistent, or does anything look out of the ordinary?
   e. Are the beams generally of a consistent size, or are they all different (leaving all different ceiling spaces under them)? Refer to Section 7.4.2, "Ceiling Spaces (Conflicts)," for a related discussion.
   f. Where are the large beams?
   g. Where are the unusual designs? Is enough information given (dimensions, layout, etc.) for proper evaluation and shop drawing preparation? Refer to Section 7.4.12, "Inadequate Level of Detail," for a related discussion.

2. Review the architectural, plumbing, H.V.A.C., and electrical plans.
   a. Are listed column line dimensions between all designs consistent? (Refer to Section 7.4.5, "Column and Beam Locations.")
   b. Are there large ducts shown crossing the large beams identified in section 1.e.? [Refer to Section 7.4.2, "Ceiling Spaces (Conflicts)."]
   c. Are there light fixtures in the areas of the large ducts? (Refer to Section 7.4.13, "Light Fixture Locations.")
   d. Where is the sprinkler main? Does it cross large beams, ducts, and light fixtures?

3. Spot-check architectural dimension strings. Add up architectural plan dimensions at random to see if they consistently add up to the same structural column line dimensions. (Refer to Section 7.4.19, "Numerous Details and Dimension Strings.")

4. Individually overlay the architectural, plumbing, H.V.A.C. and lighting plans and note:

a. All apparent discrepancies of column locations;

b. Important or critical locations of potential conflicts, such as points where large ducts pass under large beams or where large drain lines pitch while they cross several beams. [Refer to Section 7.4.2, "Ceiling Spaces (Conflicts)."]

5. Cross-check the conflict lists of each design compared to step 4 to determine the points where multiple conflicts are likely. For example, the individual comparisons of the plumbing, H.V.A.C., and lighting plans with the structural plans may collectively indicate a designed overlap of a large duct, a sanitary drain line, and a recessed ceiling light—all at the same point directly under a large beam. In this kind of situation, there may be no problem—if the owner is willing to live with a 6' ceiling height.

6. Prioritize all such real or potential conflicts and identify all available solutions.

7. Analyze each possible solution and determine:

    a. The cost of each correction; and

    b. The net impact, if any, on planned sequences.

8. Prepare a change order proposal in accordance with the recommendations in Part Four for each significant conflict. Refer to Section 6.3.5, "Sample Notification Letter to the Owner on Changes," and determine if its use may be appropriate at this time.

7.4.6 Design Change Telltales

Description

While reading through the plans and specifications, it may become apparent that specification sections have been rewritten, design details have been changed, and material and equipment added or deleted—at the last minute. While there is nothing wrong with this practice in and of itself, these eleventh-hour modifications may have been done for any number of reasons. They may involve the correction of an architect/owner oversight or simply a change in an owner's requirements. Whatever the case, the corrections are done at a time when the plans and specifications are and have been generally complete. Dimensions have been worked out, the design disciplines have been coordinated, and the final design checks have been completed. Each last-minute modification then must be recoordinated with all other areas of the documents. Each affected dimension string must be corrected. The whole recoordination process must be done after the plans and specifications have been allowed to cool and after all systematic coordination efforts have stopped. As such, the potential for oversight and error with the introduction of each last-minute design change is great. (Examples of the specific kinds of coordination problems that all this late activity can generate are detailed in Section 7.4.19, "Numerous Details and Dimension Strings.")

Some of the telltale signs to look for in the documents that will confirm the extent of these risky activities include:

1. Different style type on the plans or in the specifications.
2. Handwriting in the specifications.
3. Hand-drawn details on computer-generated plans.
4. Out-of-sequence reference marks and details.
5. Out-of-sequence or inserted pages in specifications and details.
6. Different handwriting on the plans.
7. Different use of language (different ways of saying the same thing may lead to different interpretations).

Action

As you use the contract documents, try to become aware of the form of the presentation as well as the content. Observe the care with which the documents were prepared. A large amount of these telltales may be an indication that the design coordination process or even the design itself was completed improperly or that the owner was not sure of its requirements in the first place. Either way, it can mean a poor set of documents that will eventually be riddled with the kinds of problems described throughout this chapter. As problems are discovered, consider the action steps provided for each situation.

1. Review Section 7.4.12, "Inadequate Level of Detail," and Section 7.4.19, "Numerous Details and Dimension Strings," for a discussion of the most common types of problems generated by last-minute contract modifications. These sections also include a summary of the suggested action steps.

7.4.7 Design Discipline Interfaces

Description

Some of the most significant areas in which the risks of error and oversight are extremely high are the points where the various design disciplines interface with each other. Different project objectives determine the relative priorities of each design component. However, all systems structural, mechanical, electrical, site, architectural, and many specialty systems, must satisfy their own respective requirements while keeping within the grand scheme of the project. They must do so without introducing interferences or contradictions with each other. It is for this reason that the table of contents for Section 7.4 references specific examples of the design coordination issue.

Action

If a particular item is suspect, following the steps given in Section 7.4.5, "Column and Beam Locations," the same general process of systematic coordination will expose all potential areas of conflict resulting from lack of design coordination.

7.4.8 Duplication of Design

Description

Duplications in design can occur in the plans, specifications, or both. Duplications can manifest themselves in several forms, including:

1. Specifications for the same or similar items are found in two different sections.
2. Specifications for different items intended for the same function are included in two different specifications.
3. Descriptions for different items intended for the same function are included in the same specification.

When a duplication is discovered, the owner will invariably take the position that he or she is absolutely entitled to the work in question. After all, the work is specified *twice*. Additionally, if the work is substantial, expect a request for a credit proposal.

The larger the dollar value of the work in question, the more important it becomes to understand the concepts of Chapter 3, Proven Strategies for Applying Construction Contracts. By understanding and applying the rules of construction contract law, one may be able to channel owner approval directly to the *preferred* specification or method. In certain cases, it may be possible to demonstrate that even though the work is in fact described in two places, it was in reality left out of the bid altogether due to the contract structure and/or rules of contract interpretation.

Remember the rules regarding:

- Reasonable expectations (Section 3.3.2).
- Ambiguities (Section 3.3.3).
- Interpretation (Section 3.3.4).
- Rules of precedence (Section 3.3.5).
- Trade custom (Section 3.3.6).
- "Reasonable Review" (Section 3.4.7).
- "Intent" vs. "Indication" (Section 3.4.8).
- "Performance" and "Procedure" Specifications (Section 3.4.9).

Action

1. Determine all details of the apparent duplication.
    a. Is each description complete?
    b. Do the descriptions in different specification sections involve the responsibility of two or more contractors?
    c. Are the duplications in the same specification the responsibility of the same contractor?
    d. Is the same work specified twice, or are different treatments called for to accommodate the same function?

e. Of the different items and/or methods called for in the specifications, is one preferable over the other? Is one subcontractor preferred or better suited to perform the work described in the duplication? Which item or description fits within the constraints of the construction schedule?

2. Review all relevant contracts, plans, and specifications.
   a. Are any or all of the documents contracts of adhesion (refer to Section 3.3.3, "Ambiguities Resolved Against the Drafter")?
   b. Are any or all subcontractors owner-selected (refer to Section 7.3.2, "Named Subcontracts")?
   c. Are the respective subcontracts "per plans and specs," or are the scopes segmented?
   d. Are there any modifications or qualifications in any subcontract?
   e. Are the rules of precedence outlined in the specification or do you need to rely on applicable contract law (refer to Section 3.1, "Strategic Interpretation," and Section 3.3, "Rules of Contract Interpretation")?
   f. What plans are specifically the responsibility of each potentially affected subcontractor? Which are designed as being "related"? Refer to Section 7.4.24, "Specification Section 'Scopes.'"
   g. Do the "Scope of Work," "Work Included in This Section," or "Related Work" portions of the specifications for each potentially affected subcontractor help or hurt your objective(s)?

3. Analyze the duplications as objectively as possible.
   a. List the reasons why each subcontractor *should* and *should not* have included the work in question in its bids.
   b. Should any affected subcontractor have been *aware* of the work, but reasonably construed it to be accommodated by another trade?
   c. Did anyone contact the owner prior to bid with a request for clarification? If so, is there any documentation with regard to the request or the owner's response?
   d. Is each duplication clear in and of itself, or is its use strained or otherwise incomplete? (Refer to Section 3.3., "Rules of Contract Interpretation," and Section 3.4, "Applying Construction Contracts Without Resistance.")

4. Determine your preferred solution.
   a. Does any solution involve commitment of your own money, time, and effort, or are they all strictly involving subtrades?
   b. What are the dollar estimates of each solution?
   c. Is the time of each respective solution particularly good or bad?
   d. Will any of the potentially affected subcontractors be more inclined to accept the extra work without a battle?
   e. Does any solution make more sense, given design, cost, schedule, and quality considerations?

5. Determine if grounds exist to convince the owner that the duplication was reasonably not included in the general bid and that a price increase is appropriate.
   a. Reread the Chapter 3 sections noted under "Description."
   b. Considering your answers to action step 2, should the advice of an attorney be sought?
6. If the analysis under step 5 does not justify the submission of a request for an owner-approved change order, and your contract structure supports the action, submit a request to the owner for interpretation and direction.
   a. Refer to Section 7.3.2, "Named Subcontracts," Section 7.3.3, "Sample Letter to the Owner Regarding Obligation to Determine Responsibility for Questionable Work," and Section 7.3.4, "Sample Letter to Subcontractor Regarding Owner's Decision Directing Work."
   b. Modify the first paragraph of the letter to the owner to include the specific areas of the plans and specifications at issue. This will make the owner's research of the documents "easier" and guide the owner in the desired direction.

   For example, "Section 04200 3.4 requires the mason to provide the welded portion of masonry anchors attached to columns. Detail 4/S3 indicates the same work to be performed by the structural steel contractor."

   If for whatever reason it is advantageous to have the work done by the mason, the paragraph could be continued with: "General Conditions Article 4 states that specifications take precedence over plans." If it is preferable for the owner to decide that the work is the responsibility of the steel contractor, omit the foregoing line altogether. If additional technical items can be found to support the desired position, state them. If not, include references to trade custom (refer to Section 3.3.6) or anything else that supports the preferred position. Examples might include references to schedule, job practicalities, and so on.
   c. If one position is more favorable than the other interpretation, the contractor may want to make the owner's move to a particular decision more probable by listing the specific references as described above, emphasizing the references that support the desired position, including the interpretation right in the letter and by requesting *the owner's confirmation*. Upon the contractor's receipt of such confirmation, it then becomes the *owner's* decision. Follow through with the Section 7.3.4, "Letter to Subcontractor Regarding Owner's Decision Directing Work."
7. If the results of step 5 reveal that a request for change can be readily substantiated, prepare a detailed change order proposal in accordance with the recommendations in Part Four. Back up the contentions in the notification and change order proposal with specific references to the plans and specifications. Be sure to include the following in the request:

a. All technical plans and specification descriptions and details that support the request.
b. References to conflicting details.
c. The contract provisions that dictate the manner in which ambiguities are to be resolved (General Conditions entitled "Rules of Precedence" or "Correlation of Contract Documents," or something similar) or lacking such a provision, a reference to contract law and trade custom and usage may be sufficient (refer to Section 3.3, "Rules of Contract Interpretation," and Section 3.4, "Applying Construction Contracts Without Resistance").
d. A summary or conclusion logically deciding the issue in favor of a change order.

7.4.9 Sample Letters to the Owner Regarding Design Duplications

The Sample Letters to the Owner Regarding Design Duplications are coordinated with the discussion in Section 7.4.8, "Duplications of Design."

- Letter 1 is designed for use with Section 7.4.8, step 6b.
- Letter 2 is designed for use with Section 7.4.8, step 6c.
- Letter 3 is designed for use with Section 7.4.8, step 7.

In actual use, the content of each letter must be modified so that it is specific to each individual situation. Accordingly, the letters here are samples following through on the recommendations in Section 7.4.8 and those given below:

1. Describe the conflicting details *concisely* and by specific reference. See Section 7.4.8, action steps 6b, 6c. and 7, for examples.
2. Describe any contract mechanism that exists to decide the issue formally. For additional information, refer to:
   - Rules regarding ambiguities (Section 3.3.3);
   - Interpretation (Section 3.3.4);
   - Rules of precedence (Section 5.2.4);
   - Trade custom (Section 3.3.6); and
   - Reasonable review (Section 3.4.7).
3. Either request definite owner direction, or guide the owner to the desired decision by making the determination for the owner and requesting "confirmation." Make a logical deduction based on the reasons highlighted in Part Two.
4. Require action by a specific date and be clear that any delay will result in project delays, interferences and increased costs.

SAMPLE LETTERS TO THE OWNER REGARDING DESIGN DUPLICATIONS

LETTER 1

LETTERHEAD

(Date)

To: (Owner)

RE: (Project no.)
(Project description)

SUBJ: Welded portions of masonry anchors: Subcontractor responsibility

Mr. (Ms.)   :

    Section 04200 3.4 requires the masonry contractor to provide the welded portion of the masonry anchors attached to steel columns. Detail 4/S3 indicates the same work to be performed by the structural steel contractor. Both contractors have refused to perform the work as part of their contracts.

    Please confirm which party is responsible for performing the subject work. Your response by (date) is required to avoid (minimize) job delay and associated charges.

Very truly yours,

Project Manager

cc: Architect
    Masonry Contractor
    Steel Contractor
    Owner Site Representatives

SAMPLE LETTERS TO THE OWNER REGARDING DESIGN DUPLICATIONS

LETTER 2

LETTERHEAD

(Date)

To: (Owner) RE: (Project no.)
              (Project description)

SUBJ: Welded portions of masonry anchors: Subcontractor responsibility

Mr. (Ms.)   :

Section 04200 3.4 requires the masonry contractor to provide the welded portion of the masonry anchors attached to steel columns. Detail 4/S3 indicates the same work to be performed by the structural steel contractor. Both contractors have refused to perform the work as part of their contracts.

General Conditions, Article 4, states that the specifications take precedence over information in the plans. Accordingly, please confirm that it is your intention to have the mason contractor complete the work as part of Section 04200.

Very truly yours,

Project Manager

cc: Architect
    Masonry Contractor
    Steel Contractor
    Owner Site Representatives

SAMPLE LETTERS TO THE OWNER REGARDING DESIGN DUPLICATIONS

LETTER 3

LETTERHEAD

(Date)

To: (Owner) RE: (Project no.)
 (Project description)

 SUBJ: Concrete encasement at underground electrical ducts

Mr. (Ms.)   :

Section 16006 2.12 of prefiled subbid* No. 18 indicates that the concrete encasement at the underground electrical ducts is to be installed "in accordance with the requirements of Section 03300." Neither the site plans nor Section 03300 incorporates this work.

Accordingly, we request your approval of a change order in the amount of $15,000 to complete this work. The detailed proposal with all substantiating documentation is attached for your review.

Your approval is required by (date). If we do not receive a response by this date, the progress of the project will be delayed and we will add the costs attributable to the delay to the amount of this change order.

Very truly yours,

Project Manager

cc: Architect
 Masonry Contractor
 Steel Contractor
 Owner Site Representatives

*Type of owner-selected subcontract. Refer to Section 7.3.2, "Named Subcontracts."

7.4.10 "Fat" Specifications

Description

"Fat" specifications develop over a period of time at an architect's office. Every year more projects are completed, more battles are won and lost, more arbitrations occur, and more change orders are reluctantly given up. Each subsequent specification then attempts to "benefit" from the experience. With each success and failure, another clause, more boilerplate, and additional disclaimers are added to the specifications.

At first glance, a contractor who's confronted with a Fat Specification may become uneasy. After all, the specification with all its elaborations makes no effort to hide the fact that the specification's objective is to nail the contractor to the wall. It seems so ridiculously stacked in favor of the owner and/or architect that it is hard to see how any businessperson in his or her right mind would sign off on such a document.

There is, however, a significant difficulty for the owner and architect with a Fat Specification. It is based on the proposition that all the extra baggage which was added in a piecemeal fashion was not coordinated properly with the existing documents. In all probability, the extra language was added as each individual piece was dreamed up. Minimal or no effort may have been expended to see if the issue was already accommodated in some other fashion. Another kind of problem results from the new language being added to the specifications outside of the design context. For whatever reason, the intent of the additional provision is different from the intention of the persons who prepared the design.

The net result is a complicated, cryptic, and confusing specification that is riddled with overstatements, contradictions, ambiguities, and impossibilities. While the architect thought he or she was building a thick protective coating, the architect actually increased the probability of a contractual error and additional change orders.

Action

It is generally easy to spot a Fat Specification, particularly on a smaller project. The contractor should review the entire specification and should be intimately familiar with the project's requirements. The contractor must not be intimidated by the length of the specification or the architect's verbose writing style. Set forth below are several action items to keep in mind when reviewing a Fat Specification:

1. Quickly review the specifications for the following indicators:
   a. General Conditions, Special Conditions, etc., that are longer than the technical specifications;
   b. Extensive duplication in the general provisions;
   c. Descriptions and instructions that are much longer and more labored than necessary (i.e., using a paragraph to describe a responsibility, when a sentence would be sufficient); or
   d. The presence of many clauses describing requirements that are not normally encountered in specifications for this type of project.

2. Try to evaluate whether the architect, engineer, or owner seem to know what they are doing, or have prepared an overambitious specification to hide something:
3. Review the excess language for clues as to the attitude or perspective of the owner or the architect.
4. If the initial review of the specification reveals one or more of the indicators of a Fat Specification, make every effort to become intimately familiar with the specification's content so that the duplications, ambiguities, and contradictions may be spotted quickly.
5. When duplications, ambiguities and/or contradictions are encountered, immediately proceed with the recommendations included in this chapter for the quick resolution of the respective items.

7.4.11 Finish Schedule vs. Specification Index

Description

A comparison of the the Finish Schedule with the Specification Index can expose potential duplications and/or omissions. It can indicate whether or not the design process seems to have been completed in a coordinated manner, or if it was performed by different individuals who did not speak with each other. Any discrepancies discovered in this review will be fairly obvious.

Action

The comparison of Finish Schedule to the Specifications may be as simple as comparing the schedule's headings against the categories included in the technical specifications. During the comparison, the contractor needs to:

1. Confirm that each item is accommodated in the technical specifications, and that each item is only included once.
2. Upon the discovery of a discrepancy, duplication, or oversight, proceed with the respective recommendations included in this chapter.

7.4.12 Inadequate Level of Detail

Description

Providing an inadequate level of detail in a design necessary to complete a particular construction is an attempt by a designer to:

1. Avoid spending the proper amount of time necessary to complete the design; or
2. Shift the burden of unresolved or incomplete design issues to the contractor under the guise of "coordination."

Missing design information can take many forms, and can range from obvious to subtle. Some examples include:

1. Missing mounting or fastening details. (Do you want stainless steel brackets or rubber bands holding up the limestone?)
2. Not enough dimensions to allow even an elaborate calculation to locate the work properly.
3. Incomplete descriptions. (Is the blocking to be continuous, 24" on center, or eliminated altogether?)
4. Vague descriptions of special shapes, angles, and so on. (What is the *exact* angle of the spandrel glass?)
5. Imprecise, incomplete or inadequate layout information. (What is the radius of the curved stone in the front of the building? Do you have to lay it out in the parking lot to find out?)
6. Shop drawings are returned with "By GC" or "GC to Coordinate" notes on them instead of the dimensions and details requested.
7. Job meeting after meeting go by without definite and complete resolution of the deficient design issues.

These and other areas of deficient design eat into a contractor's anticipated profits because they require disproportionate amounts of time to resolve. Much time and effort is usually needed to convince the architect and the owner that the missing information is the designer's responsibility in the first place.

The most common effects of an inadequate or incomplete design are:

1. Field time is wasted each time a situation is encountered.
2. Unnecessary design liability is assumed by contractors who take it upon themselves to determine design and dimension solutions to allow the work to move forward.

Both of these effects cost the contractor time and money. Further, as mentioned previously, the assumption of design responsibility exposes the contractor to unnecessary risks.

Action

When faced with a situation involving an incomplete or an inadequate design, the contractor must:

1. Remember that the word "coordination" means *moving* information from one place to another in a timely manner and that coordination does not mean *inventing* information—that is design.
2. *Always* get the missing design information from those who are responsible to provide it, even in the seemingly most obvious circumstances.
3. Immediately confirm the completed design in writing if the architect has failed to document the corrected design and be sure to mention that it was the *architect* who ultimately provided the missing design information, no matter who originally came up with it.

4. Scrutinize "supplementary information," "clarifications" or whatever the architect chooses to call the document which contains the missing design information.
5. Determine if the supplementary information or the clarification includes any additional work or involves any extra costs.
6. Not be overwhelmed by documents, calculations or details.
7. Upon the determination that the "clarification" includes additional work, research the issue within the contexts of Chapters 2 and 3 and apply any applicable recommendations in Chapter 7.
8. Complete the process per the instructions in Part Four with the submission of the complete change order proposal.

7.4.13 Light Fixture Locations

Description

It is very common for light fixtures to conflict with other improvements at the project. This is because light fixtures:

1. Are in every area of the facility.
2. Are competing for room in confined ceiling spaces.
3. Must allow all other improvements in the ceiling areas to be located in their required locations (sprinkler heads, registers, diffusers, smoke detectors, etc.).
4. Must be located to conform to function, building code, and aesthetic requirements.

The first area in which the light fixture locations should be coordinated is between the electrical lighting plans and the architectural reflected ceiling plans. This simple overview alone will often reveal discrepancies in plan locations for the light fixtures. Many inconsistencies result from the fact that the reflected ceiling plans were prepared in the architect's office, while the lighting plans were prepared by the office of the electrical engineer and no one took the time necessary to properly coordinate the light fixtures, emergency lights, exit lights, switch locations, and the related work shown on the lighting plans.

Meanwhile, the mechanical plans, which show the H.V.A.C. registers, grilles, and diffusers, are prepared by the mechanical engineer. In many instances, the mechanical plans are schematic, incorporating some reference to coordinate their locations with the locations indicated on the reflected ceiling plans. If the persons who prepared the reflected ceiling plan took every precaution to be sure of the actual physical dimensions of each piece of equipment to be located in the ceiling, there is a good chance that the conflicts will be minimal. However, experience demonstrates that there are almost always conflicts.

Add sprinkler heads, smoke detectors, architectural shapes (soffits, recesses, etc.), any design that may be in the ceiling tile itself and the ceiling grid pattern, and it becomes easy to see the design coordination fiasco that can develop in the ceiling. If the ceiling design was properly completed, all these items will fit neatly in the horizontal plane of the ceiling.

In addition to the need to fit these improvements into the horizontal plane of the ceiling, there are even more structures that must be located above the ceiling and below the roof or the bottom of the floor above. These structures include structural members; water, sanitary, vent and sprinkler pipes; and ductwork. Invariably, the architect located ceiling recesses in spaces that looked clear. Those same spaces were also used by the plumbing, fire protection, and H.V.A.C. designers. So everything ends up being located on the individual drawings all in the same space. Refer to Section 7.4.2 for additional discussion of ceiling spaces.

Action

As with ceiling spaces in general, begin with the assumption that conflicts among light fixture locations and surrounding architectural, plumbing, H.V.A.C., structural, and architectural equipment and systems exist. Then take the following steps:

1. Overlay the lighting plans on the reflected ceiling plan and check the location of each:
   a. Ceiling light fixture;
   b. Emergency light;
   c. Soffit light;
   d. Exit light; and
   e. Undercabinet light.
2. Overlay the architectural plans and check for discrepancies that will affect light locations, ceiling patterns, and exit light layouts.
3. Overlay the H.V.A.C. plans and check the location of each register, grille, and diffuser. Confirm that the equipment's actual sizes are accommodated in the layout and that everything misses the lights.
4. Overlay the sprinkler layout and confirm that the heads miss the light fixtures.
5. Overlay the electrical power plans to make sure that the smoke detectors miss the light fixtures and everything else.
6. Catalog all conflicts discovered by this process and develop potential solutions for each conflict.
7. Assign recommended solutions to each conflict. Keep all conflicts on the list, even if the proposed solution does not add any additional time or money. Be sure that the project records document the correction for the following reasons:
   a. Although the owner's mistake has been fixed, all liability associated with the redesign should stay with the owner.
   b. As conflicts begin to pile up, they will ultimately interfere with the work. Even "no-cost" items take time and resources to resolve.

c. Since the contractor is doing the design coordination for the architect, the contractor should get paid for all this extra effort. A complete and accurate list can be used in any dispute resolution procedure to demonstrate the design team's failure to coordinate the work and the effort expended by the contractor.

8. Given the number, cost, and complexity of the changes discovered through this process, determine the best way to proceed in a manner consistent with the recommendations in Part Four. Refer to the Notification Letter to the Owner on Changes in Section 6.3.5 and consider whether the letter should be used at this time. Refer to Section 7.4.2, "Ceiling Spaces (Conflicts)," for additional related discussion.

7.4.14 Match Lines and Plan Orientations

Undeniably, match lines are necessary on many large projects. Fortunately, however, it seems as though many designers will minimize their use because of the increased drafting time and design coordination effort required to ensure that split plans are accurate. Stated simply, divided plans are difficult to read, check, and coordinate and the use of match lines provides one more possibility of an error or omission.

If match lines must be used, good design practice dictates that they should be placed precisely in the same location on every plan. If they are not, the plans will be even more confusing. In addition to increasing confusion, match lines placed in varying locations may be an indication of the general state of carelessness in the preparation of the plans altogether.

The plan orientations of all drawings should be identical. The north arrow should be in the same place, and the building outline should be consistent on each drawing. Again, it is very confusing to have different orientations among the mechanical, electrical, structural, and architectural plans. If these most basic design procedures are not followed by the design team, the contractor can expect problems at every level. Careless match line and plan orientation practices are indicative of a careless design attitude which will likely permeate all other design considerations as well. In other words, if the designers have not taken care of the most basic design/layout considerations, the contractor can expect that the more complicated responsibilities were probably completed improperly, if at all.

Action

If any match lines are present on the drawings, the contractor should analyze the plans to determine if the match lines were necessary.

1. Compare the match lines of all floors and of all drawings (architectural, structural, mechanical, electrical):
   a. Are they in the same location *every* time?
   b. Do they include the same information?
   c. Is anything missing?
   d. Are they complete and to the same extent on every plan?

2. Confirm that the north arrow is in the same place on each drawing (varying locations may quickly disclose changes in plan orientations).

3. Determine if plan orientations are the same for all drawings.

4. Negative answers to any of the foregoing considerations will expose improper use of match lines and/or otherwise deficient drafting practices. Upon such determination, reserve time to perform a detailed search of the plans to expose other areas in which the design process may have been compromised.

5. Return to the beginning of Section 7.3, and follow the action steps for each category of problem discovered during this review. Under these circumstances, it is almost a certainty that the plans will contain multiple flaws. It is critical to expose these flaws and their effects as early as possible to minimize the job impact and to secure prompt approval and payment for resulting change orders.

7.4.15 Mechanical, Electrical, and N.I.C. Equipment

Description

During the design phase, all mechanical and electrical equipment specified for the project should have been coordinated with and accommodated by the respective building systems. The obvious ones are simple: The front-end alignment machine needs a pit, and the track for the overhead hoist needs a steel beam for support. The equipment itself needs certain provisions: $1/2''$ gas line, 220-volt three-phase power, a 4'' drain, 9' head room, heated space, and so on.

One of the objectives of the shop drawing submission and approval process is to review these respective requirements:

1. To be sure that specified requirements are met;
2. To coordinate the information with all those who require it;
3. To compare the ultimate equipment configuration with the provision reflected in the facility design.

The physical characteristics of the actual equipment will determine whether:

  a. The equipment can be installed in accordance with the requirements of the current construction schedule (will the tanks fit through the overhead door, or does the roof have to be left off to get them in?); or

  b. The surrounding construction must be modified. For instance, does the $1/2''$ gas line shown on the plans need to be changed to $3/4''$ to accommodate the equipment's needs? Is the required concrete base shown on the plans? Will the architectural screen really cover the cooling tower?

The need for proper and timely coordination is obvious. The potential for a devastating interference increases with the quantity and complexity of the equipment. Hospitals, trade schools, and factories require a much more intense and sustained coordination effort than will warehouses.

Owner-furnished equipment and N.I.C. (Not in Contract) items are another source of potential confusion. Sometimes these items are indicated in the documents, but more often than not, the reference is inadequate. For instance, the reference to the N.I.C. equipment may be nothing more than a dotted outline of the designated item. As far as coordination is concerned, the contractor must evaluate whether to install the cabinets as shown, because the dotted lines designating the N.I.C. refrigerator fit in the space, or make the extra effort to confirm the equipment's actual dimensions with the owner before fabrication and/or installation. Refrigerators may not seem that serious, but what about gas burners, X-ray machines, film processors, built-in safes, and front-end alignment machines? These kinds of items requiring complicated and detailed provisions can either be accommodated properly during construction, or they can be left for dramatic and costly interferences when the project is about to be turned over to the owner. If the contractor fails to inquire about the characteristics of the N.I.C. equipment, any resultant problems will not necessarily become just the owner's problem (particularly if large dollar consequences result). The contractor should still get paid for technical changes to the contract; however, the payment may be reduced and/or changed on account of the various issues that need to be addressed and resolved in this situation. Some of these issues include:

1. Could the required change have been completed more quickly and cheaply had either the contractor or the owner coordinated the item earlier?
2. Who is primarily responsible for the delay in the equipment installation and use?
3. Are the problems attributable to the owner for failure to provide necessary information, or to the contractor for failure to coordinate? (Refer to Section 2.3 relating contract responsibilities of the owner, architect, and contractor.)
4. Use the Letter to the Owner Regarding N.I.C. Equipment Coordination included in Section 7.4.18 to either avoid this problem altogether or keep the responsibility with the owner.

Action

The equipment coordination begins with the shop drawing submission and approval process for contractor-supplied items. By contract, it is normally each respective contractor's responsibility to coordinate his or her own equipment. (Refer to Section 2.3.3 regarding contractor responsibilities and Section 3.4.3, "The Pass-Through Clause.") Included in this responsibility is the obligation to highlight clearly *all* differences in the actual items supplied from that indicated in the contract documents. This applies whether the item in question is specified equipment or an alternative product. The requirement is normally stated very clearly in the contract.

As part of the equipment coordination effort, the contractor should:

1. Send the Letter to Subcontractors Regarding Material and Equipment Coordination (see Section 7.4.16) to all trades.

2. Recognize that different companies respond to the change-highlight requirement in varying degrees if left entirely on their own. The contractor should not be required to bring this responsibility to each subcontractor's attention; however, responsible project management dictates that the contractor take the initiative and specify that all contract requirements are expected to be met properly. Each contractor must be made to understand that problems resulting from their failure to fulfill this responsibility properly will remain with them.

3. Coordinate equipment submittals with the applicable contract provisions. Note that:

   a. Changes in the contract provisions that become necessary due to differences in the actual configurations of specified equipment are the owner's responsibility and should be the subject of an owner-acknowledged change order (See Section 4.4.1).

   b. Changes in the contract provisions that become necessary as a result of differences in a contractor-proposed equal or substitution are the responsibility of the subcontractor requesting the change (refer to Section 7.4.16, "Letter to Subcontractors Regarding Material and Equipment Coordination").

4. If steps 2 or 3 expose additional work that is the owner's responsibility, submit a detailed change order proposal in accordance with the recommendations in Part Four.

5. Once all mechanical and electrical equipment submittals have been processed, any changes should have been identified and resolved. If it is a complicated equipment project, such as a trade school, consider Section 7.4.17 and the use of the Letter to the Architect Regarding Contract Equipment Coordination. This letter, when accompanied with appropriate attachments, presents a final opportunity for the designers to review all the mechanical, electrical, and other equipment that has been approved to this point and double-check that adequate provisions have been made for this equipment in the design of the facility. It is a final effort to discover any hidden coordination problems that may have been missed.

6. The final equipment coordination effort involves the items that are being furnished by the owner. It is not enough simply to be aware that the equipment is coming. The responsible contractor must know what it is, where it goes, what gets attached to it, where the provisions originate, and so on. More important, the contractor must know all these things in the stride of the current construction schedule. If not, the project will be delayed, disrupted or built in a deficient manner relative to the owner's equipment. The Letter to the Owner Regarding N.I.C. Equipment Coordination presented in Section 7.4.18 will set the wheels moving to finalize this important coordination.

7.4.16 Sample Letter to Subcontractors Regarding Material and Equipment Coordination

The Sample Letter to Subcontractors Regarding Material and Equipment Coordination is designed for use in 7.4.15, action step 1. Specifically, the letter is intended to:

1. Direct the subcontractor's attention to the subcontractor's obligation to:
    a. Coordinate all phases of its own work;
    b. Clearly highlight *all* differences and deviations in any submittal from the characteristics and configurations originally incorporated in the facility design.
2. Notify the subcontractor that all costs resulting from the subcontractor's failure to complete such coordination and highlighting will be the *subcontractor's* responsibility. (Refer to Section 3.4.3, "The Pass-Through Clause," for additional support for this contention.)
3. Emphasize that the subcontractor will be held responsible for any interference, disruption, and delay resulting from improper and/or untimely action.
4. Require proper action on *each* submittal by a specific date and in accordance with the requirements of the current construction schedule.

SAMPLE LETTER TO SUBCONTRACTORS REGARDING MATERIAL AND EQUIPMENT COORDINATION

LETTERHEAD

(Date)

To: (Subcontractor 1)

RE: (Project no.)
(Project description)

SUBJ: (Specification section or subcontract number)
Material and equipment coordination

Mr. (Ms.)   :

General Conditions (or other appropriate reference) Article ( ) requires that you call specific attention to *any and all* differences and deviations in the materials and equipment to be provided under your subcontract no. ( ) in their respective submittals. This requirement applies whether the items in question are submitted as specified, as equal, or as substitutions.

Your approval submissions are required on or before the dates indicated by the current construction schedule. It is your responsibility to be aware of the current schedule and to comply with it in every respect.

Please be advised that your failure to submit shop drawings in proper form and in a timely manner, as well as your failure to bring all deviations directly to the attention of (insert name of your own company) and the architect in your submittals may result in interference, disruption, and delay. Your company will be charged for all costs which result from your company's failure to comply with your contractual obligations.

Thank you for your cooperation.

Very truly yours,

Project Manager

cc: Owner
    Architect

7.4.17 Sample Letter to the Architect Regarding Contract Equipment Coordination

The Sample Letter to the Architect Regarding Contract Equipment Coordination is designed for use in Section 7.4.15, step 5. This letter is designed to:

1. Provide a convenient catalog of all mechanical and electrical equipment that has been approved to date, with the mechanical, electrical, structural, and unusual requirements for each.

2. Give all designers a final opportunity to review *their* contract plans and specifications to confirm that each piece of equipment has been properly accommodated.

   By utilizing the Sample Letter to the Architect Regarding Contract Equipment Coordination, the contractor is shifting a significant burden with regard to coordination to the design team. Even if the contractor does not receive any response to the letter, the designers will have assumed a significant amount of responsibility for the equipment coordination, by default.

3. Emphasize the need for speed and the idea that time is of the essence in regard to designer response, if the seriousness of any problem is to be kept to a minimum.

4. Notify the owner (by copy) as well as the designers that construction is continuing with the best information that is available at the moment. If extensive correction becomes necessary because the designers took their time to analyze the situation, it will not be the contractor's problem.

SAMPLE LETTER TO THE ARCHITECT REGARDING CONTRACT EQUIPMENT COORDINATION

LETTERHEAD

(Date)

To: Architect

RE: (Project no.)
(Project description)

SUBJ: Material and equipment coordination

Mr. (Ms.)   :

    Attached is the list of all mechanical and electrical equipment which has been approved to date and which will be incorporated into the project, along with their respective connection and support requirements.

    Please confirm that the present design of the facility will properly accommodate each piece of equipment as listed. If you should discover a discrepancy, please advise this office of any necessary corrective action immediately.

    Construction is progressing based on the information included in the contract documents and in accordance with the current construction schedule. To minimize the impact of any possible change at this late date, your immediate review and response is required.

    Thank you for your prompt response.

Very truly yours,

Project Manager

cc: Owner w/attachment
    Electrical Engineer w/attachment
    Mechanical Engineer w/attachment
    Structural Engineer w/attachment

7.4.18 Sample Letter to the Owner Regarding N.I.C. Equipment Coordination

The Sample Letter to the Owner Regarding N.I.C. Equipment Coordination is designed for use in Section 7.4.15, action step 6. The letter:

1. Requests that the owner provide all necessary information regarding all owner-furnished material and equipment for proper coordination of the work.

2. Requests that the information be provided in a timely manner relative to the requirements of the current construction schedule. As with Section 7.4.17, "Letter to the Architect Regarding Contract Equipment Coordination," this letter emphasizes the necessity for speed in owner response time.

3. Notifies the owner and the architect (by copy) that construction is continuing with the best information available at the moment. If extensive correction becomes necessary because of the owner's failure to furnish a timely response, the owner will remain responsible for all resulting costs.

SAMPLE LETTER TO THE OWNER
REGARDING N.I.C. EQUIPMENT COORDINATION

LETTERHEAD

(Date)

To: Owner

RE: (Project no.)
(Project description)

SUBJ: N.I.C. equipment coordination

Mr. (Ms.)   :

    Please provide all information regarding owner-furnished equipment and material upon your receipt of this letter so that this equipment can be properly coordinated with the work. We need this information, as well as the equipment and material, in sufficient time to allow incorporation without disruption or interference. Your attention is directed to the specific requirements set forth in the current construction schedule.

    As you know, construction is progressing in accordance with the information included in the current contract documents. We request your immediate response to minimize the impact of any possible change.

    Thank you for your prompt response.

Very truly yours,

Project Manager

cc: Architect
    Owner Field Representatives
    Any agency that you know of supplying
      equipment for the owner

7.4.19 Numerous Details and Dimension Strings

Description

One mark of a designer who is aware of the dangers of document changes is that the right information is shown the least number of times. If the design can be completed by providing all required information only once, it is to everyone's advantage. In contrast, if an item is detailed too many times, by the same or different design offices, two types of problems can occur:

1. As the number of details and incidence of repeated information increases, so does the probability of a discrepancy. (Where there are multiple and unnecessary dimension strings to the same end points, for example, there's always a good possibility that they will not add up.)
2. Design changes that occur during the latter stages of the design process have to be accounted for in every detail and dimension location that is affected by the design change.

Refer to Section 7.4.8, "Duplication of Design," for a related discussion of specifications as well as specific direction to take upon discovery of a conflict.

Action

It would be very difficult (if at all possible) to perform a systematic analysis of the plans to find all cases of multiple details and too many dimension strings. The problem is compounded by the increase in the number, size, and complexity of the plans. As such, the appropriate action for the discovery and correction of the resulting conflicts in this category involves the development of an awareness of good design practices versus bad. As the contractor becomes more familiar with the plans, the contractor should develop an almost reflex-like response that springs into action when the contractor observes duplication in the contract documents. The responsive reflex causes the contractor to double-check the precise information included at each location. The following action steps will help to develop this kind of awareness and action response:

1. As you observe repeated design details, make note of them. Too many details is the opposite of the kind of problems discussed in Section 7.4.12, "Inadequate Level of Detail." Each time a repeated detail or piece of information is discovered, check it against the previous details that were used for coordination.
2. When multiple dimension strings are apparent, spot-check them as often as convenient. Randomly add up duplicating lines of dimension strings to see if they are consistently correct.
3. Always double-check. Establish and maintain the discipline to immediately research any detail or notation that appears unusual at first glance. Confirm that the information is correct and that your first questioning impression was right. Do it now, or it will be forgotten until it affects job profitability.

4. Upon confirmation of an error, return to the action steps Section 7.4.8, "Duplication of Design." Consider the suggestions outlined there and proceed accordingly.

7.4.20 Performance and Procedure Specifications

Description

Section 3.4.9 described both performance and procedure specifications in detail. Refer to that section for description, application, and cautions surrounding these types of provisions. This section discusses the problems created when both a performance *and* a procedure specification are given for the same item.

If a product or material is specifically described with sufficient particularity, a contractor should be able to prepare a responsible estimate. If a performance specification is also given for the same material, the contractor is not obligated to verify that the precisely described and dimensioned material meets the stated performance requirements. That much was up to the designer.

In the most common situation, the difficulty arises because the performance requirement dictates the need for a material that in some way is superior (and more expensive) than the product or material that was specified. Whether thicker, more dense, smoother, straighter, or whatever, the designer will likely claim that the performance specification takes precedence over the procedure specification. Unless there is specific contract language which decides the issue, this does not have to be the case.

Action

When the contractor encounters a performance and procedure specification for a single item, the contractor should:

1. Analyze both the performance and procedure specifications completely.
    a. Are they mutually exclusive?
    b. Can they be reconciled or interpreted to be compatible?
    c. Which is more expensive?
    d. Which did the company estimate?
    e. Has one been included in the approved Schedule of Values?
    f. What will it cost to achieve both?
    g. Which specification does the owner/architect prefer?
    h. Is time or material availability a factor?
    i. Which is the more complete or otherwise more appropriate specification?
2. Determine the specification that is preferred from the standpoint of the contractor, given the circumstances.
3. Assemble the details and arguments that will best support the contractor's position. Consider:

a. Section 2.3, owner, architect, and contractor responsibilities.

b. Section 3.3, rules of contract interpretation.

c. Section 3.4, applying construction contracts without resistance.

Remember that a contractor has the right to choose a valid, albeit competitive, interpretation from among the alternatives.

4. After considering action steps 2 and 3, determine whether an acceptable function can be achieved while presenting a change order to provide the superior detail. Use action step 3 to support the "competitive" interpretation.

5. Assemble a complete change order proposal to provide the more expensive option. Follow the recommendations in Part Four, and be sure to state precise time requirements for owner action.

6. If a favorable response is not received by the stated date and project interference is imminent, consider notifying the owner that you are proceeding per your original interpretation unless a stop order is received. Alternatively, the contractor could offer to proceed "under protest" and with a complete reservation of the contractor's right to receive additional compensation as the result of the owner's failure to acknowledge the contractor's reasonable interpretation of the ambiguous specification. This last approach is disfavored because the contractor loses any advantage associated with performing the work and is required to finance this aspect of the project until the issue is resolved.

7.4.21 Proprietary Restrictions (Public)

Description

Specifications for public projects cannot unreasonably restrict sources of supply in a competitive situation. Except possibly for very esoteric, one-of-a-kind items, the requirements of a given product cannot be so narrow as to prevent or narrow competition unnecessarily. The common specification practice is to name a minimum of three acceptable manufacturers for a given product. In addition, the words "or equal" are normally included to allow a contractor to propose still more manufacturers of competing products giving the same or superior function, performance, and quality levels.

If, for example, one metallic floor hardener is specified (on a public project) without any mention of an allowance for an "equal" product, the specification should appear strange and should be investigated. The contractor's investigation should start with a review of the estimate for the project and a discussion with the estimator. The questions to be considered include:

1. Was the estimate based on the specified product or an alternate product?
2. If the estimate was based on an alternate product, is it equal to the specified product?

   a. Was the alternate product manufactured with precisely the same ingredients?

b. Are the installation instructions the same?

c. Do independent testing laboratory results confirm identical performance?

3. Is the specified product more expensive?
4. Is the owner insisting on the use of the specified product?

Under these conditions, the contractor should be entitled to a change order to provide the material specified in a proprietary manner.

Action

If a specification on a public project names fewer than three acceptable manufacturers, and/or fails to include the words "or equal," it may be an unlawful proprietary specification. Review the specification with legal counsel. Additionally, the contractor should consider the following approach:

1. If the proposed bid includes competing products, try to be as specific as possible in the proposal. If allowed by the bid documents and bid procedure, try to condition the bid on the use of the competing products. Since the rules governing public projects are designed to afford each bidder an equal opportunity to receive the contract and to avoid subjective evaluations, this procedure will not be available on traditional "hard" dollar public works. If the procedure is permitted and the competitive bid is accepted, the acceptance will include the "equal" product. If the bid documents allow no modifications, refer to Section 3.3.3, "Ambiguities Resolved Against the Drafter," and Section 3.3.4, "Right to Choose the Interpretation," for related discussion regarding adhesion contracts.

2. If using an "equal" product in a proprietary bid situation, be as clear as possible in the approval submission with regard to the right to use the equal.

   a. Comply with *all* submittal requirements in *every* respect. Do not leave any opportunity for the owner or architect to create delays and difficulties on account of technicalities.

   b. Indicate in the submission precisely why and how the material or item submitted meets or exceeds all function, performance, and quality criteria of the specified item. Be sure to highlight *all* differences. Address and resolve each difference in favor of the alternate product. Try not to give the owner or architect any opportunity to come to their own conclusions regarding any differences.

   c. If the price differential justifies the action, consult with an attorney concerning the legality of the proprietary specification. If the specification is unlawful, be sure to point out the problem to the owner.

3. The owner's response to action step 2 will be to:

   a. Approve the "equal" submission.

   b. Approve the "equal" submission, but only with a credit change order.

   c. Reject the "equal" submission, and insist upon the specified item or material.

4. If the owner accepts the "equal" only with a credit change order, the contractor can either:
   a. Evaluate the requested credit and determine whether it is acceptable given the time, trouble, and expense associated with keeping the issue alive. Be sure to weight this consideration against the backdrop of maintaining credibility in future similar actions.
   b. Reject the action. If step 2 involves a material that truly is equal, and the material was submitted against an illegal proprietary specification, the contract has every right to reject the owner's demand. Consider notifying the owner that the equal product will be provided and incorporated into the project without any adjustment to the contract price since the owner's action has confirmed the acceptability of the product and the only unresolved issue involves whether the requested credit is unwarranted.

   Refer to Section 7.4.22, "Sample Letter to the Owner Regarding Equal to Proprietary Item," for one way to consider treating the issue.

5. If the owner rejects the "equal" and insists upon the specified material or item, reiterate the arguments set forth in step 2. If the decision is to provide the specified material, advise the owner that the material is being provided in accordance with the owner's demand and with the expectation that a change order will be issued to cover the increase in price. Be certain that the notice mentions the owner's unwarranted action and that there is not any contractual support for the owner's position.

Refer to Section 7.4.23, "Sample Letter to the Owner Regarding Rejection of Equal for Proprietary Item," for an example of the letter described in step 5. Follow up this notification with a complete change order proposal prepared in accordance with the recommendations in Part Four.

7.4.22 Sample Letter to the Owner Regarding Equal to Proprietary Item

The Sample Letter to the Owner Regarding Equal to Proprietary Item is provided as an example of the procedure outlined in Section 7.4.21, step 4. The letter:

1. Acknowledges that the owner has constructively approved the "equal" item or material and that it is in fact acceptable for use in the project.
2. Rejects the owner's demand for a credit change order as being outside the owner's contractual rights.
3. Notifies the owner that the "equal" product will be provided without any change in price.
4. Advises the owner that if the owner interferes with the work, the owner will be expected to pay for resulting costs.

**SAMPLE LETTER TO THE OWNER
REGARDING EQUAL TO PROPRIETARY ITEM**

LETTERHEAD

(Date)

To: (Owner)

RE: (Project no.)
(Project description)

SUBJ: Equal approval of proprietary item (description of item)

Mr. (Ms.)   :

    The subject item (material) was submitted for approval against an improper proprietary specification. Your decision to approve this item on the condition of a credit change order is not in accordance with the terms of the contract and, as such, cannot be accepted.

    Because you have acknowledged the product to conform to the function, performance, and quality criteria of the specification, the use of this product has been approved. Accordingly, please be advised that the product will be furnished at no change in price.

Very truly yours,

Project Manager

cc: Architect

7.4.23 Sample Letter to the Owner Regarding Rejection of Equal for Proprietary Item

The Sample Letter to the Owner Regarding Rejection of Equal for Proprietary Item is provided as an example of the procedure outlined in Section 7.4.21, step 5. The letter is designed to:

1. Acknowledge that the owner has approved the "equal" item, and that it is acceptable for use in the project.
2. Reject the owner's action as being outside the owner's contractual rights.
3. Notify the owner that the illegally specified item will be provided but only with the expectation of a corresponding increase in the contract price.
4. Require a definite owner action by a specific date.
5. Advise the owner that:
    a. Failure to respond by the indicated date constitutes approval and acceptance of the change order.
    b. Any inappropriate action which interferes with the work will delay the project and all resulting costs will be submitted to the owner for payment as part of this change order.

SAMPLE LETTER TO THE OWNER
REGARDING REJECTION OF EQUAL FOR PROPRIETARY ITEM

LETTERHEAD

(Date)   Certified Letter

To: (Owner)   RE: (Project no.)
                                           (Project description)

                                        SUBJ: Rejection of equal
                                                      for proprietary item.
                                                      (Description of item)

Mr. (Ms.)   :

    The subject item (material) was submitted for approval against an improper proprietary specification. For the reasons stipulated in (refer to submission correspondence in Section 7.4.21, action step 2) the (equal product) meets or exceeds all specified requirements. Your decision to force the use of the specified proprietary item (material) is not in accordance with the contract and, as such, cannot be accepted.

    Accordingly, the (name the specified item) will be provided for the additional sum of $(   ), per the detailed change order proposal, attached.

    Your approval of this change order is required by (date). Please be advised that your failure to respond by this date constitutes acceptance of the change order by default. Any subsequent action which results in additional costs due to project delays and interferences will be added to this change order price.

    Please take notice.

Very truly yours,

Project Manager

cc: Architect
      Owner Field Representatives

7.4.24 Specification Section "Scopes"

Description

At the beginning of each technical specification section, there are usually two sections entitled "Work Included in This Section" and "Related Work Included in Other Sections" or something similar. The potential for error that exists is similar to the problem described in Section 7.4.1, "As Indicated." That is, the "Work Included in This Section" may only be "included" in the table of contents. Additionally, the items listed in the "Related Work Included in Other Sections" may have been assumed to have been placed there by someone else, and missed because of a breakdown in the design coordination process (refer to Section 7.4.7, "Design Discipline Interfaces"). This will be particularly true where the "Other Sections" involve another design discipline altogether, such as structural, mechanical, or electrical.

If precise specification section references are included in the "Other Sections," the chances are good that the design coordination process was completed properly. However, the only way to be sure is to check.

Action

It is not practical to perform a complete and detailed check of each specification section cross-reference prior to starting work on a project. However, the contractor should review each section quickly to determine if the design coordination process does or does not appear to have been done effectively. If anything appears unusual at this point, by all means check it.

In all likelihood, the opportunity or necessity to check individual cross-references will not present itself until some conflict requires resolution. At that time, detailed research will become absolutely necessary. In the event that specification sections are subsequently found to be missing or inadequate, refer to Section 7.4.12, "Inadequate Level of Detail," and Section 7.4.1, "As Indicated," for a detailed discussion of related problems and specific direction.

7.5 SITE

7.5.1 Introduction

Discovering change orders on the site boils down to comparing in precise detail what is really in existence at the site to the condition which is represented in the plans and specifications. Before any existing site conditions are disturbed, it is wise to verify and document properly the exact state of affairs. Taking these important precautions can protect the contractor from exposures as simple as repairing existing pavement cracks or as significant as removing hundreds or even thousands of extra cubic yards of earth from the site.

7.5.2 Grades, Elevations, and Contours

The new site contours are normally prepared by either a registered land surveyor or a professional engineer. At times, the existing data are verified before the new designs proceed, but often they are not. For the most part:

1. Grade contours are lifted from existing land surveys;
2. Manhole locations and storm and sanitary line invert elevations are provided by the city engineer's office or other authority;
3. Telephone and power line locations are verified with the respective utility companies; and
4. Transformer pad locations and configurations are coordinated with the power company.

During the bid preparation, the contractor has no choice but to rely upon the accuracy of the data. This information has direct effects on the estimates for:

1. Earthwork cuts and fills;
2. Demolitions and alterations;
3. Trench excavation lengths, widths, and depths;
4. Sequence of site activities;
5. Planned locations of material stockpiles, staging areas, and field offices;
6. Provisions for temporary power, phone, and water;
7. Surface water control;
8. Time and extent of road excavations and tie-ins;
9. Pavement cutting and patching;
10. Excavation shoring; and
11. Treatment of rock.

The difficulty, of course, occurs when there is a discrepancy between the actual site configuration and the condition depicted in the documents. As set forth above, the source of the discrepancies can be from any number of origins, but the procedure for prompt resolution will be similar for most of them.

Refer to Section 7.2.6, "Interferences from Utilities Not Properly Shown," and Section 7.4.4, "Changed Existing Conditions," for related discussions and action steps.

Action

With each new project, the contractor should:

1. Arrange to photograph the entire site before *anything* is done. Make a detailed film record showing the actual existing conditions. Photograph *all* areas within the contract limit line, whether or not work is being done everywhere. Film is cheap. Arguments are not.

2. Before the site engineer begins any portion of the building layout:
    a. Spot-check existing grades at several easily identifiable locations. If the spot-checks prove accurate, proceed with the building layout as planned. If any discrepancies are discovered in the spot-checks, continue with additional checks.
    b. Check the locations of manholes, markers, or any available indicator to verify the locations of existing telephone, water, sewer, and gas lines and fuel tanks.
    c. Open a few manholes and spot-check actual invert elevations. Compare the actual elevations with the information shown on the plans.
    d. Check the locations of telephone poles, street signs, pole guys, and any other existing facilities and confirm that the facilities don't interfere with roads, walks, pavement, excavations, or other site improvements.
    e. Check the actual horizontal distances between telephone poles, light poles, manholes, drainage structures, and so on. Compare the distances to the plans. Significant or multiple discrepancies may indicate an overall lack of quality in the plans and may be the basis for additional investigations.
3. If definite discrepancies are discovered, the owner must immediately be notified. Depending upon the extent of the discrepancy and the value of the correction, stopping the work in the affected area needs to be considered. This, however, is a very serious action and should not be implemented unless the preservation of the existing differing site conditions is necessary for prompt and/or equitable resolution of the problem. (Remember that many contracts and some laws provide that a differing site condition should not be disturbed until the owner or the owner's agent has had an opportunity to inspect the condition. See Section 5.6, "Latent Conditions (Defects)." If the dollars and time are or have the potential to become significant, the contractor's actions should be coordinated very closely with the advice of an attorney.
4. Prepare precise documentation of all relevant characteristics of the existing condition. If certified and stamped engineered drawings and calculations are required, make immediate arrangements for their preparation.
5. See Section 7.5.3, "Sample Letter to the Owner Regarding Changed Site Conditions," for an example of the kind of notification discussed in action step 3.
6. Regardless of whether the format suggested in Section 7.5.3 or another approach is selected based on the specific circumstances, follow up with a detailed change order proposal in accordance with the recommendations in Part Four.

In many situations, it may be difficult to generate the precise information necessary to calculate accurate estimates at the early notification stage. If this is the case, confine the owner notification to the facts that:

a. A discrepancy exists.
b. It will cost the owner money.
c. It is interfering with the work.
d. Detailed calculations are being developed now.
e. A change order will be submitted for approval and payment as soon as the detailed calculations are completed.

7.5.3 Sample Letters to the Owner Regarding Changed Site Conditions

The Sample Letter to the Owner Regarding Changed Site Conditions—Letter 1 is an example of the notification discussed in Section 7.5.2, "Grades, Elevations, and Contours," action step 3, and Section 7.4.4, "Changed Existing Conditions," action step 6. This letter notifies the owner that:

1. Discrepancies exist in a specific area.
2. All substantiating calculations are complete, and the costs are included in the attached detailed change order proposal.
3. Calculations are based upon established contract procedures.
4. Owner approval is required before the changed work can proceed.
5. Approval is required by the specified date.
6. All rights are reserved to claim costs and damages resulting from unforeseen effects.

SAMPLE LETTER TO THE OWNER
REGARDING CHANGED SITE CONDITIONS

LETTER 1

LETTERHEAD

(Date)

To: (Owner)

RE: (Project no.)
(Project description)

SUBJ: Changed existing conditions

Mr. (Ms.)   :

    Drawing SU-1 indicates the pipe invert elevation at M.H. #2 to be 35.5'. The actual invert elevation is 25.5'. The corresponding increase in the contract price is $(   ), computed in accordance with the procedures of Earth and Rock Excavation Article 4. Refer to the detailed change order proposal attached.

    Your approval is required by (date) to allow the work to proceed without additional interruption. The effects of the change activity sequences and durations on the construction schedule are now being reviewed. When the analysis is complete, you will be notified of the additional costs.

    Please take notice that we reserve all rights to claim all damages resulting from effects which we cannot anticipate at this time.

Very truly yours,

Project Manager

cc: Owner Site Representative(s)
    Architect w/att.

PROSPECTING FOR CHANGE ORDERS AND THEIR COMPONENTS

The Letter to the Owner Regarding Changed Site Conditions—Letter 2 is designed as an example of the notification discussed in Section 7.5.2, action step 5. This letter notifies the owner that:

1. Discrepancies exist in a specific area.
2. Research is continuing, and engineered data will be provided.
3. All costs resulting from this condition, including costs related to the effect of this condition on activity sequences and durations, will be calculated and, when enough is known, will be submitted for payment.
4. *Direction* is required from the owner to stop or continue work by a precise date.
5. If the owner has not responded to the notification by the specified date, the work will proceed immediately following the completion of engineered data necessary to document the condition.

SAMPLE LETTER TO THE OWNER
REGARDING CHANGED SITE CONDITIONS

LETTER 2

LETTERHEAD

(Date)

To: (Owner)

RE: (Project no.)
(Project description)

SUBJ: Changed existing conditions

Mr. (Ms.)   :

It has been discovered that existing site grades are significantly different from those indicated on drawing L-1. Per my conversation with Mr. (Name), we have made arrangements with a registered land surveyor to immediately complete a corrected layout and calculation. When the data is complete, you will be advised of all costs associated with this change.

We anticipate the survey will be complete on (date). Please confirm whether we should either *stop work* or proceed with the additional work by (day before "date" above) to minimize the interferences caused by this change.

The effects of the changed activity sequences and durations on the construction schedule are now being reviewed. When this analysis is complete, you will be notified of the additional costs.

Please take notice that we reserve all rights to claim all damages resulting from effects which we cannot anticipate at this time.

Very truly yours,

Project Manager

cc: Owner Site Representative(s)
    Architect w/att.

7.6 CHANGE ORDER DISCOVERY CHECKLIST

Sections 7.1 through 7.5 cataloged many of the most common sources of changes that occur in the various stages of every construction project. As with the Change Order Research Checklist in Section 6.3.3, the Change Order Discovery Checklist summarizes the concerns of the entire section. It is designed to provide a convenient mechanism for the identification, review and clear consideration of all potential effects of a particular change. It is an organized summary of Chapter 7 that catalogs the precise areas to consider when performing a systematic and complete search of the contract documents. Performing the analysis at the earliest possible time will expose all real and potential interferences *before* they have an opportunity to do their worst damage. Completing the detailed checklist will help guide the responsible project manager through all the considerations mentioned in this chapter, from the most basic to the complex, without the risk of oversight.

When the components of an apparent change are discovered, refer to the appropriate Chapter 7 section for detailed discussion and action steps. Reread Chapter 6, Using the Change Order Process to Your Maximum Advantage. Refer then to Section 6.3.3, "Change Order Research Checklist," and complete *that* process for each potential change discovered.

Each section of the Change Order Discovery Checklist is coordinated with that corresponding section in this chapter. When your checklist answers seem to point to an apparent source of a potential change order, refer back to the respective section for a more detailed review of the concepts and related action steps.

CHANGE ORDER DISCOVERY CHECKLIST

A. PREDESIGN   YES   NO

1. 7.2.1 Adjacent Properties
   a. Have all properties adjacent to the site perimeter been reviewed in detail?   ____ ____
   b. Are there:
      - Seasonal watercourses?   ____ ____
      - Heavy traffic patterns?   ____ ____
      - Other independent construction activities?   ____ ____
      - Other _____   ____ ____
      - Other _____   ____ ____
2. 7.2.2 Boring (Subsurface) Data
   a. Are boring depths inconsistent?   ____ ____
   b. Are boring locations erratic or unusual?   ____ ____
   c. Are boring locations relevant to construction?
      - Are borings provided outside the area?   ____ ____
      - Are gaps left within the building area?   ____ ____
   d. What time of year were the borings taken? ____
3. 7.2.3 Building Code Compliance
   a. Have any violations of the building codes been observed by any building official when the building permit was applied for?   ____ ____
   b. Do any portions of the design appear out of the ordinary?
      - Headroom?   ____ ____
      - Entrances/exits?   ____ ____
      - ADA provisions?   ____ ____
      - Fire separations?   ____ ____
      - Lighting?   ____ ____
      - Ventilation?   ____ ____
      - Other _____   ____ ____
      - Other _____   ____ ____
4. 7.2.4 Easements/Rights of Way
   a. Are there designated easements?   ____ ____
   b. If so, will they adversely affect the construction operations?   ____ ____
   c. Do local traffic patterns restrict access?   ____ ____
   d. Are there parking areas, traffic patterns, business, etc., at the contract limit line that will restrict operations in any way?   ____ ____
   e. If the answer to 4.a is yes, are there conditions on the easement?   ____ ____
   f. If a restriction to the construction operations is evident, has the estimate accommodated it in some way?   ____ ____

g. If the answer to 4.f is no, should a reasonable prebid site investigation have disclosed the condition? ____ ____
5. 7.2.5 Special Agency Approvals
   a. Does any portion of the site encroach on an inland wetlands or any other protected area? ____ ____
   b. If so, are all appropriate approvals in place? ____ ____
   c. If required approvals are not apparent, have you requested the confirming information from the owner? ____ ____
6. 7.2.6 Interference from Utilities Not Properly Shown
   a. Have the characteristics of all existing utilities been verified with each respective company? ____ ____
   b. Has each company representative reviewed the details at the site? ____ ____
   c. Is anything different from that represented on the plans? ____ ____
   d. Are the current utility charges for the various tie-ins the same as those given at the time of bid? ____ ____
7. 7.2.7 Plan Approvals (Building Permit)
   a. Has the building permit been applied for at the earliest possible time? ____ ____
   b. Were there *any* problems? ____ ____
   c. Were any notes or corrections made on the plans? ____ ____
   d. Has the permit been delayed in any way? ____ ____
   e. Is a permit required (and a Certificate of Occupancy necessary) for temporary field offices? ____ ____
8. 7.2.8 Temporary Utilities—Availability Within the Contract Limit Lines
   a. Have the anticipated conditions at the time of bid been confirmed? ____ ____
   b. Are conditions adequate? ____ ____
   c. Are site conditions now different? ____ ____
      - Are additional telephone/power poles needed? ____ ____
      - Is power available at all (without generating equipment)? ____ ____
      - Is previously anticipated use of existing facilities now prevented? ____ ____
      - Are temporary heat and protection now required due to owner-caused delay? ____ ____
      - Is water available in sufficient amounts for construction? ____ ____

B. THE CONTRACT AND BID DOCUMENTS
1. 7.3.1 Award Date
   a. Has an extension in the contract award date been requested? ____ ____

b. If so, is there any basis upon which to ask for an increase in the contract sum?
  - Will acceleration be necessary?
  - Will a portion of the project now be placed into unfavorable weather conditions as a result of the start-up delay?
c. Can more favorable contract terms be required at this time?
  - Is your bid substantially lower than the next bidder's?
  - Can you complete the facility in less time than your competitors?
  - Were you involved in design development?
  - Is the owner tied to you in any way?

2. 7.3.2 Named Subcontracts
   a. Are there owner-selected subcontracts on the project?
   b. Does any disclaimer exist that limits the owner's liability for subcontractor selection?
   c. Are the subcontract agreements themselves owner-defined?
   d. Is any specific procedure in place to resolve disputes between two owner-defined subcontracts?
   e. Will the owner in fact make decisions (or will there be constant attempts to drop the responsibility on the general contractor)?

3. 7.3.5 (Price/Bid) Allowances
   a. Are there allowances anywhere in the contract?
   b. If so, have all allowance items been bid or rebid yet?
   c. Have or will all allowance items been awarded in time to prevent schedule interruption?

4. 7.3.6 Contract Time
   a. Did the first schedule draft drastically exceed the allowed contract time?
   b. Did subsequent schedule drafts incorporate unusual or excessive compressions and accelerations?
   c. Did any long-lead-time purchases dramatically exceed the originally anticipated deliveries?
   d. If so, were they for specified items?
   e. Had the contract award date been extended?
   f. Had the site start date been extended for an owner-caused reason?
   g. If the answer to either 4.e or 4.f is yes, was the schedule logic affected?

h. Did extra work result? ____ ____
i. Can clear cause–effect relationships be demonstrated to justify more contract time? ____ ____

C. PLANS AND SPECIFICATIONS
1. 7.4.1 "As Indicated"
   a. Are notes without specific references common (such as "As Indicated," "See Specs," "See Plans," and so on)? ____ ____
   b. Has each vague reference been researched to confirm that completing details do in fact exist? ____ ____
   c. If so, were there any incomplete, conflicting, or missing references? ____ ____
   d. If so, has each instance been cataloged for individual consideration? ____ ____
2. 7.4.2 Ceiling Spaces (Conflicts)
   a. Is there a contract clause clearly noting that the sub- or trade contractors are responsible for the coordination of their work? ____ ____
   b. Have all areas of potential conflict in the ceilings been properly coordinated?
      - Is there enough room to pitch all pipe? ____ ____
      - Do pitched lines miss all steel and concrete beams? ____ ____
      - Can all ducts pass below beams at all locations shown? ____ ____
      - Do too many items occupy the same space in any area? ____ ____
      - If so, can enough space be made, or can anything be moved? ____ ____
      - Are there large ducts shown to cross large beams and/or other significant obstructions? ____ ____
      - Will all light fixtures fit in the remaining spaces:
         Height? ____ ____
         Plan? ____ ____
      - Are there elaborate architectural, structural, or special shapes continuing into the ceiling? ____ ____
      - If so, do other building systems or equipment penetrate any part of them? ____ ____
      - If so, has the *actual* size of each part or item been confirmed? ____ ____
3. 7.4.4 Changed Existing Conditions
   a. Has the estimate been reviewed for:
      - All site work considerations? ____ ____
      - Any interferences with existing structures? ____ ____
      - Any noted conditions of existing structures? ____ ____

- Locations, extent, makeup, and conditions of existing utilities?
- Traffic patterns and site access?
- Anticipated storage and staging areas?
- Parking and security arrangements?
   b. Have the estimators visited the site to review all items in (a) with the project management team?
   c. Have any changes between conditions existing now and those existing at the time of bid become apparent?
4. 7.4.5 Column and Beam Locations
   a. Have the structural drawings been reviewed in detail?
      - Are column layouts erratic or unusual?
      - Are there any unusually long spans requiring relatively large structural members?
      - Are there unusual shapes, angles, slopes, or connections?
      - Are elevation changes strained or confusing?
      - Are beam sizes all different (with different ceiling spaces below them)?
      - Have the locations of all large beams been reviewed?
      - Are there unusual designs?
      - If so, is enough information included for proper shop drawing preparation the first time around?
   b. After reviewing the architectural, plumbing, H.V.A.C., and electrical plans:
      - Are listed column line dimensions between all designs consistent?
      - Are there large ducts shown crossing large beams?
      - Are there light fixtures in the areas of large ducts?
      - Does the sprinkler main cross large beams, ducts, or light fixtures?
      - Do random spot-checks or architectural dimension strings reveal any discrepancies?
5. 7.4.6 Design Change Telltales
   a. Is there a large number of apparent last-minute design changes?
      - Different styles of type or handwriting in the specifications?
      - Incomplete erasures?
      - Out-of-sequence reference marks or inserted pages in the specifications?

- Different handwriting on the plans? ____ ____
- Different use of language for the same or similar remarks? ____ ____

6. 7.4.7 Design Discipline Interfaces
   a. Has any review to this point revealed any problems at the points where design disciplines cross each other? ____ ____

7. 7.4.8 Duplication of Design
   a. Have any duplications been observed? ____ ____
   b. If so:
      - Is each description complete? ____ ____
      - Are the descriptions in different specification sections with different contractors involved? ____ ____
      - Are the duplications included in the same specification? ____ ____
      - Is the same work specified twice? ____ ____
      - Is different work specified for the same function? ____ ____
      - Is any of the available options preferred? ____ ____
   c. In a review of relevant contracts, plans, and specifications:
      - Are any or all contracts adhesion contracts? ____ ____
      - Are any subcontractors owner-selected? ____ ____
      - Are the affected subcontracts "per plans and specs"? ____ ____
      - Are there modifications to any contract? ____ ____
      - Are the rules of precedence outlined in the specification? ____ ____
      - Are all affected plans noted to be the responsibility of the affected subcontractor(s)? ____ ____
      - Do the descriptions of work in the affected and related specification sections help establish responsibility for affected work? ____ ____
   d. Objectively analyze each duplication:
      - Have all the reasons why each subcontractor should and should not have carried the work in their bids been considered? ____ ____
      - Should any contractor have reasonably construed it to be included by another trade? ____ ____
      - Did anyone request clarification from the owner prior to bid? ____ ____
      - If so, is the request and/or response documented? ____ ____
      - Is each duplication clear and complete in itself? ____ ____
   e. Is there a preferred solution?
      - Does any solution involve your own time or money? ____ ____

- Are the dollar estimates of each solution a consideration?
- Is the timing of any solution particularly good or bad?
- Is any potentially affected contractor more inclined to accept the extra work?
- Does any solution make more sense?
  f. Do grounds exist to convince the owner that duplicated work is in fact not included anywhere?
8. 7.4.10 "Fat" Specifications
   a. Does a review of the documents reveal:
      - An unusually fat "front end"?
      - Extensive duplication in the general provisions?
      - Long and/or labored descriptions and instructions?
      - "Catch-all" phrases and boilerplate not specifically applying to project conditions?
9. 7.4.11 Finish Schedule vs. Specification Index
   a. In a comparison of the Finish Schedule to the Specification Index:
      - Is each item accounted for?
      - Is each item included only once?
10. 7.4.12 Inadequate Level of Detail
    a. If enough design information has not been originally provided:
       - Will the architect respond now with the complete information?
       - Is it confirmed in writing?
       - Are there any additional cost implications?
11. 7.4.13 Light Fixture Locations
    a. In overlaying the lighting plans on the reflected ceiling plans, are there conflicts in:
       - Ceiling light fixtures?
       - Emergency lights?
       - Soffit lights?
       - Exit lights?
       - Undercabinet lights?
    b. In overlaying the architectural plans, are there conflicts in walls, soffits, or cabinets?
    c. In overlaying the H.V.A.C. plans:
       - Are there conflicts in register, grille, or diffuser locations?
       - Are the actual equipment sizes accommodated?
       - Does everything miss the lights?
    d. In overlaying the sprinkler plans:
       - Do the heads miss the lights?

- Do the heads fall in the center or quarter center of the ceiling tile?
- Is there an architectural pattern in the ceiling tile that will change location preference?
  e. In overlaying the electrical plans:
  - Do the smoke detectors miss the lights (and everything else)?
12. 7.4.14 Match Lines and Plan Orientations
    a. Are match lines present?
    b. If so:
       - Are they necessary?
       - Are they in the same location *every* time?
       - Do they include the same information?
       - Is anything missing?
       - Are they complete and detailed to the same extent on every plan?
    c. Is the north arrow in the same place on each drawing?
    d. Are the orientations the same for each plan?
13. 7.4.15 Mechanical, Electrical, and N.I.C. Equipment
    a. Are differences highlighted in *all* approval submissions?
    b. Has the Section 7.4.16 Letter to Subcontractors Regarding Material and Equipment Coordination been sent?
    c. Has the Section 7.4.17 Letter to the Architect Regarding Contract Equipment Coordination been sent?
    d. Has the Section 7.4.18 Letter to the Owner Regarding N.I.C. Equipment Coordination been sent?
14. 7.4.19 Numerous Details and Dimension Strings
    a. Have repeated design details been observed?
    b. Are there many instances of multiple dimension strings?
    c. If so, have spot-checks uncovered errors?
15. 7.4.20 Performance and Procedure Specifications
    a. Are there any instances in which both performance *and* procedure specifications occur for the same item?
    b. If so:
       - Are they mutually exclusive?
       - Can they be made to be compatible?
       - Is one or the other more expensive?
       - Is one preferred over the other?
       - Has one been included in the Schedule of Values?

- Is it cost-prohibitive to accomplish both?
- Is time or material availability a factor?
- Is one more complete or otherwise more appropriate?
       c. Have all the details and arguments supporting the preferred position been assembled?
   16. 7.4.21 Proprietary Restrictions (Public)
       a. Does the specification being considered:
           - Name fewer than three acceptable suppliers?
           - Include the words "or equal"?
       b. Do you intend to use an "equal" product?
       c. If so, does the owner want a credit change order?
       d. If so, has the Section 7.4.22 Letter to the Owner Regarding Equal for Proprietary Item been considered and sent?
       e. Has the owner rejected the "equal" submission?
       f. If so, has the Section 7.4.23 Letter to the Owner Regarding Rejection of Equal for Proprietary Item been considered and sent?
   17. 7.4.25 Specification Section "Scopes"
       a. Does the design coordination process appear to have been done correctly?
       b. Are specific cross-references included?
       c. Does the Scope section appear to be complete?

D. SITE
   1. 7.5.2 Grades, Elevations, and Contours
       a. Has the entire site been photographed *before* any work has begun?
       b. Have the existing grades been spot-checked for accuracy?
       c. If so, have any discrepancies been discovered?
       d. If so, has a detailed check been arranged?
       e. Have the locations of existing telephone, water, sewer, and gas lines been verified?
       f. Have the manholes been opened to spot-check actual pipe invert elevations?
       g. Have the locations of telephone poles, street signs, pole guys, and any other constructions been checked to avoid interference with the site improvements?
       h. Have the actual horizontal distances between telephone poles, light poles, manholes, drainage structures, etc., been checked for accuracy?
       i. Have any discrepancies discovered been documented in the most accurate and unquestionable manner available?

Part Four
Change Order Proposal, Preparation, and Presentation

8

Designing and Constructing Effective Change Order Proposals

8.1	Change Order Components	183
	8.1.1 Introduction: The Three Costs	183
	8.1.2 Direct Costs	184
	8.1.3 Indirect Costs	184
	8.1.4 Transforming Indirect Costs into Direct Costs	185
	8.1.5 Direct Project Management and Administrative Costs FORM	186
	8.1.6 Consequential Costs (Damages)	188
	8.1.7 Practical Management of the Three-Cost Approach	189
8.2	Developing the Change Order Proposal	190
	8.2.1 Change Order Identification/Notification	191
	8.2.2 Sample LETTER TO THE OWNER Regarding Pending Change Order	193
	8.2.3 Assembling Component Prices	195
	8.2.4 Sample Change Order General Conditions CHECKLIST and ESTIMATE SHEET	196
	8.2.5 Assembling Subcontract Prices	198
	8.2.6 Sample LETTER TO SUBCONTRACTOR—Request for Change Order Quotation	199
	8.2.7 Sample LETTER TO SUBCONTRACTOR—Change Quotation, Second Request	201
	8.2.8 Sample LETTER TO SUBCONTRACTOR Regarding Change Order Price by Default	203
	8.2.9 Sample Change Order Telephone Quotation FORM	205

	8.2.10	Sample LETTER TO SUBCONTRACTOR Confirming Telephone Quote	207
	8.2.11	Determining Schedule Impact	209
8.3	Finalizing the Proposal	212
	8.3.1	Introduction	212
	8.3.2	Proposal Format and Timing	213
	8.3.3	Sample Change Order Proposal COVER LETTER	214
	8.3.4	Sample LETTER TO THE OWNER Regarding Change Order Cost Escalation Due to Untimely Action	217
	8.3.5	Presenting Change Order Components	219
	8.3.6	Presenting the Total Change Order Price	221
	8.3.7	Presenting the Effects on Contract Time	221
	8.3.8	Requiring Approval Action	222
	8.3.9	Additional Terms and Conditions	223

8.1 CHANGE ORDER COMPONENTS

8.1.1 Introduction: The Three Costs

Chapter 7 outlined the most common areas in which to uncover contract changes. If not actively sought out, these conflicts can have a serious detrimental impact on the project schedule, the cost of the work and the contractor's overall performance. Early detection, accurate determination of scope changes, and firmly pinpointing responsibility for the added costs should facilitate the prompt approval of requested changes. Even though the facts giving rise to the requested change can be confirmed through proper measurement and good bookkeeping, sometimes securing the owner's agreement that extra compensation is warranted is another matter entirely.

Before change order proposals can be designed and constructed, an understanding of their components must be developed. As with all construction contracts, change orders have three basic components:

1. scope;
2. time; and
3. costs.

Three cost categories will add up to the maximum amount of compensation justified for the contractor on the subject change. The three cost categories are:

1. Direct costs;
2. Indirect costs; and
3. Consequential costs (damages).

As was first discussed in Section 3.4.10, "Equitable Adjustment," their application to a particular change is largely a matter of degree. Consequential costs on a very small change, for example, may be so small as to not justify any lengthy debate. On another change, however, the consequential costs may dwarf the actual direct cost portions, due to extensive interference, delays, and disruptions. The application of each cost category must, therefore, be individually considered for each change proposal. If it is then decided to omit a particular cost category, it will be done after proper consideration and by design. Even in this case, however, the omitted category should be accommodated in some manner so as to leave the possibility of a further request open if circumstances change. A clear understanding of the cost categories and their relationship to one another and to the change order proposal will enable the responsible project manager to determine whether all three costs categories should be included in a particular proposal. This chapter will develop the methods for developing and analyzing each cost category. Additionally, this chapter will discuss the presentation technique that will accomplish the contractor's objectives.

8.1.2 Direct Costs

Direct costs are usually thought of as the "hard" costs of a change. They include those items that are specifically and uniquely attributable to the modified work. Although many times direct costs are thought of as the on-site expenses, in reality they are usually much more. All direct costs share the following common characteristics:

1. They are clearly assignable to an event; and
2. They can easily be documented.

By their nature, they also facilitate accurate accounting. Because of their clear cause–effect relationships, they are readily justifiable as change order cost items. They include such things as:

1. Labor;
2. Material costs;
3. Site supervision;
4. Off-site material carrying costs;
5. Shipping costs;
6. Restocking and/or cancellation charges;
7. Additional performance and payment bond premiums;
8. Temporary protection;
9. Temporary heat, light, and power;
10. Additional material handling costs;
11. Safety equipment, staging, scaffolding, and lights; and
12. Any other item directly and specifically attributable to the change.

The keys traits of direct costs are that they must be easily and clearly identifiable and they must lend themselves to complete, comprehensive accounting.

8.1.3 Indirect Costs

Indirect costs (the soft costs) are those items that, although precipitated specifically by the change, are less obvious and may be more difficult to recognize. If recognized, they are also harder to apportion to a specific event, and, as such, are more difficult to count. Most indirect costs can be demonstrated to apply in principle without too much trouble once their relationship to the change proposal is understood. The amount that is to be applied to the change in question, however, is the issue that will be debated. Most arguments revolve around the question of whether the item can be considered as a specifically attributable cost, or whether it is simply part of the contractor's "overhead." Additional bond premium costs, for example, are often caught up in this argument. Indirect costs include such items as:

1. Home office overhead;

2. Off-site supervision;
3. Change order preparation, research, negotiation, and associated travel;
4. The effects of project interference and disruption;
5. Time delays; and
6. Increases in guaranty and warranty durations.

The keys to keeping an indirect cost item clearly identified as a specific change order expense in both principle and absolute value are clear documentation and accurate, verifiable records. By establishing clear change order accountability the contractor can transform an indirect cost item into a direct cost. This transition must be accomplished in the mind of the owner and, if done, the contractor will be well on his or her way to maximizing the total cost amounts against which the contractor will then apply the overhead and profit (or fee) percentages. If this conversion is not accomplished, these kinds of "soft" costs will be dumped into "overhead," if allowed at all.

8.1.4 Transforming Indirect Costs into Direct Costs

In order to change an indirect cost into a direct cost the contractor must clarify and confirm its applicability to the change proposal. The contractor must demonstrate to the architect and the owner that the cost item in question both is caused directly by the modification and has been accurately measured. Demonstrating the cause–effect relationship will require skills commensurate with the cost of the item and the complexity of the relationships. If the cost is high and the interaction complex, a comprehensive schedule analysis showing all related effects may be required to convince the owner that it is in fact the change (and not some peripheral activity) that is giving rise to the added costs. In other cases, the relationship will be more clear and it will be easier to show that the cost is a direct result of the modification.

In any event, agreeing to the applicability of the cost item in principle is the first part of the problem. The second, usually more difficult, part is assigning a value to the item in a convincing manner. The cost assignment must be verifiable to the greatest extent possible. This goal may be accomplished by establishing a procedure that will first demonstrate that keeping track of the respective costs is a normal part of the business. The contractor must be able to demonstrate that the activities are routinely accounted for in the manner presented. As part of the procedure, the costs must be cataloged and clearly assigned to the change activity. If all this can be accomplished, the cost category can be transformed from the vague classification entitled "overhead" to a direct item against which the contractor should be able to add the overhead percentage. "Overhead" will rightfully be left to only those items that still cannot be appropriated directly to a specific event.

An example of the kind of procedure that will help make the transition from indirect to direct cost more smooth is the Direct Project Management and Administrative Costs Form in the next section.

8.1.5 Direct Project Management and Administrative Costs Form

The Direct Project Management and Administrative Costs Form is an example of the effort described in Section 8.1.4 to transform otherwise indirect cost items into direct, measurable charges. It is designed to take the "Change order preparation, research, negotiation, and associated travel" item in Section 8.1.3, "Indirect Costs," and create a direct, billable item. If this can be accomplished, the real cost of the item for the respective change will become clear. Overhead will then be applied to this cost as with any other direct charge. By utilizing the procedure set forth in the Administrative Costs Form, the contractor can identify certain indirect costs as being attributable to a particular change and can prevent these costs from being absorbed into "overhead."

The form gives the impression that cataloging and keeping track of the respective cost items is a normal part of the business. It gives the message that there is nothing unusual about segregating the real charges that are caused by a particular change. It makes it clear that the contractor expects to be compensated for all costs associated with or arising from the change (refer to Section 3.4.10, "Equitable Adjustment," for more on this concept).

Finally, it is important to note that it is not necessary to come up with *all* charges on the form each time that it is applied. For instance, the contractor may indicated "N/A" (not applicable) in the line items not being used at this time. Treating the noncharged items in this way will accomplish two things. First, it will confirm to anyone reviewing the change proposal at a later date that the cost item in question was considered and that its omission was specifically intended. Second, it demonstrates the same thing to the owner. It communicates that although it has been decided that the cost item will not be included on *this* change (presumably due to its small size), it can and will apply on later changes if its size increases significantly. It may even apply to *this* change if owner inaction increases required management attention. This idea will be developed further in Section 8.1.7, "Practical Management of the Three-Cost Approach."

SAMPLE CONTRACT CHANGE ORDER
DIRECT PROJECT MANAGEMENT AND ADMINISTRATIVE COSTS

Project: _____   Contract Change No.: _____

Owner No.: _____   Owner Change No. _____

(Company) No.: _____   Description: _____

DIRECT COSTS
(Attach additional schedules if required)

1. CHANGE ORDER PREPARATION (Costs incurred as of Proposal submission date)
   a. Research _____ Hrs @ $____/Hr $_____.__
   b. Design _____ Hrs @ $____/Hr $_____.__
   c. Design review/verification _____ Hrs @ $____/Hr $_____.__
   d. Site inspection _____ Hrs @ $____/Hr $_____.__
   e. Travel (a., b., c., d.) _____ Hrs @ $____/Hr $_____.__
   f. Coordination of trades _____ Hrs @ $____/Hr $_____.__
   g. Coordination of Utilities _____ Hrs @ $____/Hr $_____.__
   h. Clarifications _____ Hrs @ $____/Hr $_____.__
   i. Phone calls _____ Hrs @ $____/Hr $_____.__
      Phone co. charges (attach copies)
   j. Facsimile charges _____ @ $____/Ea $_____.__
   k. Proposal construction _____ Hrs @ $____/Hr $_____.__
   l. Secretary _____ Hrs @ $____/Hr $_____.__
   m. Photographs _____ @ $____/Ea $_____.__
   n. Estimating _____ Hrs @ $____/Hr $_____.__
   o. Postage/shipping/handling _____ Hrs @ $____/Hr $_____.__
   p. _____ _____ @ $____ $_____.__
   q. _____ _____ @ $____ $_____.__
   r. _____ _____ @ $____ $_____.__

2. Prosecution of the work (costs incurred as the work progresses)
   a. Field coordination _____ Hrs @ $____/Hr $_____.__
   b. Site inspection _____ Hrs @ $____/Hr $_____.__
   c. Travel (a., b.) _____ Mi @ $____/Mi $_____.__
   d. Tests _____ Hrs @ $____/Hr $_____.__
   e. Photographs _____ @ $____/Ea $_____.__
   f. Phone calls _____ Hrs @ $____/Hr $_____.__
      Phone co. charges
   g. Postage/shipping/handling _____ Hrs @ $____/Hr $_____.__
   h. _____ _____ @ $____ $_____.__
   i. _____ _____ @ $____ $_____.__
   j. _____ _____ @ $____ $_____.__

3. Total direct costs $_____.__

Prep. by: _____   Date: _____

8.1.6 Consequential Costs (Damages)

The third cost category of every change is consequential costs. These costs present the contractor with the most difficulty when trying to secure recognition from the owner that the cost is a legitimate item. The kinds of things that fall into the consequential cost category are most often thought of as damages—the stuff that lawyers seek to recover in lawsuits. They are costs resulting from the variety of impacts and effects caused by a change. A common example is freezing temperatures. Why should winter conditions—snow, ice, and cold—be anybody's fault? After all, everybody knows that they occur in varying severity, at about the same time every year. What makes it compensable in the change order context involves whether the winter would have been avoided entirely (or at least during the construction of the critically exposed work) if not for a particular change. If it is the change that put the work into the winter, the subsequent accommodations for the winter conditions become part of the change cost.

Other kinds of consequential costs include:

1. Strikes.
2. Interference and disruption.
3. Project delay costs.
4. Approval delays that alter the originally anticipated sequences or conditions.
5. Delay in retainage release (carrying costs).
6. Delay in project close-out (keeping capacity tied up—opportunity costs).
7. Delay in contract work (cash flow interruption and opportunity costs).
8. Canceled contracts.
9. Lost profit.
10. Acceleration.

On private projects, a good client relationship is more difficult to preserve in cost increase circumstances. A valued relationship may be placed in jeopardy if an aggressive posture is assumed in pursuing damages. On public and other low-bid jobs, however, the approach can be quite different.

Applying the three costs of every change order will help to accomplish three objectives. First, it will afford the contractor the opportunity to collect all the soft costs that would otherwise have been lumped into "overhead" and minimized (or have been left out of the calculation entirely). Second, the approach reserves the contractor's right to claims that may come up at a later date. Third, in a private contract situation, it will allow the contractor to present to the owner a catalog of all the cost items for which the contractor is entitled to receive compensation, but which are not being pursued at this time because of the ongoing "good" relationship. That knowledge can make the costs of the first two types (direct and indirect) much more palatable to the owner.

8.1.7 Practical Management of the Three-Cost Approach

We have already discussed how the three-cost approach helps to make the collection of a change's "soft" costs easier, and how it can be used to reserve the contractor's right to future claims. As the change proposal moves through the indirect costs and into the consequential cost category, the difficulty in gaining the owner's acceptance of all charges increases substantially. The real extent and responsibility for delays, for example, are areas that will continue to feed attorneys for a long time to come. By their fundamental nature, changes inevitably create friction. The temperature generated by the friction will rise with each additional cost item. Add the fact that the change will most probably expose someone's mistake to justify payment, and it becomes easier to see that a thoughtless, blunt application of the three-cost approach will not solidify friendships.

This is not to say that the approach cannot be used effectively on both private and public projects. To be effective, however, the contractor must recognize the owner's position and must adopt a thoughtful and professional approach to the requested change. The contractor must present the basis for his or her entitlement to the change, while at the same time trying to keep the relationship intact.

On low bid, public projects, it will be easier to apply the three-cost procedure because there is minimal risk to future business. In any case, the best way to implement the technique is to use the format on *every* change order proposal. The format itself must become a matter of operating policy. If this is accomplished, it will remove a large degree of the personal animosity that would otherwise be directed to the project manager or the project management team, if the owner thought that the form had been dreamed up on the particular job solely for his or her benefit. If, on the other hand, the form can be demonstrated to be an established company procedure, the owner, architect or construction manager will find it more difficult to blame the "person" presenting the request for change. The person presenting the request will be able to demonstrate that "It is a matter of company policy, my hands are tied." Some owners may just respond to this impression and begin to work *with* the project manager in a team approach, almost as if it were "against the procedure" rather than against a person. It's like going in to fill out a loan application at the bank. People get hypnotized by the form and not necessarily the information that it requests. In this situation many people tend to fill out all the information, even if it is completely irrelevant or redundant. Refer to Section 12.40, "Power of Legitimacy," for a more complete discussion of this effect.

It is important to understand that even on very small change proposals, and on those changes that do not have a cost associated with one or more of the categories, that the three components should still be listed on the proposal form. Where a value is not appropriate, simply indicate "Not Applicable" or "No Change." This must be done every time if the credibility of the procedure is to be maintained.

An example of the format applied to the typical change order proposal is included in Figure 8-1 to clarify the three-cost approach:

```
A.  Direct costs
    1. Labor (per breakdown, attached)              $7,000
    2. Material (per quotes, attached)               3,000
    3. Other direct costs (per schedule, attached)   2,000
            Total                                              $12,000

B.  Indirect costs
    1. Project management costs (per schedule,
       attached)                                    $  900
            Subtotal                                           $    900
            Overhead (10%, A + B)                                 1,290
            Subtotal                                           $14,190
            Profit (10%)                                          1,419
            Subtotal                                           $15,609

C.  Consequential costs
    1. Job delay
       (per schedule analysis and cost calculation,
       attached)
       4 days @ $600/day                            $2,400
            Subtotal                                           $ 2,400
            Total change order                                 $18,009
```

FIGURE 8-1 THE THREE-COST APPROACH

8.2 DEVELOPING THE CHANGE ORDER PROPOSAL

As mentioned previously, change order proposal development begins with the recognition of a difference in the work from that represented in the contract documents. Nearly every change has the potential to become significant, no matter how small its origins. The gravity of this effect can become severe through either improper handling or unreasonable owner reaction. Effective change order presentations begins with this fundamental realization. All project changes are, therefore, given full attention, and every important consideration is given the opportunity for complete development, regardless of the size of the change.

The change order proposal process should be a systematic procedure that is applied consistently and to the same extent to each change situation. The development of a systematic procedure for change order preparation, presentation and follow-up lends itself to the use of form letters, easily managed follow-up activities and comprehensive checklists. Using the forms and following the outlines in the remainder of this section will promote complete documentation. Each proposal will be assembled, prepared, and finalized in the fastest manner possible for the given set of circumstances. The contractor will be left in the best possible position to negotiate and finalize the outcome. The inclusion of relevant terms, conditions, and disclaimers will maximize the contractor's options if conditions of the change or its approval become different.

For each and every change, the change order proposal procedure may be summarized as:

1. Identification of extra/changed work;
2. Notification to the owner of possible or definite costs and project effects;
3. Assembling all cost components from outside vendors and contractors;
4. Compiling all internal costs as they apply to the three categories of costs identified in Section 8.1;
5. Evaluating schedule impact; and
6. Determining appropriate terms and conditions to maintain maximum protection in the future.

8.2.1 Change Order Identification/Notification

Complete and correct change order identification is the first step in the proposal process. As simple as it sounds, unnecessary confusion (resulting in procrastination, reexplanation, and so on) will result if it is overlooked. The description should be short and accurate and use the same terms as those indicated in and around the work affected by the change. If the work changes "Section 08331, Rolling Counter Shutters," for example, do not refer to the work as an "overhead door." There is enough confusion on today's large complex projects. Avoid adding to this confusion through the inconsistent use of language.

Company policies and procedures regarding the information which must be included in the reference section of each letter vary widely between contractors. Further, some specifications include provisions which govern the content of the information in this section. At a minimum, each change order should be identified first with the project reference. Immediately thereafter, the "Subject" should include a change estimate file reference, along with a concise but accurate description.

Example:
 RE: RTC Towers
 SUBJ: (009)
  Changes to Beam and Plate Lintels

Refer to Part Five, Change Order and File Administration, for comprehensive instructions on change order file arrangement and use.

Once properly identified, owner notification of the change becomes the next critical consideration. Every well-written construction contract will contain some provision requiring that the owner be notified in writing of any possible modification giving rise to a change (whether an increase or decrease) in the contract price and/or time. In addition, the contract will normally require such written notification within a precise number of days of the realization of its existence. The application of this notice requirement, however, is similar to the requirement that was first mentioned in Section 7.3.6, "Contract Time." That is, the clock will start ticking from the point at which it was first realized that the condition would have a definite effect on the contract price and/or time—not necessarily at the point where it was realized that there was a change. This is an important distinction which needs to be understood by project management.

The first thing, then, is to be aware of any notice requirement in the contract and to comply with it. The failure to provide the requisite notice may be used to evade, delay or confuse an otherwise proper change order request. The contractor will be required to devote time and energy to an explanation of the notice afforded the owner. This explanation will detract from even a solid request. In contrast, if the requirement has been complied with, there is no problem; and the parties can move on to the next issue. Do not risk losing an otherwise valid claim because of a technicality.

If the change proposal is straightforward and easy to assemble, do so in accordance with the recommendations of this chapter. Include the notice of the change order as part of the proposal's introduction. Follow this recommendation only if the complete proposal can be delivered within the time requirements noted.

On large and/or complex proposals, plain logistics usually demand more time for proper assembly and presentation. In this case, owner notification must still be accomplished on time, independently of the final proposal. A short letter *concisely* defining the issue will suffice. This notification should be sent even if the contractor is not certain that the different condition will result in extra costs or the need for additional time. A notification advising the owner of the possibility of a cost or time impact will comply with most notice provisions. A sample of such a notification is given in Section 8.2.2, "Sample Letter to the Owner Regarding Pending Change Order."

A proper notification letter is a very powerful tool. Set forth below are some guidelines that will help the contractor to maximize the advantages to be gained by issuing proper notification:

1. Always establish the earliest possible date of owner awareness. If the issue was first discussed at a job meeting two weeks ago, refer back to that date as the point of reference. Likewise, an early telephone conversation can be pinned down as the notification date.

2. Notify everyone who might possibly require notice, either directly or by copy. The persons interested in the notice include:
    a. The owner,
    b. The owner's field representatives,
    c. The architect, and
    d. All definitely and *potentially* affected subcontractors and suppliers.

3. Establish the change order file immediately. Include the file identification on the first written correspondence, if possible. If the first documentation is dated prior to the change order realization, dig it out now and reiden-

tify it. Refer to Section 6.3.1, "Establishing the Change Order File." Be sure that the file is absolutely complete in every respect. This attention alone can save hundreds to thousands of dollars in research time later, not to mention sidestepping the risk of losing a portion of the claim because of losing a document. There is just no excuse for a poor file.

Refer to Section 6.3, "Preparing the Change Order," for an additional discussion of change order notification and important action.

8.2.2 Sample Letter to the Owner Regarding Pending Change Order

The Sample Letter to the Owner Regarding Pending Change Order is one example of the important owner notification described in Section 8.2.1, "Change Order Identification/Notification." Specifically, it notifies the owner that:

1. A change has occurred that is beyond the scope of the original contract.
2. It appears that the change will have an effect on the contract price and/or time.
3. All component prices and a detailed analysis of all project effects are being assembled and when all such items are known, a complete proposal will be submitted.
4. The contract notice provision has been satisfied, without any question.
5. The earliest possible date of notification is established by reference to the original discussion (where possible).

This letter fulfills the technical notice requirement while affording the contractor the time necessary to complete the proposal properly. Strategies and negotiating positions can be evaluated in a less hurried manner as well; refer to "Quick Deals" (Section 12.52) and "Acceptance Time" (Section 12.2) for more on this idea. Before this kind of letter becomes part of an established change procedure, it is a good idea to have its intent and language reviewed by an executive management attorney to confirm that it is appropriate in light of the contractor's business objectives, local law and customs and the work undertaken by the contractor.

SAMPLE LETTER TO THE OWNER
REGARDING PENDING CHANGE ORDER

LETTERHEAD

(Date)

To: (Owner)

RE: (Project no.)
(Project title)

SUBJ: (Change file number)
Welded portions of
masonry anchors

Mr. (Ms.)   :

As reviewed at the Job Meeting on (date), the welded portion of the masonry anchors is not specified in either Section 04200 Masonry or Section 05100 Structural Steel.

At this time, it appears that there will be an increase in the contract sum to cover the cost of this additional work, and possibly an increase in the project time as well. We are proceeding with the assembly of all component prices and will complete a detailed analysis of all project effects. Upon the completion of our analysis, a change order proposal will be submitted.

Please consider this your notification in accordance with the requirements of General Conditions, Article 12.

Very truly yours,

Project Manager

cc: Owner Field Representatives
    Architect
    Mason Contractor
    Steel Contractor

8.2.3 Assembling Component Prices

Assembling change order component prices may be divided into two categories: internal and external. The internal company charges are again split among the direct, indirect, and consequential costs described in Section 8.1, "Change Order Components." External costs are from subcontractors (trade contractors in construction management contracts) and suppliers. Figure 8-2 clarifies the relationships.

The determination of direct internal costs is the responsibility of the contractor's estimating group. Since there may be as many methods to estimate the direct portion of a change order as there are contracting companies and because estimating is beyond the scope of this manual, estimating procedures will not be treated here in depth. The important consideration, however, is to be absolutely certain that the changed work has been reviewed and analyzed in intimate detail. Section 8.3.5, "Presenting Change Order Components," will develop the reasons for certain types of approaches.

Section 8.2.4 presents the Change Order General Conditions Checklist and Estimate Sheet. The use of this checklist will help to ensure that no important and justifiable cost items are overlooked.

After all internal direct costs have been assembled, recall the discussion in Section 8.1.4 on converting indirect to direct costs. Review and use the Section 8.1.5 Direct Project Management and Administrative Costs Form to catalog these charges to the greatest extent possible. From there, focus on Section 8.1.3 and indirect costs. Review all the additional costs which were incurred (or which will be incurred) as a result of the change and which are usually lumped into "overhead." Try to pinpoint these cost items, and the amounts associated with each item, as precisely as possible.

The external portion of change order component prices involves the assembly of the costs for all affected subcontractors and suppliers. The proper treatment of these items is critical to smooth acceptance of the final change order proposal. We discuss this topic in detail in Sections 8.2.5 through 8.2.9.

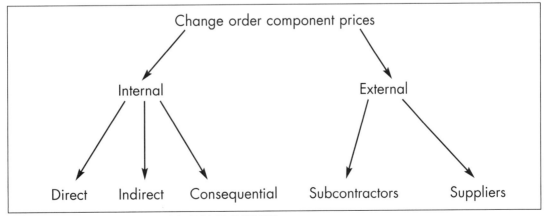

FIGURE 8-2 CHANGE ORDER COMPONENT PRICES

8.2.4 Sample Change Order General Conditions Checklist and Estimate Sheet

The Sample Change Order General Conditions Checklist and Estimate Sheet is provided to allow fast, comprehensive consideration of the monetary impact of the change order work on project administration. Some categories on this form may be specifically excluded by contract on a particular project. Even if the itemization of these administrative costs is not limited by contract, gaining owner acceptance of certain items as direct costs and not overhead can be a challenging and uphill battle. However, the contractor will not be in a position to request and argue for the recovery of these costs if the contractor is not aware of their existence. Utilization of the Sample Change Order General Conditions Checklist and Estimate Sheet will facilitate the consideration of these costs on every change order. Remember that failing to secure approval of a particular cost item on a certain change with a specific owner does not mean that the cost will not be accepted under different circumstances. The success of the contractor's request for these administrative costs as a direct project cost is dependent on a variety of factors including the complexity of the change, the disposition of the owner and the contractor's ability to identify the subject costs and demonstrate that the costs are a direct result of the change. If at first the contractor does not succeed, the contractor should keep trying.

CHANGE ORDER GENERAL CONDITIONS CHECKLIST AND ESTIMATE SHEET

Project _____ # _____ Change Estimate No. _____
Owner Bulletin No. _____

	Material	Labor	Total
1. Supervision			
a. Project manager			
b. Superintendent(s)			
c. Project and Office engineer(s)			
d. Field engineer(s)			
e. Additional foremen			
f. Accountant/time keeping/ material check			
g. Home office supervision			
h. _____			
2. Temporary Facilities			
a. Field office(s)			
b. Material trailers/sheds			
c. Temporary toilets			
d. Temporary roads			
e. Safety protection/equipment			
f. _____			
3. Field Support			
a. Office/first-aid supplies			
b. Blueprinting/copying/photos			
c. Telephone			
d. Fire/theft alarm			
e. Insurance			
f. Home office expense			
g. _____			
4. Temporary Utilities			
a. Heat			
b. Light and power			
c. Water			
d. Elevators/lifting/moving			
e. Tests/inspections			
f. _____			
5. Construction Equipment			
a. Small tools (expendables)			
b. Trash removal/light trucking			
c. _____			
6. Special Conditions			
a. Winter conditions			
b. Snow removal			
c. Cutting and patching			
d. Final cleanup			
e. _____			

Total change order general conditions $_____

8.2.5 Assembling Subcontract Prices

Subcontract proposals most often comprise the major portion the external prices indicated in Figure 8-2. They can also be the most significant portion of the direct cost category. Notwithstanding the importance of subcontractor proposals, they are often received late, incomplete, and in otherwise improper form. If a subcontractor's proposal is late, it delays the entire process, thereby increasing the significance of the change on the total project. The additional costs associated with a delay caused by a late proposal will be extremely difficult, if at all possible, to collect. If a subcontractor's proposal is improper in its format, other problems quickly develop. If time pressures mandate that the incomplete or improper proposal be submitted to the owner "as is," the best that the contractor can receive is an approval with some condition that the issue with the subcontractor proposal be resolved in the future. This can present a significant problem for the contractor if he or she is unable to resolve the matter with the subcontractor. More realistically, the entire proposal will probably be rejected or returned without action pending proper submission. This risk is greater on a public project where objectives are different and format is king. However, compliance with the procedural and content requirements can be an important factor on private projects as well. Remember, the proposal should exude professionalism, competence and credibility.

In contrast, if the contractor can consistently compel the subcontractor to submit the proposal in the proper format, with all the supporting information required, the contractor will have greater control over the final outcome. This is true both in terms of owner approval and in the management of the subcontractors through the changed work. Each change order presents a marvelous opportunity to increase total control over both the owner and the affected subcontractors through the strategic manipulation of the information which is passed in both directions. Do not miss these important opportunities as they arise.

The most effective process for securing timely, complete proposals in proper form boils down to:

1. *Requesting the proposal.* Attention is carefully called to applicable contract responsibilities. All requirements to be met in the proposal submission are to be carefully cataloged.
2. *Following up on the first request.* This action is performed either if the proposal has not been submitted by the required date or it has been submitted in a manner that is in some way unresponsive to the requirements.
3. *Notification of damages (if necessary) due to delay in the subcontractor's response.*

These activities will be developed further in the following sections:

- 8.2.6, "Sample Letter to Subcontractor—Request for Change Order Quotation"
- 8.2.7, "Change Quotation, Second Request"

In some cases (often, but not always, on small changes), a subcontractor will perform work that is part of a documented compensable change, but never bother to submit even a lump-sum price. This practice can create a variety of administrative problems for the contractor, especially at project close-out, and can generate costs that will far exceed the costs of the change itself. If this happens, consider using the "Sample Letter to Subcontractor Regarding Change Order Price by Default" described in Section 8.2.8 as a tactical maneuver to close the issue one way or another.

Finally, telephone quotations will occasionally be used in an effort to expedite the general proposal. Depending upon the owner's policies and the nature of the change, they may indeed help to speed up the process. If so, consider Section 8.2.9, "Sample Change Order Telephone Quotation Form," to be sure that all relevant information is secured. The most important consideration here, assuming a useful, detailed proposal, is immediately to secure written confirmation of the quote from the subcontractor. Use the Sample Letter to Subcontractor Confirming Telephone Quote presented in Section 8.2.10 to tie the loose ends together.

In these days of Federal Express, telegrams, electronic mail, telcopy, and facsimile transmissions, there is no excuse for not producing and delivering a written document very quickly. Accordingly, double-check the advisability of utilizing a telephone quotation before wasting valuable time on one. The use of telephone quotes is also questionable in the case of a complicated proposal or where the level of trust between the parties is either low or untried. Finally, be aware that some states have specific rules governing telephone quotations. For example, California Commercial Code section 2204 provides that an oral quotation to a contractor involving a price in excess of $2,500.00 is only valid for 48 hours unless the contractor confirms the quotation, in writing, within this time frame.

8.2.6 Sample Letter to Subcontractor—Request for Change Order Quotation

The Sample Letter to Subcontractor—Request for Change Order Quotation is designed to secure subcontractor change order proposals in the fastest, most complete manner possible under a given set of circumstances. Specifically, the letter:

1. Transmits all documents and information required to price the change to the subcontractor;
2. Requests written confirmation of "no change in price" if the work does not affect the particular subcontractor;
3. Calls attention to any General Conditions or subcontract requirement regarding change order form and/or content.
4. Directs the subcontractor's attention to all relevant considerations affecting price, time, and effect on work sequence and other trades.
5. Reminds the subcontractor that substantiating records, in proper form, are required;
6. Requires a complete, correct response by a specific date.

SAMPLE LETTER TO SUBCONTRACTOR
REQUEST FOR CHANGE ORDER QUOTATION

LETTERHEAD

(Date)

To: (Subcontractor)

RE: (Project no.)
(Project title)

SUBJ: (Change file number)
(Change description)
Proposal request

Mr. (Ms.):

Attached is the (identify and date all enclosures necessary to price the change).

If the subject work does not affect your trade, submit your confirmation of no change in contract price or time.

If the changed work affects your company, please submit:

- Price to complete the changed work.
- Any applicable credit for contract work.
- All substantiating labor and material records.
- Material and equipment delivery times after change order approval.
- Time required to perform the work (separate major items).
- Work of any other trade affected.
- All conditions required to perform the work.
- Any significant weather, site, or other constraints that are beyond your control.
- All other applicable information.

Your attention is called to (insert the appropriate General Conditions or subcontract article reference) for proper format and required level of detail. If you have any questions in this regard, contact me immediately.

Your complete response, in proper form, is required by (date) to avoid unnecessary delay associated with this change.

Very truly yours,

Project Manager

8.2.7 Sample Letter to Subcontractor—Change Quotation, Second Request

The Sample Letter to Subcontractor—Change Quotation, Second Request is an example of a follow-up to the letter requesting the change quotation (see Section 8.2.6). It is to be sent by certified mail, return receipt requested, immediately after the required response date indicated in the original proposal request. Specifically, this letter:

1. Notifies the subcontractor that the required response time has passed.
2. Requires immediate, expedited action.
3. Notifies the subcontractor of its responsibility for all additional damages resulting from the subcontractor's failure to comply with the request.
4. Reserves the right to claim damages caused by the subcontractor's inappropriate response.

SAMPLE LETTER TO SUBCONTRACTOR— CHANGE QUOTATION, SECOND REQUEST

LETTERHEAD

(Date)　　　　　　　　　　　　　　CERTIFIED MAIL/RETURN RECEIPT

To: (Subcontractor named　　　　RE: (Project no.)
　　in 8.2.6 letter)　　　　　　　　　　(Project title)

　　　　　　　　　　　　　　　　　SUBJ: (Change file number)
　　　　　　　　　　　　　　　　　　　　(Change description)
　　　　　　　　　　　　　　　　　　　　Second request for
　　　　　　　　　　　　　　　　　　　　proposal

Mr. (Ms.)　　　　:

The following requests for change quotations remain outstanding:

Change No.	Description	Date Requested	Date Due

　　Your lack of response by the date required is now interfering with the completion of our proposal and with the work. Your proposal(s) is (are) now to be delivered to this office by (allow two days). In the event we do not receive proper responses, in the form required by your subcontract, your company will be charged for all costs associated with any resulting delay.

　　Please take notice that we reserve all rights to claim all damages resulting from your untimely response.

Very truly yours,

Project Manager

8.2.8 Sample Letter to Subcontractor Regarding Change Order Price by Default

The Sample Letter to Subcontractor Regarding Change Order Price by Default is not what it may first appear to be. While some subcontractors may be willing to accept what the letter threatens, the real intent is to wake up the sleeping subcontractor. It is to be sent to the subcontractor who has failed to provide a price for a small change, despite repeated requests, threats and phone calls. The letter tells the subcontractor what the final outcome is going to be. It might even jar the subcontractor into finally getting the price submitted. If not, it may be possible to simply proceed with the change in accordance with the terms of the letter.

Specifically, the letter:

1. Notifies the subcontractor that the subcontractor's repeated failure to submit a price has resulted in the price being determined for him or her.

2. Advises the subcontractor that a unilateral change order will be processed in the indicated amount, after which the issue will no longer be open for any negotiation.

3. Reserves all the rights to claim damages resulting from the subcontractor's continuing failure to act.

SAMPLE LETTER TO SUBCONTRACTOR
REGARDING CHANGE ORDER PRICE BY DEFAULT

LETTERHEAD

(Date) CERTIFIED MAIL/RETURN RECEIPT

To: (Subcontractor) RE: (Project no.)
 (Project title)

 SUBJ: (Change file no.)
 (Change description)
 Change order price
 by default

Mr. (Ms.)   :

    Your continuing failure to respond to repeated requests for change order quotations is now delaying project close-out and generating unnecessary and excessive overhead expenses. Accordingly, if your price is not received by (two days after the date above), a unilateral change order will be processed in the amount of $(   ). At that time, the file will be closed, and there will be no opportunity for further review.

    Be advised that any damages resulting from your lack of attention to this matter will be backcharged to your account. Additionally, we reserve all rights to claim all damages resulting from your untimely response.

    Please take notice.

Very truly yours,

Project Manager

8.2.9 Sample Change Order Telephone Quotation Form

The Sample Change Order Telephone Quotation Form allows the quick assembly of all relevant parts of a complete change order proposal. Filling in the categories on the form will help to ensure that important qualifications are included and that all the information necessary for proper evaluation of the quotation is complete. Once the form is completed, it can be used in the general change proposal as backup. If you intend to do this, however, be sure that the owner does not have some policy that requires the subcontractors' proposals to be signed. If this is the case, the Telephone Quotation Form will expedite the preparation of the general proposal to the owner, but will not comply with the foregoing contractual requirement. Ultimately, the contractor will be required to obtain and submit a signed subcontractor proposal. If presented with this situation and the subcontractor is unable to submit a written proposal due to a lack of administrative resources, consider having the subcontractor sign a copy of the Telephone Quotation Form.

It is important to understand the seriousness of the written confirmation of the quote. Such a confirmation must be based on a clear understanding with the subcontractor and the subcontractor must be made aware that the telephone quotation will be confirmed in writing. This should be a condition for taking the quote over the phone in the first place. The contractor should also explain the Section 8.2.10 Letter to Subcontractor Confirming Telephone Quote.

After the telephone quotation is received and the Telephone Quotation Form is completed, the contractor should read the *entire* contents of the form back to the person giving the quotation, exactly as the information has been recorded on the Telephone Quotation Form. Resolve any discrepancies in the telephone quotation before ending the dialogue or utilizing the quotation.

If you confirm the terms of the telephone quotation in this way, the submission of the written confirmation should not present a problem. Remember to confirm telephone quotations, in writing, as soon as possible.

SAMPLE CHANGE ORDER
TELEPHONE QUOTATION FORM

LETTERHEAD

Date: _____ 19____   Firm: _____
                                        _____
Project: _____   _____
   no.: _____   By: _____
         _____   _____
Change no.: _____   Phone: (____) _____

Documents Included:

Description	Date/Rev.	Description	Date/Rev.
_____	_____	_____	_____
_____	_____	_____	_____
_____	_____	_____	_____
_____	_____	_____	_____

____ Price increase: $ _____   Sales tax included? _____
____ Price decrease: $ _____   Bonds supplied? _____
____ No change

Material/equipment delivery lead times after C.O. approval: _____

Time required to complete the work (separate the major items): _____
_____

Work of any other trade affected: _____
_____

Special conditions required to perform the work: _____
_____

Significant weather, site, or other constraints: _____
_____

Other applicable information: _____
_____

Alternates, qualifications, exclusions: _____
_____

Taken by: _____

<u>CONFIRM IN WRITING</u>

8.2.10 Sample Letter to Subcontractor Confirming Telephone Quote

The Sample Letter to Subcontractor Confirming Telephone Quote transmits the telephone quotation form of Section 8.2.9 to the subcontractor and secures written confirmation by the request of an acknowledging signature. If the procedure set forth in 8.2.9 was followed, there should not be a serious problem in securing the confirmation. This statement is based on the following:

1. There was a complete understanding of the confirmation procedure as a condition for receiving the price over the phone in the first place.
2. The form was read back to the party giving the information in the exact language that it was recorded.
3. After the information was read back to the subcontractor, any discrepancy was resolved.

SAMPLE LETTER TO SUBCONTRACTOR CONFIRMING TELEPHONE QUOTE

LETTERHEAD

(Date)   CERTIFIED MAIL/RETURN RECEIPT

To: (Subcontractor)   RE: (Project no.)
                                 (Project title)

                          SUBJ: (Change file no.)
                                 (Change description)
                                 Confirmation of C.O.
                                 telephone quote

Mr. (Ms.)   :

    Per our agreement of (date of the telephone quote), attached is the subject Change Order Telephone Quotation Form. Please confirm that all information set forth on this form is accurate by signing below and returning this letter to my attention.

    Your response is requested by (date).

Very truly yours,

Project Manager

         Accepted by: _____
                 Title: _____
                 Date: _____

8.2.11 Determining Schedule Impact

Time should be addressed on every change order proposal. On simple or small changes, the time effect on a project may be negligible or nonexistent. Even so, read this section and reconsider the temptation to state prematurely that "no additional time is required."

The process of incorporating schedule impact information into your change order proposal can be summarized as:

1. Determining activity cause–effect relationships.
2. Adjusting resulting schedule logic.
3. Computing the net effect on project time.
4. Representing, substantiating, and presenting the time analysis.
5. Establishing a value for the time.

To demonstrate convincingly that additional project time is required by a specific change, the direct cause–effect relationships of the activities must be displayed. For proposals submitted prior to the performance of any work, two schedules are needed: the as-planned schedule and the adjusted schedule. The as-planned schedule outlines the sequence of construction that would have proceeded per the original contract, had it not been for the change. Even if a detailed construction schedule has not been completed for the entire project at this stage of the project, one must be finalized now for at least the work sequence affected by the proposed change. The as-planned schedule must be based on the work as originally described in the contract documents and should only be as complex or as simple as the issue itself dictates. The adjusted schedule is the document that incorporates the modified work with all its effects on logic sequences and activity durations. It is a recomputation of the entire network of construction activities, after the adjustment to the as-planned schedule for the proposed change. The adjusted schedule may not include a new project end date. The difference in end dates, if any, is the amount of time attributable to the respective change.

As complicated as all this may first appear, it does not have to be. The simplest way to complete the demonstration will ultimately be the most effective in communicating the concepts to those who must approve the result. If sufficient detail can be achieved through the use of a simple bar chart, or by merely listing obvious effects on significant milestones, then by all means leave it at that. A simple, inexpensive, but very effective method to achieve this kind of analysis is detailed in *Construction Scheduling Simplified*, Prentice Hall, 1985.

It is important to understand the effects of two concepts regarding the level of complexity when trying to develop a quick understanding in persons who need to approve a scheduling change or calculation. These concepts are:

1. The more people who are involved in the approval process, the simpler, or more developmental, the method of presentation the better.

2. The nature of the change impact and the complexity of the interactions will determine whether an easily identifiable cause–effect demonstration can be used by itself.

If the contractor is working with a single individual, who happens to have a fluent scheduling background, buzzwords and network analysis may feed his or her ego. Reams of computer printout might actually help in this case, especially if it is necessary to convince the person with the scheduling background that the analysis is complete. More commonly, however, the contractor will have to sell the analysis to several individuals, regardless of what the formal authority structure may be. These people will come from a variety of backgrounds and they will have interest levels ranging from intense and informed to completely uninterested and unbelieving. The first concept is not intended to insult anyone's intelligence. It is simply recognition that (as with any presentation) there is a dramatic inverse relationship between body count and true understanding. That is, as the number of people listening increases, the overall level of real understanding decreases, given the same information and effort.

With regard to the second concept, if the network interaction is complex, it may be advantageous to consider a presentation that simply begins with a statement of fact—that so many days have been added to the completion date by the change. Beyond that, the backup analysis, in all its detail, can be attached.

Whether or not the calculations are enclosed with the change order, understand that a *personal* presentation of the effects may be necessary, particularly if the additional time is significant. Resist the urge at the proposal stage to display your brilliance with sophisticated (but confusing) talents. Boil down the explanation to a simple and straightforward presentation of the cause and the effect. Be certain not to compromise the accuracy of the information during this simplification process.

A number of circumstances may improve your ability to use the simplest form of schedule analysis right on the proposal itself, or to incorporate it into the supporting documents. These circumstances include:

- Owner/architect predisposition;
- Complexity of the activity relationships;
- Accountability for all effects; and
- Level of owner sophistication vis-á-vis construction scheduling.

If conditions are right, a simple analysis might just be appreciated by the persons who must ultimately do the hard work of reviewing and substantiating the claim. Such an analysis is presented in Figure 8.3.

The time to complete the changed work is as follows:	
1. (Date) ready to perform contract work	0 W.D.
2. Finalize change order/submit proposal	11 W.D.
3. Owner approval of proposal, required by (date)	5 W.D.
4. Fabricate, deliver, and install the C.O. work	10 W.D.
5. Complete contract (affected work)	<u>10 W.D.</u>
6. Total time to complete this change if approved by (date in 3)	36 W.D.
W.D.—Working days.	

FIGURE 8-3 SIMPLIFIED SCHEDULE IMPACT ANALYSIS

The analysis is technically accurate for a change that impacts the critical path directly with no resequencing of schedule logic. In any other situation, a scheduling purist may object to the logic of this presentation. However, many owners will appreciate (and approve) this simplified format

Finally, when it becomes time for the road show, try to assemble *all* the people in the owner's camp who will have input into the owner's final decision. Just as a consumer product salesperson makes every effort to ensure that both spouses (the decision-making *team*) are present together for the sales pitch, the contractor should make sure that the decision-making authority is in attendance for the presentation of the change. The reasons for this are twofold.

First, the contractor's management team is the most qualified and will be the most persuasive in explaining the position. Do not leave such an important responsibility up to the owner's field representative or the architect to explain it to the decision maker. Even if that person accepts the position taken in the presentation, most of the underlying facts will be forgotten, and important justifications will be lost. Furthermore, the owner's representative and the architect will not be able to "go to bat" for the proposal in the face of objections. All they can do is their best to explain the position taken in the presentation, and only so far as they understand it. Second, as any good salesperson knows, it is the contractor who will best field objections and move the owner directly to buying the reasoning behind the presentation and the desired result. If all the authority and its support functions are present, it will be more difficult for the owner's team to come up with some "valid" excuse for not finalizing the agreement now. If possible, keep them there until an agreement is reached on at least something. Remember the old adage: The "good" salesperson only leaves after hearing "no" four to six times and the "great" one does not leave until hearing "no" eight to *twelve* times.

Once the number of days attributable to the change has been established, the final component is cost, or price per day. On-site expenses are the most straightforward to compute. All that is necessary is the total of personnel, equipment, facilities and supplies, and the application of your historical costs. The only things that may complicate this portion of the analysis are escalation computations if the impact is anticipated to extend well into the future. The more nebulous portion of the calculation is that of home office overhead. Over the years, contractors, owners, attorneys, accountants and the courts have tried to develop a methodology for the allocation of the contractor's home overhead to one of several projects. One such methodology was developed in *Eichleay Corp.*, (1960) ASBCA No. 5183, 60-2 B.C.A.. This formula takes a portion of the contractor's actual historical overhead as it applies to the current operation and prorates it to the project in question. It considers the value of the job and its original planned operations. It then weighs those factors against the company's total output for the period and its related overhead expenses. The result assigns a specific dollar amount to each day on the particular project. (Refer to Section 9.6, "Historical Cost Records," for an important related discussion in the practical use of this technique.)

The actual process is not as difficult as the description first makes it appear. However, since the key to the contractor's success is establishing believability in the calculation's accuracy, the calculation should be prepared by the contractor's accountant and should be based on information that can be substantiated with proper documentation. An example of a formula used to calculate home office overhead allocable to each day a project is delayed is contained in Figure 8-4.

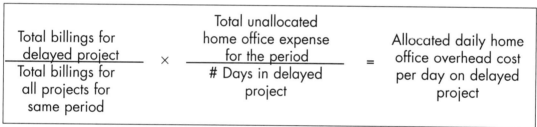

FIGURE 8-4 SAMPLE HOME OFFICE OVERHEAD DAILY RATE CALCULATION

8.3 FINALIZING THE PROPOSAL
8.3.1 Introduction

To this point, the reader has seen where to discover changes in every corner of the project, developed an understanding of the components of every change order proposal and learned how to lay the groundwork for the actual presentation. This section will finalize the assembly of all components. We will review each type of cost component and analyze the best method of its presentation for a given set of circumstances. The presentation of the "final" change order price will be considered, along with the important qualifying remarks to afford the contractor the maximum amount of protection from unanticipated changes. Once the proposal is completed, the methods for forcing prompt owner action will be detailed. Recommendations for action in the face of owner inaction will also be provided so that the work can move forward while the ball is in the owner's court. The section will close with a discussion of terms and conditions that should be a part of every change order proposal. Selected terms and conditions for special situations will also be reviewed. These terms should be used as a starting point. Local laws, as well as local customs and the contractor's business objectives, may require the use of additional or different terms.

8.3.2 Proposal Format and Timing

The change order proposal format should follow whatever overall design that will best advise the owner:

1. That a change exists;
2. The final price of the change; and
3. The terms which form the basis of the price.

Depending upon the people and the organizations who must review and approve the proposal, the proposal and supporting documents may very well come under close scrutiny by several people in different departments. Moreover, these individuals will likely be concerned with the legitimacy of only a component or two of the entire package. If the format of the proposal forces these separate review stations to get mired in disorganized detail, the proposal review will be delayed at each station. Worse yet, if each point is not presented clearly enough for a person with average understanding of the subject to grasp within a few seconds, the contractor runs the risk of losing initial approval of an issue on principle—only through the misunderstanding of an overworked clerk. Even though the proposal can be clarified and reintroduced in many instances, revisions and resubmissions waste time and cost money.

On each project, the contractor should study the process for the review, evaluation and approval of change order proposals, before a particular format is selected for use on the project. If the contractor understands who will be reviewing the respective cost components, as well as their procedures and methods of review, the contractor will be able to package the information in the best manner possible. The contractor will also be able to focus the reviewer's attention on the items which require his or her attention and away from less important matters.

No matter how the change order proposal is handled, each change order proposal should include or incorporate the following parts:

1. Notice;
2. Cost component summary;
3. Schedule impact and supporting analysis;
4. "Final" price presentation;
5. Required acceptance date; and
6. Standard and special terms and conditions.

Sections 8.2.1, "Change Order Identification/Notification," and 8.2.2, "Sample Letter to the Owner Regarding Pending Change Order," developed the concept of notice in detail. Whether the notice is part of the complete change order proposal or a separate document (as provided for in Section 8.2.2, "Sample Letter to the Owner") is a function of the speed with which the complete change order proposal can be prepared. As mentioned in Section 8.2.1, the notice must be given in accordance with the requirements of the contract. If the complexity of the change or the logistics of the project require additional time to prepare a complete proposal, send the notice first and then follow up with the complete proposal. As set forth above, the separate prior notice should be included with the complete proposal.

The balance of the proposal format will amount to presenting items 2 through 6. In most situations, the best overall change order package format will consist of:

1. A cover letter containing:
    a. Notice (if applicable);
    b. Change order total cost (lump sum);
    c. References to attached supporting documents;
    d. Additional project time required;
    e. Required change order approval/acceptance date; and
    f. Standard and special terms and conditions.
2. All required supporting data should be cataloged to correspond to the cost items contained in the cover letter.

Refer to Section 8.3.3, "Sample Change Order Proposal Cover Letter," for a sample letter which incorporates the layout described in item 1.

One other letter than can become in itself a key document in the ultimate change order outcome—particularly if the issue winds up as a dispute—is the Letter to the Owner Regarding Change Order Cost Escalation Due to Untimely Action. A sample of this letter and a discussion of its use are contained in Section 8.3.4.

8.3.3 Sample Change Order Proposal Cover Letter

The Sample Change Order Proposal Cover Letter is designed to clarify the application of the change order cover letter format detailed in Section 8.3.2. Specifically, the letter:

1. Refers to the earliest change order notification date;
2. Summarizes the change order total cost;
3. Refers to all appropriate supporting information;
4. States the additional time that the change adds to the contract;
5. Requires owner approval by a precise date;
6. Notifies the owner of consequences of failing to act by the required date; and
7. Incorporates all standard and special terms and conditions and reservations of rights.

When assembling the supporting documents for a large or complex change, consider using an index or a summary page. Additionally, try to keep the supporting information for various cost and schedule items separate. That way, if one item becomes a bone of contention, it can be segregated from the other "approvable" costs. To the greatest extent possible, the contractor should adopt a format that will facilitate the prompt review, evaluation and approval of each requested change. Separating the various cost and schedule components will allow the contractor to debate questionable items, without dragging the acceptable portions of the request into the argument.

SAMPLE CHANGE ORDER
PROPOSAL COVER LETTER

LETTERHEAD

(Date)
To: (Owner)

RE: (Project no.)
(Project title)

SUBJ: (Change file no.)
(Change description)
Submission 1

Mr. (Ms.)   :

Per the (refer to Section 8.2.2, "Letter to the Owner Regarding Pending Change Order"), our total price to perform the subject change is $(   ). All supporting documentation is attached.

Our schedule analysis has determined that the resulting change activity sequence will add (insert number) working days to the project schedule. This cost is included in the above price.

Your approval is required by (date) to maintain the above price and schedule; any approval after this date will increase the cost and time associated with this change, which additional cost and time will be added to our proposal.

We call your attention to the exclusions and qualifications detailed in the subcontractor's quotations and reserve the right to quote on all costs that may be unforeseen at this time.

Very truly yours,

Project Manager

cc: Owner Field Representatives
    Architect

8.3.4 Sample Letter to the Owner Regarding Change Order Cost Escalation Due to Untimely Action

The Sample Letter to the Owner Regarding Change Order Cost Escalation Due to Untimely Action is designed as a follow-up notification to a change order proposal that was not acted upon by the required deadline. Specifically, this letter:

1. Calls attention to the original response deadline and the consequences of failing to meet it;
2. Confirms the statement in the original proposal that the money meter is running;
3. Notifies the owner that the price set forth in the original proposal is increasing at a certain dollar amount per day;
4. Clarifies that each additional day without proper response adds additional time to the project schedule.

The letter's language is intentionally blunt. It is intended to communicate that the proposal approval delay has become a serious matter. Heightening the sense of urgency with regard to change order approval will, hopefully, force action and keep the work moving.

With respect to time, the contractor should note that the additional time necessary to complete the change may not be a simple 1:1 relationship (i.e., one additional day of contract time for each day of approval delay). The project is going forward on other fronts. Each day of approval delay may give rise to additional interferences and resequencing.

SAMPLE LETTER TO THE OWNER REGARDING CHANGE ORDER COST ESCALATION DUE TO UNTIMELY ACTION

LETTERHEAD

(Date) Certified Mail/Return Receipt

To: (Owner) RE: (Project no.)
  (Project title)

  SUBJ: (Change file no.)
   (Change description)
   Lack of response to
   Submission 1

Mr. (Ms.)   :

Change proposal no _____, dated (date) for the subject work required your approval by (date). To date, we have not received any response to this proposal, and the lack of a definitive response is now creating additional interference that was not included in the original change order price.

Each day beyond the required date is adding $(   ) per day to the change order price in administrative overhead alone. Additional items may also be added to the proposal price after your authorization to proceed is finally received. Accordingly, upon receipt of your approval, all costs resulting from the effects of the untimely response will be submitted to correct the final change order price.

We reserve the right to claim all damages resulting from untimely approval action.

Very truly yours,

Project Manager

cc: Owner Field Representatives
  Architect

8.3.5 Presenting Change Order Components

As first mentioned in Section 8.1.7, "Practical Management of the Three-Cost Approach," it is important to account for each category of direct, indirect, and consequential costs on *every* change proposal, even on small changes or proposals where they may not apply. If a category doesn't apply, indicate "Not Applicable" or something similar. There are several reasons for this:

1. It indicates that the category has been considered and that no cost was applied to the change in its current form. If conditions change, or if unforeseen effects become evident, a cost may apply.
2. It eliminates the possibility of overlooking important charges in the rush to get the proposal out.
3. It transmits the message that the cost treatment is a company standard procedure, which may help to reduce personal hostility or the risk of appearing too aggressive.
4. It gets the owner *used* to looking at it. The early novelty may give way to familiarity with the concept, and to ultimate acceptance of the cost as a legitimate item. (Refer to Section 12.2, "Acceptance Time," for more on this idea.)
5. It assures the owner that he or she is not being singled out and creates the appearance of "business as usual," which may minimize any feeling on the part of the owner that he or she is being taken advantage of by the contractor.
6. It gets everyone in the contractor's organization used to considering the costs and pushing for their recovery. The approach also broadens the awareness of all employees when considering the added expenses which are associated with changes.

Next, summarize the components and their parts in a simple, easy-to-follow schedule. Refer to Figure 8-1 as a good example. The cover letter will refer to all supporting data which is attached to the proposal, so leave the actual backup calculations, worksheets, and so on where they belong—cataloged and stacked behind the summary. All of the supporting information is available for reference when someone is interested. In contracts, working the details into a long and tedious cover sheet schedule will only increase confusion, make the whole thing more difficult to read, and ultimately delay the final outcome. Keep the summary schedule simple and keep the substantiating details in easy-to-find, manageable attachments.

It is interesting to note that there seems to be an inverse correlation between the final size and complexity of the "substantiation" and the likelihood that it will be gone through in intimate detail. That is, so long as the cost component summary is clear, the fatter and more complicated the backup, and the more minute the level of detail included in the calculations (along with references to other sources), the more likely the data will be assumed to be correct. Overworked clerks and office engineers may very well complete only a superficial review. Even if an item does not lend itself to precise calculation, it will be to the contractor's advantage to develop a system or procedure which at least gives the appearance of a precise calculation.

Finally, the contractor should be aware that it is, almost without exception, easier to secure approval for a large number of small items than it is to gain approval for a small number of large items. All psychology aside, its been proven over and over again that this process works. Additionally, remembering the idea will be worth significant monies with each new change order proposal.

The contractor must realize right from the start that nobody can visualize the complicated sequence of events in detail, with all its interferences, in the way that the contractor can. When presenting the change order proposal, the contractor must paint a detailed picture of the situation to force complete understanding. Only then can the approver sign off on the requested change with the feeling that the price is fair and reasonable. It is up to the contractor to get the others to appreciate the point of view which forms the basis of the change order proposal and justifies the change order price. The contractor must shatter any preconceptions held by the owner or architect about the change and educate them to the contractor's way of thinking.

For example, in an alteration project, if the contractor simply lists as a cost item "Remove and Replace Window—$400/Opening," the contractor can be met with outrage if the price seems high. At the very least, the contractor will have to explain the detailed sequence necessary to complete the work, with all its individual cost components. Even if the approver believes the presentation and that the change request is proper, the approver is faced with the problem of justifying the change to superiors, auditors, or whomever. In that case, the approver has got to come up with a detailed activity list.

In contrast, the contractor can avoid this situation altogether by dividing the subject of the change into several smaller activities. The same work activity could be broken out, with specific costs assigned to each component, such as:

1. Notify room occupant to relocate		$ 10
2. Move room furniture		35
3. Remove curtains and blinds		35
4. Remove window casing		70
5. Remove window and dispose in dumpster on building ground level		105
6. Distribute new window		35
7. Install new window and wash glass		140
8. Reinstall curtains and blinds		35
9. Replace furniture and clean up		70
	Total	$535

The particular example may exaggerate the effect, but the result is real. That is, the individual components themselves do not seem so unreasonable. In many cases, the final unit price arrived at will be even greater than the price contained in the lump sum estimate. Don't sell the technique short. It works. Use it.

8.3.6 Presenting the Total Change Order Price

The objective in presenting the final change order price goes beyond presenting a complete summary of all components. The total price components of the change order proposal should include the qualifying language that affords the contractor the maximum protection possible in the face of a wide range of possible owner actions (including inaction). It is an effort to keep all options open if the conditions of approval or the circumstances of the change itself move away from those anticipated and included in the price. The talent lies in the ability to appear definite and final, while actually leaving the door open to reintroduce the issue or any part of it.

The proposal qualifications that can accomplish this range from the subtle to the elaborate, and their selection depends upon the individual circumstances. At the very least, the proposal implies (if not emphatically states) that all related work indicated in the change order documents has been priced. Beyond this, the detailed cost component breakdown confirms that no stone has been left unturned. If one has, the relevant qualification to the proposal is included. So it is important to be clear that the details represent the circumstances and considerations that could reasonably be observed and/or anticipated as of the time the proposal was prepared.

Since the detailed breakdowns are attached, the proposal should state that the total price is "for work expressly described, only." If you are unsure of the owner's disposition, consider adding a more direct qualification to the effect that "all rights are reserved to submit additional costs for effects unforeseen at this time, or if conditions of the work or its approval change."

8.3.7 Presenting the Effects on Contract Time

Section 8.2.1 introduced the concept of the effect of change orders on total project time and the determination of schedule impact. Here, the final treatment of contract time in the proposal document will be outlined.

For any time extension to be valid, the proposed change must have an impact on the critical path of construction activities on the project. To demonstrate this impact, the contractor must submit some kind of schedule analysis with the change order proposal.

As part of every change order proposal, the contractor must clearly identify any changes to the logic of the current as-planned schedule (i.e., the original as-planned schedule, with adjustments on account of previous changes). After the changes are identified, the contractor must decide whether to include all the calculations as backup in the proposal package, or present the main points in summary form. In either case, the number of days that can be linked to the change order should be listed directly in the cover letter. This correspondence should be similar to the sample contained in Section 8.3.3,

"Sample Change Order Proposal Cover Letter." If the back analysis is clear, straightforward, and sound (subject to relatively minor challenge), then by all means include it. If, however, the nature of the effects is complex and the reasons for certain logic are not clear by themselves, including the detailed analysis may raise more objections than it puts to rest. In this situation, a personal presentation that demonstrates the reasoning in a clear, logical fashion will be much more effective. Opportunities can be presented to air disagreements and satisfy all questions with complete explanations.

If this maneuver makes sense for a certain proposal, do not include the schedule backup with the proposal. State the time required, and attempt to arrange a meeting as soon as possible to resolve the issue. As mentioned previously, the contractor should try to control the information content as well as the manner in which it is finally transmitted.

8.3.8 Requiring Approval Action

It is essential for a proposal to require a definite owner response by a specific date. Several factors will affect the manner and timing of the contractor's efforts to move an owner to a final action. They include:

1. The dollar size of the change.
2. The potential for significant impact on the contract work.
3. The existing relationship with the owner.
4. The history of owner action on past changes.

In today's business environment, it is an unfortunate fact that deadlines are usually the earliest date by which one can hope for the required occurrence. People target their activities toward stated objectives. If they have planned properly, have understood the requirements, experience no setbacks or unforeseen difficulties, and are responsive in a complete way, the deadline will be met. Contrast this scenario with the process that happens daily in the owner's organization. Add to that the procrastination factor that can accompany the action because the owner really does not want to let the change order go, and the probability of the original deadline being met plummets.

For these reasons, it may be naive to expect the greater portion of the required approval deadlines to be met in the first round. Therefore, document all critical dates, establish the resulting effects, and ultimately assign costs to the continuing delay. It will be the perceived ability and intention to follow through on these positions (or at least to pursue them) that will prod the owner to act.

One way to accomplish this is to consider assigning a cost per day beyond the response deadline in the change order proposal cover letter. If this is done, however, it must not be a frivolous notification. The contractor must be able to substantiate the number, just as with any other amount.

Another possibility is to consider a kind of "default clause." That is, notify the owner in the proposal that certain things will or will not happen if no response is received by the deadline date. These and other treatments will be considered further in Section 8.3.9, "Additional Terms and Conditions."

Know who has the formal approval and document signing authority. It may appear obvious at first, but it may not always be what it seems. Confirm the approval procedure before starting the change order process. This will allow the project management team to channel all correspondence directly to the decision maker(s) and avoid time-consuming sidetracks.

Finally, it is important to understand that great care must be exercised in the execution of the final change order form. Most change order forms state that the change order represents all costs associated with the change. In addition, the change order represents and the change order form may state that the final agreement voids all prior discussions, negotiations, and agreements, if not clearly mentioned in the final document. It is, therefore, imperative that the contractor have a complete understanding of the change and all its effects. It is also imperative that all reservations of rights and other special terms are in place and properly represented in the "final" change order. This is especially true for any consideration involving a future action or adjustment.

8.3.9 Additional Terms and Conditions

Besides the standard garden-variety terms and proposal qualifications discussed to this point, the last portion of the change proposal includes all terms and conditions that are the result of particular conditions, considerations or circumstances surrounding a specific change. All circumstances that are unique to the specific change must be carefully considered. For example, will the change prevent project close-out for an excessive period of time? If so, the release of the retainage amount on the entire contract sum will be delayed. In this circumstance, a precondition of the change may be the release of a significant portion of the retainage amount on time, without waiting for the change order work to be completed.

Another example arises when the individual subcontractor quotations contain creative clauses. Care must be taken to be sure that these clauses are not lost in the final change resolution with the owner. If these special subcontractor conditions are lost during the contractor's negotiation with the owner, the contractor may be responsible for those considerations as far as the respective subcontractors are concerned. In this situation, the following language may be included in the contractor's proposal cover letter:

> Your attention is called to the qualifications, exclusions, and conditions included in the respective subcontractors' quotations. These will remain in effect after the execution of the final change order unless otherwise directly addressed and resolved.

In any event, be certain that all the efforts taken to protect contractual rights and project momentum are not lost in the execution of the final change order document. If the owner breaches a condition of the change order approval, follow through on the consequences that were stated in the change order proposal. This must be done firmly and definitely to give legitimacy to future notifications. All statements *must* be believable, if they are to be taken seriously by the owner. Above all, be sure that the execution of the final document does not void prior negotiations and agreements by the fine-print inclusion of boilerplate disclaimers that may cancel such transactions.

9

Substantiating Change Order Prices: Settling Arguments Before They Begin

9.1	Introduction	225
9.2	Lump-Sum Prices	227
	9.2.1 Sample LETTER TO SUBCONTRACTOR Regarding Improper Proposal Submission	228
9.3	Detailed Cost Breakdowns	231
9.4	Time and Material	233
	9.4.1 Sample LETTER TO SUBCONTRACTOR Regarding T & M Submission Requirements	235
9.5	Unit Prices	237
9.6	Historical Cost Records	238
9.7	Industry Sources	239
9.8	Invoices—Records of Direct Payment	239
9.9	The Schedule of Values	240

9.1 INTRODUCTION

Chapter 8 was concerned with preparation of the change order proposal cover sheet. The discussion included detailed treatment of the three major cost categories of every change. This chapter is concerned with generating the supporting information ("backup") that will substantiate the cost components summarized on the change order proposal form.

The key to fast, total acceptance by the owner of any claim for additional costs and time is the definitiveness with which the respective issue can be presented. The contractor must strive to remove as much uncertainty from the process as possible. The contractor must also try to eliminate every opportunity for subjective input and evaluation. Just as the indirect cost portions should be converted into direct costs by way of precise records (refer to Section 8.1.4, "Transforming Indirect Costs into Direct Costs"), the backup cost breakdowns should be well researched, comprehensive, and in sufficient detail to satisfy every skeptic concerned with the review and approval process.

Proper substantiation begins with the justification of the entitlement to additional costs. If the contract allows the contractor to bill for a certain cost category, refer directly to it. In evaluating the change order request, the reviewer is going to have to find the contract provision anyway. By referring to the provision, the contractor saves the reviewer time and trouble and demonstrates that he or she has read the contract and is aware of the terms. In the final analysis, the extra effort will cut valuable time off the approval process. Make it as easy as possible for the reviewer and the approver to arrive logically at the proper decision—approval of the requested change.

Beyond the approval of a cost item in principle, the remainder of the backup effort will be devoted to establishing beyond argument the value of each component. Again, the overriding objective is to present as much detailed information as possible so that each cost component in the change order request is substantiated and above reproach. From crew size, to production output rate, to activity sequence, to individual wage rates, and so on, each cost must be shown to come from an authoritative and accepted reference. Show that no assumptions or contingencies have been introduced to inflate the total price unnecessarily. If assumptions (or qualifications) are required, be certain to identify the assumption as well as the reason for the assumption. Remember the discussion of allowances in Section 7.3.5. Consider whether an allowance should be used in the request rather than an assumption.

The point is to provide detailed supporting information, arranged in a format that walks the reviewer/approver through the requested change in the order in which the contractor wants the change to be considered. This will make it easier for the reviewer/approver to understand the contractor's point of view and accept the change. A logical, step-by-step approach to the requested change will facilitate the reviewer's justification of the change to the decision maker. Everything required to evaluate and approve the request should be included in the proposal and the supporting information.

Even more important than the backup itself, the properly substantiated change order request will make the reviewers look good to their superiors because it will give the appearance that they have done so much hard work evaluating all aspects of the change. By providing the reviewer with proper supporting information in a well-organized and easy-to-understand format, the contractor will encourage the prompt review and approval of change order requests. The properly prepared proposal is easy to justify and makes the reviewer look good to the decision-making authority. This strategy will make the contractor look good too.

In the evaluation and preparation of each of the three major cost categories of every change—direct, to indirect, to consequential costs—the nature of the charges moves from the concrete to the more abstract. As this occurs, the difficulty in coming up with hard substantiation for the softer-type costs can increase dramatically. Be creative. Learn to look at the "same old problems" in a new light. Do not allow yourself to be confined by routine, average thinking.

Good supporting information is an essential part of the prompt and efficient evaluation and approval of changes. It is also important because it:

1. *Ties related files together.* Multiple change orders in a given area can build upon each other fast. They can combine to compound interference and delay effects until they are well out of proportion to the actual direct costs associated with the changes. If each change is considered and priced independently, important interactions will be overlooked. The contractor may miss the cumulative impact of several related changes on the project and the owner may wind up with a bargain (at the contractor's expense) as a result of the contractor's oversight.

2. *Maintains important history and project record.* The more aggressive a change proposal, the more likely that certain charges will be denied by the owner at the time of formal change order approval. If the detailed costs for every category have been prepared at the time of the original change proposal, their believability in terms of their accuracy will be at their highest. The charges will appear much more legitimate if included in a later claim than charges that are calculated at the time of the claim. Small denied charges on many changes add up to substantial sums. If the details are individually identified and substantiated as part of the change order procedure throughout the project, a simple tabulation will be all that is needed at the time of the claim. This will substantially reduce claim preparation costs, while providing the most powerful documents available.

3. *Provides comprehensive accounting.* If all the details are itemized, reevaluations are easy in the event of the occurrence of an unforeseen condition. A good form and complete treatment will first confirm to the owner (or an arbitration panel) that the additional costs arising from the unanticipated condition or effect were not already accommodated in the approved change.

4. *Allows easy identification of trends.* Relevant summary information on change proposals and what happens to them after they are submitted to the owner will allow the contractor to spot how the overall process has been going. Chronically late owner action and repeated owner abuses in approved amounts are examples of the kind of information that can be very useful to the contractor when the contractor prepares, submits or goes to negotiate the next change.

5. *Can direct the path of negotiations.* A good change order proposal form and format can decide what will and what will not be placed on an agenda. The treatment of a given item may likewise steer attention to it and away from others. The form and format of the supporting information affords the contractor the opportunity to present the information in the light most favorable to the contractor and to manipulate the negotiations related to the change.

6. *Communicates precise attention to detail.* Elaborate, detailed attachments clearly communicate that time, attention, and hard work went into the proposal preparation. Everything is there for a reason. All the statements are serious and well thought out. Nothing is frivolous. Good records make it clear that the contractor knows what he or she is doing and that he or she intends to conduct the entire process in a professional manner.

7. *Can intimidate the owner or architect.* A well-presented and fully supported change proposal will clearly demonstrate that the contractor is serious and is not interested in wasting time. The proposal reflects that the contractor knows what he or she is due and that the contractor fully expects to receive it. Bluffing and frivolous tactics can be avoided and arguments over incomplete or unsubstantiated proposals will be eliminated. In certain instances, the owner and/or architect may even be intimidated by the realization that the contractor is more prepared than they can ever hope to be.

9.2 LUMP-SUM PRICES

On balance, lump-sum price submissions are more of a hindrance to the approval process than a help. While the lump-sum pricing option is contained in nearly every contract and subcontract standard form and it may appear to be the cleanest and most rapid form to submit, an analysis of approved changes reveals that lump-sum prices can delay change order approval and even reduce the amount that a contractor will ultimately receive for an additional cost item. If the applicable contract allows the lump-sum pricing option, the pricing method can be used in situations where the price is very small and obviously reasonable.

When receiving lump-sum prices from a subcontractor, several problems can arise if the price is not given the attention that it deserves. Lump-sum prices are difficult to evaluate and difficult to justify to the owner. If the subcontractor does not provide some meaningful basis against which to check the price, it is very difficult to determine if the price is fair and reasonable.

The only way to properly evaluate a subcontractor's lump-sum price is to generate some kind of breakdown, however informal, that will translate the lump sum into a form that makes sense. The net result of this evaluation exercise is that the responsible contractor winds up doing a lot of work for the benefit of the subcontractor. If the evaluation verifies the reasonableness of the lump sum, there is no problem and the contractor will be able to justify the subcontractor's price to the owner. If, however, the analysis uncovers that the lump sum was exorbitant, notifying the subcontractor will only cause him or her to come up with a breakdown.

One way to help avoid these time-wasters is to eliminate the language in subcontracts that allows the submission of a lump-sum price. Be clear in the terms and conditions that change order prices are to be submitted in sufficient detail to allow efficient checking of each significant component of the change. In addition, indicate that the costs for any delay in the proposal and subsequent change order approval caused by the subcontractor's improper submission will be backcharged to the subcontractor. To emphasize this requirement, include the cost associated with the review of an improperly prepared change proposal as part of the backcharge amount. Even if the collection of these costs is not pursued, the inclusion of the costs in correspondence to lackadaisical subcontractors will motivate many to submit all future proposals in the correct format. Consider utilizing Section 9.2.1, "Sample Letter to Subcontractor Regarding Improper Proposal Submission," to induce a problem subcontractor to become more responsive to the contractual requirements for proposals and to establish a basis for charging the subcontractor for the improper performance.

When completing the pricing for the change order proposal, be certain to evaluate the disadvantages associated with submitting a lump-sum price, even if the lump sum price is only part of a larger proposal. As mentioned in the previous discussion regarding subcontractor proposals, lump-sum prices are difficult to evaluate because they do not provide the reviewer with an adequate basis for analyzing the price. The use of lump-sum prices will usually prolong the review process. In this situation, the delay is the contractor's fault because the contractor made an improper submission.

Finally, do not overlook the approach detailed in Section 8.3.5, "Presenting Change Order Components," that outlines the advantages of submitting a large number of small costs instead of a small number of large ones. It is a very important strategic consideration that should not be underestimated.

9.2.1 Sample Letter to Subcontractor Regarding Improper Proposal Submission

The Sample Letter to Subcontractor Regarding Improper Proposal Submission is designed to compel the subcontractor who has submitted an unusable proposal to correct and resubmit a proper proposal in a timely manner. Specifically, the letter:

1. Returns the improper submission to the subcontractor for correction.

2. Calls attention to the contract or subcontract provision requiring the correct format.
3. Requires a resubmission in proper form by a specific, expedited date.
4. Advises the subcontractor that excessive time required to research, evaluate, and process the improper submission will be charged to the subcontractor if the problem is not corrected immediately.
5. Notifies the subcontractor that the subcontractor will be held responsible for all delays and additional interferences caused by the subcontractor's improper action (this is the primary reason the letter is sent by certified mail).

Note that the letter begins by confirming the discussion as of the date of the letter. This is important for two reasons. First, it removes any doubt as to the earliest possible notification date concerning the issue. Second, and perhaps more important, it confirms that there was every opportunity for the subcontractor to review the requirements and clarify any misunderstandings by way of personal conversation. If the resubmission is not in any better shape than the first submission, the contractor has every justification to follow through with the threats in the letter.

SAMPLE LETTER TO SUBCONTRACTOR REGARDING IMPROPER PROPOSAL SUBMISSION

LETTERHEAD

(Date)
To: (Subcontractor)

CERTIFIED MAIL-RETURN RECEIPT
RE: (Project no.)
     (Project title)

SUBJ: (Change file no.)
       (Change description)

Mr. (Ms.)   :

Per our conversation earlier today, your proposal for the subject change order is being returned for correction. Article (insert appropriate contract or subcontract provision) requires a detailed breakdown, properly itemized to allow evaluation.

Please correct your proposal and resubmit it in the proper format. The corrected submission is required by (date of two working days after conversation date). Your lack of action after that date will result in interferences that will be your responsibility.

Please be advised that excessive time required to evaluate and process improper proposal submissions will be backcharged to your company and that you will be held responsible for delays and additional interferences resulting from improper action.

Please correct and resubmit your proposal immediately.

Very truly yours,

Project Manager

9.3 DETAILED COST BREAKDOWNS

Too many subcontractors assume that because a general contractor or a construction manager get a percentage markup on subcontractors' proposals, those proposals will simply be bundled up and sent through to the owner. While this arrangement may have something to do with the way certain prime contractors choose to do business, contractors who get their changes approved quickly have a different approach. Each change order proposal will incorporate details and breakdowns which are specific to the particular change. However, there are several common areas which occur with sufficient regularity to allow the development of a systematic review procedure for almost all subcontractor change proposals. These common areas of review include:

1. *Work sequence and crew size*
    - Does the work incorporated in the proposal fit into the overall concept of the change? If not, is there sufficient information in the proposal to clarify the activities included, their sequences, and their interfaces with contract work?
    - Does the crew size make sense? Are support functions properly accounted for?
    - Are supervisor wage rates properly applied, or does the proposal include a charge for superintendents which is not allowed by the particular contract?
2. *Wage rates*
    - Are wage rate schedules included?
    - If union, are the rate schedules clear as to local?
    - Are overhead and profit figures consistent with those allowed by contract?
    - If union, do the wage rates compare identically with the published trade agreements?
    - Do insurance and tax figures appear to be in line with industry percentages?
    - Are premium time charges allowed? If so, is there an agreement in place regarding the overhead and profit applicable to these charges?
3. *Material quantities and prices*
    - Do rough, spot-check estimates generally confirm quantities listed?
    - Do prices appear to meet those expected?
    - Do random telephone spot-checks to suppliers confirm unit price accuracy? (Note that it will be necessary to check just a few prices to confirm the accuracy of the subcontractor's pricing.)

- Should the proposal be given to an estimators more experienced or familiar with the type of work included in the proposal for review? Will this step save time?
- Have unit prices or allowances been incorporated into the original bid package or subcontract?

4. *Overhead and profit application*
   - What is the percentage allowance in the prime contract?
   - Does the Pass-Through Clause (see Section 3.4.3) allow application of those rates to subcontractor prices?
   - Are any specific percentages included in the subcontract agreement?
   - Are overhead and profit allowed on the premium time portion?
   - Are overhead and profit applied to the "fringes" portion? (Many government contracts, for example, allow overhead and profit to be applied to the base rate, but not to the fringe benefits. Only overhead may be applied to the fringe benefits.) Be sure to check this requirement. Remember, the contractor must know the terms of his or her contract. These types of mistakes delay the change order process and tarnish the contractor's credibility.

When incorporating a subcontractor's proposal into the proposal to the owner, several ideas will help in the overall presentation.

1. *Verify and "correct" subcontractor's proposals visibly.* Mark up the subcontractors' proposals with *all* the corrections and changes. Include the marked-up sheets as backup to show all the hard work that went into the verification and correction process. Let the owner know that the proposal has been reviewed and analyzed. Show the owner (and management) that the proposal has been checked and that errors and improper charges have been corrected. Remember, the person who is willing to do his or her homework is more believable than are persons who are not. The marked-up sheets will give legitimacy to the contractor's review procedures and will facilitate a quick review and acceptance by the owner's representative (with a certain amount of relief because of the work that the representative will not have to do). In many instances the "corrections" will be assumed to be right, whether they are or not.

2. *Include any notes on material checks, references, and verifications.* Material prices submitted to the owner without checks and references will only be checked by the owner. Material prices that include backup showing that the prices have been checked will be assumed to be substantially correct.

3. *Overhead and profit application.* Check what is allowed by the prime contract before preparing the proposal. Do it right the first time. At the very least, it will communicate to the owner that the contractor has read the contract and is familiar with the contract terms. This fact alone may relax the owner's representative. In subsequent situations where the representative is not exactly familiar with a contractual requirement, the contractor's track record of correct application of contractual provisions will lead

the representative to assume that the contractor has handled the situation correctly. Do not underestimate the value of establishing a track record of doing things properly.

4. *Consider using a standard format.* If the owner has unusual and/or confusing change order format and requirements, consider sending a supply of either owner-supplied forms or forms designed to comply with the owner's requirements to all subcontractors to use in their change order pricing. This will speed all subsequent proposal reviews by reducing the process to a routing procedure. Billable categories will not be overlooked, and all costs will be presented and applied properly. It will also facilitate current and later research, particularly if someone else has to figure it all out at a later date.

9.4 TIME AND MATERIAL

Time and Material (T & M) is another pricing option normally contained in the Changes Clause of both the general contract and subcontracts. As a means to control the final outcome of a change, this option is considered to be the worst. On balance, a T & M arrangement with a subcontractor requires supervision and review that are disproportionate to the amount of work in question in that the contractor has a certain responsibility to manage the work properly (particularly in a construction management-type arrangement). T & M also eliminates the need for a subcontractor to be as closely concerned with crew size and productivity, since the subcontractor intends to pass the final cost on to the owner. For these reasons, it may be far more cost-efficient to hash out all of the details and agree on a final price prior to the start of the changed work. A hard dollar price component provides an important check on runaway subcontractor costs.

There are, however, situations in which implementing the T & M option with a subcontractor is unavoidable. In these cases, there are precautions that can be taken to improve the control over an otherwise nebulous situation. These precautions include:

1. *Daily means daily.* A 2" stack of T & M tickets presented near the end of the altered work is impossible to analyze and evaluate with any real degree of effectiveness. Even if the contractor has maintained detailed records against which to check the subcontractor's records, the process will be long, tedious, and error-prone. From the outset, the contractor must emphasize that the daily submission and review requirements of all T & M tickets will be strictly enforced. Requiring the subcontractor to present tickets every day will facilitate the prompt review and verification that needs to be done on a current basis. Notify the subcontractor that T & M tickets that have not been signed by the contractor *on the day that the work was performed* will not be recognized as an additional expense. Section 9.4.1, Sample Letter to Subcontractor Regarding T & M's Requirement Submission, will help fulfill this purpose. As a practical matter, it

may be difficult to enforce this condition without a significant battle. Clear language in the subcontract agreement, coupled with the contractor's use of the Section 9.4.1 notification, can form the basis for such an action. The real purpose of the letter, however, is simply to prompt the subcontractor's organization into an acceptable mode of operation right at the start. After all, all that the contractor really wants is T & M tickets to be submitted daily so that the subcontractor costs can be monitored and verified as the work proceeds.

2. *Hourly verification.* If multiple items are being billed on a time and material basis, or if the same individuals are being used concurrently on contract work, check the total number of hours billed for a given person on a given day. Do the hours add up to twelve hours for an eight-hour day? If so, such abuses can and should be detected on the first occurrence. Early discovery and prompt notification can straighten out this situation.

3. *Labor classification verification.* Is the subcontractor billing a 50% apprentice at the journeyman's rate? Is a high wage rate from another local being added to the stack of forms?

4. *Overhead and profit application.* On straight-time T & M, are the overhead and profit rates being billed in accordance with the terms specified in the subcontract agreement? In a premium-time situation, has overhead and profit been improperly applied to the premium-time portion of the bill?

5. *Photographs.* Film is cheap. In *every* situation involving *any* question, the contractor should take a photograph of the item at the very least. If the problem is significant, consider arranging for a professional photographer or video to document the facts. Refer to Section 10.5, "Photographs—What, When, and How," for additional guidance in this area.

6. *Early evaluation of production rates.* As soon as is reasonable after the start of the time and material work, begin reviewing the actual production rates against historical rates or rates that were reasonably expected. Small liberties taken by the subcontractor at the start can be put back into perspective quickly. Significant chronic abuse, however, must be dealt with swiftly and decisively. If T & M tickets are deemed to be excessive, try to resolve the issue in the field. Confront the individual presenting the ticket with the dispute over his or her claim of actual production.

When confronting a subcontractor suspected of taking liberties with T & M reports, improper reductions will be loudly protested. On the other hand, appropriate criticisms will be met with sidestepping remarks and less direct responses. The differences will be easy to spot. Loud protestations, with proper explanations, produce confidence.

If efforts to resolve the problem in the field are unsuccessful, protest immediately to the next higher authority in the subcontractor's organization. While it is generally not advisable to go over anyone's head until all options at a given level have been exhausted, progress up the authority ladder one rung at a time until someone in the subcontractor's organization will meet at the site to evaluate the problem.

When presenting time and material tickets to an owner, it is important to understand that the owner has the same concerns with the contractor's tickets as the contractor has when reviewing and evaluating tickets presented by subcontractors. Owners experience the same feelings of loss of control. The tickets are difficult to check, the work has to be watched closely, and the totals *always* seem to add up to more than originally expected. Many owners believe that the contractor is operating without any risk, and therefore without any motivation to complete the work efficiently and at the best possible price. The results of this situation can range from simple owner bad feelings about the whole arrangement to delays in change order approval and disputes over time and material components.

A final consideration with regard to the use of the time and material procedure with the owner involves those instances where the responsibility for certain work is not clear, or where the contractor has been directed to proceed at no additional cost with work that the contractor definitely believes is an "extra." If circumstances mandate that the contractor proceed with the work pending a final decision, or in anticipation of a later claim, use T & M tickets to keep accurate, indisputable records. The use of the T & M tickets in this manner is detailed extensively in Chapter 10, Using Project Records to Discover, Define, Support, and Track Change Orders and Claims. Section 10.9.2, "Sample Letter to the Owner Regarding Acknowledgment of Actual Work Performed," can also support this effort. Refer to Section 10.9, "Time and Material Tickets," for additional related discussion.

9.4.1 Sample Letter to Subcontractor Regarding T & M Submission Requirements

The Sample Letter to Subcontractor Regarding T & M Submission Requirements is designed to follow through with the recommendations in Section 9.4 and prompt the subcontractor to submit proper time and material tickets on time. Specifically, it notifies the subcontractor that:

1. *Daily* authorization and acknowledgment of work performed are absolutely required.
2. T & M tickets not approved on the day that the work was performed will not be recognized as an additional cost.
3. Labor and material charges require detailed breakdowns and substantiation for all costs.
4. Signatures by field personnel only confirm the fact that certain work was performed with certain forces on that day. Field personnel do not acknowledge or agree in any way that the work itself is extra or that the rates charged are acceptable. Other persons in the contractor's organization are responsible for determining if the work is an extra and if the rates are appropriate.

As mentioned in Section 9.4, the real objective of the letter is simply to prompt the subcontractor to utilize the T & M process correctly. If the proper routine is established at the start, the entire process will move much more efficiently.

SAMPLE LETTER TO SUBCONTRACTOR
REGARDING T & M SUBMISSION REQUIREMENTS

LETTERHEAD

(Date)

To: (Subcontractor) RE: (Project no.)
  (Project title)

  SUBJ: (Change file no.)
  (Change description)

Mr. (Ms.)   :

On (date), your company was directed to proceed with the subject work on a time and material basis, in accordance with the provisions of (insert appropriate contract or subcontract reference). The conditions of this arrangement are as follows:

1. T & M tickets are to be signed *daily*. Tickets that are not signed on the day that the work was actually performed will not be recognized as an additional expense item.

2. Precise labor classifications are to be noted on each ticket.

3. Material invoices are to be attached.

4. Signatures by field personnel only confirm the fact that certain work was performed with certain forces on that day. Field personnel do not acknowledge or agree that the work is in addition to the contract or that the rates charged are acceptable. Both of these matters are subject to further review in accordance with the terms of the (insert appropriate contract or subcontractor references).

Thank you for your cooperation.

Very truly yours,

Project Manager

9.5 UNIT PRICES

Unit prices can be included in the prime contract, or the general contractor can incorporate them into the respective subcontract agreements. Unit prices are included to remove at least a portion of the uncertainty attached to an item's definition, the method of calculation, or the price itself for a unit of work. Typical situations are described, and prices are cataloged for variations in quantities.

The objective in including unit prices in a subcontract is to add certainty to the consideration of possible additions and/or corrections that have a good probability of materializing during construction. The contractor should try to think through as many of these situations as possible in the bid package preparation stage. For example, an electric bid package for interior improvements to shell space may include unit price quotations for items such as:

- Light switches, including wiring.
- Duplex outlets, including wiring.
- Each type of light fixture provided.
- Exit lights.
- Emergency lights.
- Junction boxes with conduit to ceiling space for later telephone or computer wiring.

Having these prices established at the time of the original bid will make it very easy for the contractor or the owner to evaluate the cost of adding these items if the scope of the contract increases. Similarly, each other bid package should anticipate as many potential change situations as possible, and attempt to cover their eventuality via the unit price. If this can be achieved, valuable time will not be wasted on "price negotiation" when additional costs fall into these categories.

Unit prices in the owner's contract may, however, be a different matter. Contracts on some projects may allow quotes on unit prices in a manner similar to the situation just described for subcontractors. Other contracts are not so reasonable. On many public projects, for example, unit prices for additional work may actually be included in the bid documents for the project. A common category of this type is earth and rock excavation. An elaborate specification may be detailed for the variety of circumstances that can occur when subsurface or other conditions result in a change order. Each situation is then accommodated with its unit price. These prices are extremely difficult to change during construction.

Fortunately, specifications with unit prices usually include some language qualifying the unit price application. They may, for example, note that the contractor can request that the owner consider a modification to the unit price if circumstances can be demonstrated to be extreme or unusual. For such an effort to succeed, the contractor must clearly demonstrate to the owner that the unit price does not apply to the situation. The work in question must be redefined so that it is out of the descriptions included in the unit price specification. In practice, unit price changes will not occur very easily, if at all. Even so, documenting the effort and the supporting arguments will strengthen the contractor's position in any later dispute.

9.6 HISTORICAL COST RECORDS

Historical cost records—the actual expenses for certain items as recorded as part of your standard company procedure—are probably the best method available to substantiate the overhead amounts that are applied to the extra work. Types of costs included as part of these records are:

1. Administrative overhead rates.
2. Insurances and taxes on labor.
3. Payment and performance bond premiums.
4. Detailed job cost records.

As mentioned previously, the contractor's first task is to identify and accurately assemble all of the costs which arise from or relate to a particular situation. Then the contractor should try to convert as many of the costs typically included in the "indirect" cost category into "direct" costs (refer to Section 8.1.4, "Transforming Indirect Costs into Direct Costs," for more on this idea). Direct costs are easy to tabulate and verify. The uncertainty associated with an indirect expense must be removed. The only way to accomplish this result is through some kind of "precise" calculation (or at least one that looks precise).

Establish a job cost system throughout the organization that will track these "overhead" kinds of expenses in the same way that the more conventional direct costs are tracked. This system will involve indoctrinating administrative personnel into the habit of "charging" their work activities to the applicable cost item on a daily basis, as the work is being performed. Field personnel have been (or should have been) tracking their time in this manner as part of the contractor's project records. Getting the administrative staff to do the same thing may take some time, but it is not difficult.

If *all* costs are recorded and tracked, there is no uncertainty. If, on the other hand, conventionally "indirect" costs are not tracked specifically, the best that the contractor will be able to present is established and accepted accounting formulas that allocate a portion of total overhead to an individual project. Even though the use of these formulas might look like a clean way to accomplish the objective without having to spend all the time and effort keeping individual records, the contractor is doing himself or herself a serious disservice. The accounting formulas provide calculations that average the overhead application over all the projects. The formulas prorate the overhead amount across the large and small projects, allocating proportionate amounts to each. Because they average the expenses, they do not allocate the proper amount of overhead costs to the problem job. A project that is experiencing multiple changes and delays requires disproportionate amounts of management time and attention. Extended field and home office administration expenses, opportunity costs, reduced bonding capacity, and so on all contribute to an increased percentage of the total overhead expense *really* allocable to the problem project. This consideration is ignored by conventional accounting formulas. If the contractor can maintain direct records with regard to overhead costs, the contractor will be in a better position to recover such costs.

9.7 INDUSTRY SOURCES

Estimating quantities of materials is straightforward and easily verifiable. Estimating crew sizes, labor classifications, and production rates is a more subjective determination. These estimates are based on experience, research, anticipation, and sometimes hope. Correspondingly, these estimates are more difficult to substantiate with hard calculations than are simple material estimates. Reliance on historical records may help in this regard, but they can still be disputed. Historical records may not precisely fit the descriptions of the changed work, and as such, any conclusions drawn about the rates of production will be open to debate.

The use of established manuals, productivity analysis, equipment rental rate guides, and similar publications can go a long way in solidifying the work description and production rates. Estimating guides will often list optimal crew configurations for certain types of work. If contractor can locate these kinds of schedules, all that becomes necessary is a correct application of the formulas required to tailor the information and prices to the specific work item in the subject location. Nationally recognized publications and/or schedules provide the reviewer with the needed basis for the prompt approval of the contractor's estimate. Find out which industry sources are familiar, commonly used in the project's location, and acceptable to the owner. Researching this information before expending time and effort on the preparation of an estimate will save the contractor time and money.

A final point to be made in the face of criticism by the owner of anticipated sequences, crew sizes, and production rates is to indicate firmly that the estimates are based on actual experience. Remind the owner that the industry demands that estimates be accurate or the business fails. If required and appropriate, demonstrate the use of industry estimates on successful bids and/or completed projects. From that point, work to find ways to reduce the uncertainty in the mind of the reviewer, and make the presentation more convincing.

9.8 INVOICES—RECORDS OF DIRECT PAYMENT

These are probably the best kinds of records to produce to substantiate costs, assuming no unit price application. Material invoices are easily verifiable. In most cases, the review of material invoices is limited to being certain that the price is within reason and that the invoice does not include additional items that are not part of the change.

One major stumbling block to the quick review and approval of direct cost records involves quotations from a subcontractor's subcontract. The same care must be exercised in evaluating the sub-subcontractor's proposal as is necessary for the subcontractor's quote (as detailed in Section 9.3, "Detailed Cost Breakdowns"). Beside all those considerations, be careful to review the sub-subcontractor's overhead and profit percentages. As archaic as it seems, many government contracts do not recognize the sub-subcontractor's right to include overhead and profit in a change order proposal, even in the presence of a clear Pass-Through Clause (see Section 3.4.3). In this situation, the sub-subcontractor's overhead and profit needs to be built into the costs so that this procedural snag can be avoided. It is for these kinds of reasons that an intimate knowledge of the contract provisions, as well as their practical application, is absolutely necessary to speed every change proposal through as fast as possible.

9.9 THE SCHEDULE OF VALUES

The Schedule of Values is a document that itemizes the individual components that make up the entire project. It can be broken down into quantities and unit prices and is used expressly for preparing and evaluating progress (partial) payment requests. The information contained in this schedule normally cannot be used as a basis for an adjustment in the contract price, because the unit prices contained in the schedule normally include overhead and profit and the distribution of costs may not be entirely accurate. This inaccuracy results from some contractors trying to "front-load" the schedule; that is, including a larger percentage of overhead and profit (and possibly some cost items) in the items that will be billed for early in the schedule sequence. This will help a contractor with early cash flow on a project, if the schedule is approved by the owner. For these reasons, the schedule of values is not often used as a basis to establish prices in a change situation.

Occasionally, it may be attractive to try to get the high unit prices in the Schedule of Values accepted as part of a change order proposal. If this occurs, consider it closely. If this procedure is accepted by the owner, it can set a dangerous precedent to confirm the method of unit price establishment on future change orders. In later proposals involving cost items with less favorable (or downright unattractive) unit prices, the schedule may be forced on the contractor by reason of the prior change.

10

Using Project Records to Discover, Define, Support, and Track Change Orders and Claims

10.1	Introduction	243
	10.1.1 Active Working Files	243
	10.1.2 Item Completion and Close-Out	244
	10.1.3 Archives	244
10.2	Establishing Dates in the Correspondence	244
10.3	Daily Field Reports	245
	10.3.1 Sample Daily Field Report FORM	247
10.4	Payroll Records	251
	10.4.1 Sample Field Payroll Report FORM	251
	10.4.2 Sample Monthly Administrative Time Sheet	252
10.5	Photographs—What, When, and How	255
	10.5.1 Introduction	255
	10.5.2 Photograph Layout Requirements	256
	10.5.3 Sample Photograph Layout FORM	256
10.6	Construction Schedules	258
	10.6.1 As-Planned, As-Built, and Adjusted Schedules	258
	10.6.2 Six Requirements for Presentable Evidence	259
10.7	Using Job Meetings to Establish Dates, Scopes, and Responsibilities	261
	10.7.1 Introduction	261
	10.7.2 Job Meeting and Minutes Guidelines	262
	10.7.3 Sample LETTER TO SUBCONTRACTORS Regarding Mandatory Job Meeting Attendance	264

	10.7.4 Sample LETTER TO SUBCONTRACTORS Regarding Lack of Job Meeting Attendance	266
	10.7.5 Sample Job Meeting Minutes FORM	268
10.8	Shop Drawings and Approved Submittals	271
	10.8.1 Approval Responsibility	271
	10.8.2 Approval Response Time	273
	10.8.3 Treatment of Differing Conditions	273
	10.8.4 Absolute Contractor Responsibility	273
10.9	Time and Material Tickets	274
	10.9.1 Introduction	274
	10.9.2 Sample LETTER TO THE OWNER Regarding Acknowledgment of Actual Work Performed	275
	10.9.3 Sample T & M FORM	277

10.1 INTRODUCTION

Complete and accurate records of all significant project events are an absolute necessity in today's construction industry. Projects and issues are increasingly complex. They involve elaborate contract requirements, specific submittal and approval procedures, sophisticated interpretations, and detailed procedures for resolution of disagreements. Beyond all this, commitments from dozens of subcontractors and suppliers, timetables, and performance requirements must be established, documented, and monitored. In any kind of hindsight analysis (as in a claim), good, comprehensive, and properly correlated records are essential for clear visibility and understanding. In a dispute resolution situation, the only real proof will be the project records.

Good records are absolutely essential to:

1. Fully document historical facts, including all significant events and information;
2. Correlate issues with appropriate contract provisions to provide the justification for action;
3. Catalog the chronology of an issue and the sequence of important events;
4. Simplify complex relationships and interactions by highlighting and organizing significant information;
5. Arrange relevant information in each file location that requires or may require the information during the project cycle or afterward in a claim or other dispute situation;
6. Provide status summaries of key dates, transactions, decisions, and results which can function as checklists, to be sure that all items begun are finished, and to identify important trends;
7. Form the basis of correct and timely management action; and
8. Ultimately provide organized, convincing proof to support arguments and substantiate claims.

The records systems themselves are designed to function in three capacities: active working files, item completion and close-out, and archives.

10.1.1 Active Working Files

A system of documentation is needed that correlates important information with all issues that may require the information now or in the future. Issues in process need to be tracked and researched quickly and on a continuing basis. A mechanism must be introduced to keep hot topics on the front burner. It is this *visibility* that will keep everyone's attention on the issue and will bring the matter to the fastest resolution. Finally, in-process files will coordinate administrative activities (such as shop drawing submission) with project timetables as established by the construction schedule or by a change order.

10.1.2 Item Completion and Close-Out

Issues in construction can last from days to over a year before final resolution. Dispute resolution activities (hearings, arbitration, litigation) can occur during a project, immediately after, or well after project completion. Comprehensive files are the keystone of accurate issue reconstruction and substantiation.

10.1.3 Archives

When a project is finally closed out, it may not be necessary to keep *everything* as historical records. A good file storage system will allow fast future reference if the historical information later becomes necessary or desired. Future management or new items of interest may require a look back into the past. If the historical files are stored properly, the information will be there.

Good records maximize the contractor's options, opportunities, management capability, and absolute protection. It is the contractor with the best records who will consistently convince opponents, win arguments, and make money. It has been said on many projects that the contractor does not even have to be right—the contractor just has to have the best records.

10.2 ESTABLISHING DATES IN THE CORRESPONDENCE

Sections 8.2.1, "Change Order Identification/Notification" and 8.3.2, "Proposal Format and Timing," discussed the necessity for proper and timely notification as a technical contract requirement. The objective should be to remove any question as to whether the obligation has been fulfilled. As such, the overriding principle that every contractor should adopt is that there cannot be too much notification. When in doubt, send it out, in writing.

The important point of this section is to clarify that it is not necessarily the date of the first written correspondence that qualifies as the notification date. Notification has been achieved when there is an understanding in the mind of the recipient. The written confirmation is there only to document the fact. The notification, then, can actually be said to occur in a wide variety of communications. Among these are:

1. Telephone and face-to-face conversations;
2. Job meetings;
3. Other letters;
4. Speed memos;
5. Shop drawings;
6. Design change drawings; or
7. Any document that either expressly or implicitly acknowledges an event.

Personal and telephone conversations are probably the most neglected forms of notice, perhaps because they do not have the appearance of a project record notification. Similarly, job meeting discussions are too often disregarded as proper notification of an issue. Do not lose track of these valuable early dates. A few simple guidelines will help to ensure that the ultimate written notice will take every advantage available:

1. *Reference the earliest appropriate notification date in the written notice.* Whether a telephone conversation, shop drawing, or whatever, refer to the earlier date in the first part of the written notification. An opening remark as simple as "Confirming our conversation of (date)…" or "Per your direction of (date) …" will tie the events together.

2. *Specifically name the people involved in the prior notification.* If, for example, the written notification refers to a discussion with someone else in the addressee's organization, a "confirming my conversation with your office…" statement is too weak. Even if you had left a message with a party's secretary, include the *name:* "Per my notification to your Mr. (Ms.) _____ on (date)… ."

3. *Confirm all significant discussions immediately.* Any piece of information, understanding, confirmation, promise, and so on that has or may have an effect on the work should be recorded in writing. At a minimum the written record should be contained on a daily report or job diary of some kind. Even if it begins only as a plain note in a file, the note can form an important basis for a more formal notification if necessary. A speed memo or a short but clear note in the minutes of a job meeting is even better because these records are provided to others.

4. *Be concise, but be clear.* If a reference is made to a prior letter, the reference itself is all that is needed. Elaborate descriptions of the referenced document add bulk and confusion. For example,

    In response to your (date) letter regarding the subject,

is better than:

    In response to your (date) letter regarding the changes to the mezzanine concrete slab support wherein you directed the sitework subcontractor to provide an additional twelve inches of bank run gravel under the north portion of the slab…

The reference in the first case is enough, if the referenced document is defined clearly enough so as to avoid any possible confusion with any other document. In this case, the recipient simply needs to find a copy of the referenced document to see exactly what it says in all the detail.

10.3 DAILY FIELD REPORTS

The Daily Field Report is the fundamental document that records actual job progress, together with all conditions that affect the work. The report represents the contractor's standard process for recording all causes and effects on the work at the point of production. If actually prepared daily, they are generally considered to be one of if not the best sources of job information. This is because they are prepared as the information is being generated, with no appreciable time lapse. The inclusions, descriptions, and circumstances are fresh in everyone's minds. This facilitates complete details of the work. In addition, the reports are prepared by those with authority and responsibility for the work. They have an interest in the accuracy of the information and, therefore, have a significant incentive to maintain the report's accuracy and completeness. These are the people who actually witness the work and, as such, are recognized to be the most qualified to describe the facts.

The contractor's reports should be detailed, accurate, and complete records of all jobsite events. The Daily Field Reports will become the cornerstone used to support all contentions through the presentation of the legitimate, believable actual data. The reports are invaluable to the contractor because of the wealth of information recorded in summary form. Such information customarily includes:

1. Work and activity descriptions, separated as to the physical location;
2. Labor force, broken down by subcontractor, locations, makeup and major activities;
3. Equipment used and stored by the various contractors;
4. Site administrative staff and facilities;
5. Weather conditions and temperatures at key times of each day;
6. Change order work accomplished, with relevant details;
7. Photographs taken during the day;
8. Visitors to the jobsite;
9. Inspections, accidents and other key events; and
10. Meetings, discussions, commitments, and conversations.

Because it is impossible to tell in advance what will become important, all categories must be meticulously maintained. Further, it is up to the central office to regularly police the reports to be certain that all required information is being properly recorded and that the reporting process is not being abused (i.e., criticizing the contractor's own organization). Finally, the central office must stress the requirement that the reports be completed *daily*. If the reports are allowed to be lumped together for completion even weekly, the accuracy of the information quickly degenerates. The completeness of the data is also dramatically reduced. When on site, management should refer to the reports often, during conversations, issue research, and so on. Be sure that the project manager, superintendent and foreman know that the information is depended upon and used. The minute the field staff begins to get the feeling that no one is looking at the information, the data will become very thin. Have the reports prepared in duplicate, and require the field office to send the daily report to the central office every day. Keep copies of the reports in the field office for quick reference and use.

Finally, as hard as it is to believe, many sizable construction projects today are still being constructed without any such daily reports. Daily reports are an essential management tool. A contractor's failure to maintain

CHANGE ORDER PROPOSAL, PREPARATION, AND PRESENTATION 247

a daily record in these litigious times is a clear indicator of irresponsible project management. If field reports are not being used, turn immediately to Section 10.3.1, "Sample Daily Field Report Form," and begin using it *today*.

10.3.1 Sample Daily Field Report Form

The Sample Daily Field Report Form is to be filled out every working day by the project superintendent and signed by the project manager if that is an on-site position. In the absence of an on-site superintendent, the report must be filled out by the responsible project manager—every day.

As set forth above, the report is designed to provide relevant information regarding:

1. Work and activity descriptions, separated as to the physical location;
2. Labor force, broken down by subcontractor, locations, makeup, and major activities;
3. Equipment used and stored by the various contractors;
4. Site administrative staff and facilities;
5. Weather conditions and temperatures at key times of each day;
6. Change order work accomplished, with relevant details;
7. Photographs taken during the day;
8. Visitors to the jobsite;
9. Inspections, accidents and significant events; and
10. Meetings, discussions, commitments, and conversations.

The following recommendations will help with efficient completion of the report forms:

1. *Title box.* Indicate project name, company job number, project location, and the name of the individual completing the report.
2. *Date/page.* Account for each page of the total report. Use as many report forms as necessary for a complete record of the day's events. *Never* use the back of a report to continue information; it is easy to overlook during a review and photocopying.
3. *Weather.* Include a short comment on the weather conditions of the day (cloudy, rainy, snowing, etc.). Record the temperature at the beginning and the end of each day.
4. *Equipment.* List all major pieces of equipment on the site.
5. *Visitors/problems/comments.* List all visitors to the jobsite who are not part of the workforce. Include a brief remark as to the reason for the visit. Note any special situations deserving immediate attention, such as accidents, inspection, etc. Include references to appropriate documentation.
6. *Materials.* List all materials received or requested that day only.
7. *Workforce.* Indicate the number of each type of employment classification for each type of labor involved. For example, if two minority elec-

tricians and three nonminority electricians are employed by the same subcontractor, indicate "2" in the "M" column and "3" in the "NM" columns, both beside the description "Electrician." Separate the labor breakdowns by subcontractor.

8. *Description and location.* Directly adjacent to the respective labor force information in item 7, include a short description of the type of work being performed. Also provide the location of the work, with sufficient references to allow accurate location. For example, "Continue foundation wall forming along column line A, between lines 4 and 6."

9. *C.O. No.* If the work described in item 8 applies to a change or a proposed change, insert the number of the change order or change estimate number in the column provided.

10. *Signature.* The superintendent and/or the project manager should sign the daily report.

11. *Distribution.* Send the signed report to the central office. Retain a copy at the project site.

On large projects, some of the information on the daily report can be preprinted. If the report is prepared, stored and transmitted electronically, be certain that the information is backed up so that it will not be lost in the case of a computer failure. Also, the contractor should take steps to be certain that the information cannot be altered after the report is in final form.

DAILY FIELD REPORT

Page 1 of ____

Project: _____ No.: _____ Date: _____
Location: _____ Weather: _____
Superintendent: _____ Temp: 8 AM ____ 1 PM ____ 4 PM ____

Staff

Name	Classification	Name	Classification

EQUIPMENT

Quat	Type/Size	Work/Idle	Work performed	Arrival	Departure
		/			
		/			
		/			
		/			
		/			
		/			
		/			
		/			
		/			
		/			

VISITORS / PROBLEMS / COMMENTS

_____
_____
_____
_____
_____
_____
_____
_____
_____
_____
_____
_____
_____
_____

MATERIALS

Item	Requested from	Company	Promised by

Item	Received from	Company	Time

_____   _____
Project Manager (Signature)   Superintendent (Signature)

DAILY FIELD REPORT (continued)   Page ___ of ___

FIELD WORK REPORT

Quant	Classification	Sub	Description and Location of Work	CO #
TOTAL				

10.4 PAYROLL RECORDS

Reporting payroll as a project control method is one objective. The other is to allow substantiation of actual costs as they are applied to a project. The treatment is split between the field and home office administration. Even in the field, the workforce is divided into the field labor and field office staff. In its most basic form, the actual labor force record is used for the payroll record. For those contractors with more sophisticated job cost systems, the field payroll report is broken into the specific activities performed on a given day by a particular employee. A numbering system is assigned to the possible activities as determined by the work estimates. The labor hours are reported against those activity numbers. The result is a job cost report that identifies the exact labor cost applied to the individual job components. This kind of division provides a valuable comparison to the cost estimates for the respective activities. Actual costs can be tracked against the estimates as they occur to allow identification of potential problems before opportunities for correction have passed. The kind of field payroll report form that can help to accomplish these objectives is included in Section 10.4.1, "Sample Field Payroll Report Form."

The actual field payroll report provides an indisputable account of direct costs to be applied to a particular change. Administrative staff expenses can be tracked with the same precision by utilizing the same procedure used for the field force. Administrative effort can be broken into the major activities and reported as such, particularly the time soaked up by changes. In the case of central office project executives, estimators, and accountants, who divide their time simply by project (and not by the activities within those projects), the use of Section 10.4.2, "Sample Monthly Administrative Time Sheet," will be more appropriate. Even if your particular job cost system cannot accommodate information at this level of detail, keep the records anyway. In the event that the costs become necessary, they can be assembled and tabulated for a specific situation. If these kinds of records are not available, the contractor needs to resort to those "accepted" accounting methods to allocate administrative overhead to the project in question. Utilization of the accounting formula approach will result in the inequities described in Section 9.6, "Historical Cost Records."

10.4.1 Sample Field Payroll Report Form

The Field Payroll Report Form is designed to accommodate the suggestions included in Section 10.4 for field reporting. Specifically, it provides the capability to separate each field employee's daily hours into the major activities of work. The key is the use of a job cost system that divides the estimated work into the individual activities and assigns separate numbers to each one. When the information is entered into the job cost system, the result will be separate reports. One applies the actual resources expended on the major items of work, and the other isolates those spent on the respective change orders.

10.4.2 Sample Monthly Administrative Time Sheet

Because project managers, project executives, estimators, and administrative assistants are often salaried personnel, it is technically not necessary to report their labor hours for payroll purposes. The desirability of this kind of information, however, has recently increased dramatically.

The Sample Monthly Administrative Time Sheet is designed to accommodate the recommendations in Section 10.4 as an effort to convert otherwise indirect costs to direct costs chargeable to specific projects. Its purpose is to keep a record that will provide management with accurate information regarding:

1. Where administrative overhead is spent;
2. Where the largest percentage of overhead is spent in terms of the project's life cycle;
3. The percentages of administrative overhead for a project that are spent in the field and in the central office;
4. Specific charges to be applied to changes, delays, accelerations, and reschedules;
5. What percentage of the project cost is represented by administrative or general conditions costs.

The Time Sheet is to be kept by all appropriately designated personnel. During or at the end of each day, all projects worked on (central office and field) and any other significant activity (estimating, scheduling, etc.) are listed on the form. An estimate of the amount of time spent on each activity is included. At any point during or after the life of the project, an accurate accounting can be prepared for any category desired. This information will become extremely important when converting would-be indirect costs into direct charges in any change or claim.

SAMPLE FIELD PAYROLL REPORT FORM

Project: _____ No. _____ Week Ending: ____/____/____

EMPLOYEE	SUN		MON		TUE		WED		THUR		FRI		SAT		TOTAL HOURS
	Act. No.	Hrs	Act. No.	Hrs	Act. No.	Hrs	Act. No.	Hrs	Act. No.	Hrs	Act. No.	Hrs	Act. No.	Hrs	
No. _____ Name: _____															
	Total		Total		Total		Total		Total		Total		Total		
No. _____ Name: _____															
	Total		Total		Total		Total		Total		Total		Total		
No. _____ Name: _____															
	Total		Total		Total		Total		Total		Total		Total		
No. _____ Name: _____															
	Total		Total		Total		Total		Total		Total		Total		
No. _____ Name: _____															
	Total		Total		Total		Total		Total		Total		Total		

SAMPLE MONTHLY ADMINISTRATIVE TIME SHEET

NO. _____

NAME _____
POSITION _____
MONTH _____ 19 _____ WEEK NOS. _____
REVIEWED BY _____ DATE ___/___/___

(2) WEEK PERIOD (SUNDAY TO SATURDAY)

PROJECT NAME	PROJ. NO.	S	M	T	W	T	F	S	S	M	T	W	T	F	S	TOTAL

(2) WEEK PERIOD (SUNDAY TO SATURDAY)

PROJECT NAME	PROJ. NO.	S	M	T	W	T	F	S	S	M	T	W	T	F	S	TOTAL

SIGNATURE _____

10.5 PHOTOGRAPHS—WHAT, WHEN, AND HOW

10.5.1 Introduction

The job photo record should begin with a preconstruction series. Before *any* work is begun on the site—including mobilization of temporary field offices—the entire site should be photographed in detail. All existing conditions should be included. Consider photographs of the same major areas from different directions and points of view. When in doubt, take a picture of it. Film is cheap. All that is needed for the photo record is a contact sheet of each roll of film, just to identify the areas and views included. If evidence of the exact preconstruction condition of the site becomes necessary in a change order proposal, a claim, or a defense against a claim of site damage, the contact sheets can be reviewed. The appropriate photographs can then be enlarged as required by the situation. The procedure is extremely inexpensive, but can save thousands of dollars by ending arguments or clarifying extra cost items in change order proposals.

Ongoing construction photographs are divided between regular progress photos and those required in special situations. Even if progress photographs are not required by the owner, arrange for them anyway, at least in some minimal capacity. Utilizing a procedure similar to the procedure described for the preconstruction photographs will provide good coverage of the project at a minimal expense.

In situations involving change orders, claims, backcharges, possible insurance claims, and other special situations, before-and-after photographs are advisable. In these situations, fast actions are necessary to gain the maximum possible advantage. A "precondition" photo or series of photos will be the best record to confirm the exact circumstances prior to a changed work sequence. If the duration of the activity of interest is long, progress photos showing the significant stages of its progress should be taken. The individual circumstance will determine the level of detail necessary to adequately fulfill this requirement and document the condition.

Good-quality 35mm cameras are excellent in all project photograph situations. Resist the temptation, however, to use wide-angle lenses inasmuch as they distort shapes and perspectives. Remember that the objective is to provide an accurate record. Use lenses with "normal" focal lengths, and the photographs will remain true representations of reality.

An "instant" or digital camera will also prove to be a valuable addition to the project photo effort in the field. It can be used to supplement the better quality photos or be used by itself if it turns out to be sufficient. The principal advantage is that these cameras immediately display the precise information that has been recorded in the photograph. It is therefore immediately evident if your photographs indicate everything that was intended, or if more photos and different angles are necessary. This is particularly important if conditions are likely to change quickly, leaving no time for conventional photos to be developed. Another alternative is one-hour film developing, which is now available in almost all areas. With all of these options, every contractor should be able to use photographs to record project progress, unanticipated conditions and significant events.

Finally, try to include some clear physical references in the photographs to establish perspective. For example, a photograph of a wide crack in a concrete slab will be more meaningful if it includes an open ruler beside or spanning the gap. Similarly, common objects such as coins, pencils, and cars can be used to establish size and distance perspectives in different situations.

10.5.2 Photograph Layout Requirements

If progress photographs are required by the owner, the exact requirements will normally be described in the general conditions of the contract. In any event, a number of requirements should be complied with, to ensure that the proper information is recorded and the usefulness of each photograph is preserved. Included in each photo description is:

1. Name of the project;
2. Owner and company project numbers;
3. Date of the photograph;
4. Direction of view;
5. Identification of the photographer;
6. Description of the photo subject;
7. The particular location of the photo subject at the project; and
8. References to any appropriate correspondence, or anything to tie it to the detailed project record.

The use of Section 10.5.3, "Sample Photograph Layout Form," will provide a convenient mechanism for complying with these seemingly elaborate information requirements. If the form is used consistently, its application will quickly become a routine procedure.

10.5.3 Sample Photograph Layout Form

The Sample Photograph Layout Form is designed to accommodate the suggestions of Section 10.5.2, "Photograph Layout Requirements." Specifically, it provides a convenient location to record:

1. Name of the project;
2. Owner and company project numbers;
3. Date of the photograph;
4. Direction of view;
5. Identification of the photographer;
6. Description of the photo subject;
7. The particular location of the photo subject at the project; and
8. References to any appropriate correspondence, or anything to tie it to the detailed project record.

SAMPLE PHOTOGRAPH LAYOUT FORM

Project _____ No. _____   Date ___/___/___
Taken by _____   Time ____:____
                                                AM PM

Location _____
Orientation (Indicate on Key Plan) _____
Remarks _____
_____
_____

Place photo(s) here.
Assign identification numbers to multiple photos

Draft Key Plan here

Note that the most significant feature of the layout form is a key plan of the building. If the project is a high-rise, the typical floors can be labeled. If the project has a simple layout, the key can be included at the bottom portion of the photo layout form. If this is done, the photo can be mounted on the form, with the precise location and photo direction indicated. If an 8" × 10" or even larger photo is used, simply attach the key plan to the photograph. The form is a sample. It is included to demonstrate the principles discussed. If the contractor can fit the key plan of the specific project on the bottom of the form in a manner that will leave room for the actual photograph and the other required information, so be it. If not, attach the key plan to the layout form and reference the photograph number.

10.6 CONSTRUCTION SCHEDULES

10.6.1 As-Planned, As-Built, and Adjusted Schedules

To this point, construction schedules have been referred to in the context of justifying time extensions and delays. The analysis that is used to prepare these demonstrations is performed either before the delay is actually encountered or after, in a claim situation. In the first case, reasoning is presented to convince the owner that a potential change will affect the schedule by a precise number of days. This is accomplished with the use of two separate but related schedules. The first schedule is the As-Planned Schedule. This is the schedule that indicates how the contractor intended to construct the work per the original contract. The second schedule is the Adjusted Schedule. It indicates the new sequence of work, incorporating all the influences of the change, along with the ultimate effect on the project end date. The difference between the end date depicted on the As-Planned Schedule and the Adjusted Schedule represents the time attributable to the change. If the change and adjusted schedule are accepted by the owner, the Adjusted Schedule becomes the "starting point" for determining the effect of the next time-sensitive change.

In the second situation the construction schedules are used to demonstrate where the contractor would have been had it not been for the subject change. It involves an after-the-fact analysis, often in a delay claim situation. This particular kind of analysis is most useful in cases where the project was affected by a change, but the complex and interrelated components that make up the total project confuse any clear cause–effect relationship. The dilemma is solved with the use of three schedules. First, again, is the As-Planned Schedule. Next is the historical As-Built Schedule which, through its periodic updates, incorporates *all* the complicated effects on the construction sequence. The final schedule is the Adjusted Schedule which reconstructs each update as necessary, removing the effects of the change of interest. The result is a schedule that includes all the other effects without the impact of the change being considered. The difference in the end dates between the As-Built and Adjusted Schedules becomes the amount of time attributable to the single change. A review of Figure 10-1 will clarify the relationships.

```
┌─────────────────────────────────────────────────────────────────┐
│  ┌──────────────────────────────────┐                           │
│  └──────────────────────────────────┘                           │
│   As-Planned Schedule                                           │
│   (Prior to construction)                                       │
│  ┌──────────────────────────────────────────────────────┐       │
│  └──────────────────────────────────────────────────────┘       │
│   As-Built Schedule                                             │
│   (Including all effects)                                       │
│  ┌───────────────────────────────────────────────┐              │
│  └───────────────────────────────────────────────┘              │
│   Adjusted Schedule                                             │
│   (Removing effects attributable to the change)                 │
│                                                      ╲          │
│                      Time attributable to the change ─╲→        │
└─────────────────────────────────────────────────────────────────┘
```

FIGURE 10-1 THE THREE SCHEDULES

10.6.2 Six Requirements for Presentable Evidence

The primary objective of any construction schedule is to complete the project as designed, in a systematic, coordinated manner. All this must be accomplished in the shortest possible time, consistent with material and personnel constraints, to maintain a profit. Everyone involved in today's contracting industry recognizes that changes are a normal part of the construction process. Every superintendent, project manager, and construction executive knows that in all probability, there will be changes occurring on the project that will affect the final duration and price. If these changes can be handled competently and in stride, the effects on the project will have to be endured, but responsibility and accountability for the change will remain with the proper party. Construction schedules that are constructed and updated in a clear, consistent manner with complete, thorough notes and references can facilitate keeping responsibility and accountability in the proper place. If the schedules are managed so that the information in them can be easily assembled, categorized, and supported in the event of a worst-case scenario—a claim—their strength will actually become a force to minimize serious disputes. A good schedule will provide significant, visible comparisons and convincing proof of damages, including delay, acceleration, suspension of work, inefficiencies, disruption, and interference.

Following the suggestions that form the basis for presentable evidence, it turns out, will establish the fundamentals of good records, clear accountability, and effective presentation:

1. *The schedule must be the one that was actually used to build the project.* Even if the schedule used to construct the project is substantially different from the one formally submitted and approved by the owner and/or the design professionals, the schedule will be deemed to be legitimate so long

as the schedule was actually used and depended upon by the various trades.

2. *The schedule must be periodically revised.* It is generally recognized that construction scheduling is nothing more than a plan to construct the work described in the contract documents, considering that there may be several ways to build the same project. Although contingencies may be included to allow for imprecisely defined variables, there probably has never been a schedule that anticipated every problem, coordinated every piece, and required no modifications to make it work. Changes or corrections to the as-planned schedule are inevitable. The schedule, therefore, must be updated periodically to maintain a current and accurate representation of reality. The revisions must be founded on the project record.

3. *The periodic updates must show all positive and negative influences by all parties.* No impact or influence must be singled out or deleted for the convenience of the contractor. If all updates, for example, indicate only delays caused by the owner and fail to delineate other known problems caused by other parties, it will be easy to demonstrate that a bias has been built into the document and that the schedule is not an accurate depiction of the events that transpired on the project. Failing to recognize all significant events will bring into question the validity of the entire schedule presentation.

4. *The schedule must include realistic construction logic and activity durations.* Illogical sequences or the lack of consideration of critical variables (such as indicating the installation of a 25,000-gallon tank in a building with no overhead doors to occur before the roof is constructed) will only demonstrate that the constraints in the schedule were unrealistic from the outset.

5. *The schedule must fairly represent the actual method planned to build the project.* If, for example, the schedule had been prepared primarily to cater to progress payments, or was otherwise unrelated to the actual construction of the project, it will become very clear that the schedule has very little value as a tool for managing the project.

6. *The schedule updates and analysis must be realistic and in perspective.* An overly aggressive computation of damages may hurt the validity of the entire analysis. Direct cause–effect relationships must be shown between the impact and result shown on the schedule.

In addition to the foregoing, be sure that all notes and references are correlated with the more detailed project correspondence and other records. Have all field notes, claims, commitments, deadlines, and promises detailed in the schedule documents. Include the dates and the names of the individuals in the organizations supplying information. Be prepared to justify each entry on the schedule.

10.7 USING JOB MEETINGS TO ESTABLISH DATES, SCOPES, AND RESPONSIBILITIES

10.7.1 Introduction

Job meetings are commonly set up to run on a biweekly basis throughout normal periods of construction activity. During problem periods, or through critical or accelerated work sequences, the schedule may be compressed to weekly meetings. Depending upon the type of contract arrangement, any member of the project team may be responsible for presiding over the meetings and keeping the minutes. Regardless of who is technically responsible for scheduling job meetings or maintaining the meeting minutes, the contractor cannot allow these things to "just happen." Job meeting records are critical documents that, if properly implemented, will:

1. Record history of all significant (or potentially significant) events;
2. Keep open items on the front burner until they are finally resolved, or filed for future reference;
3. Force action;
4. Clarify accountability;
5. Provide a basis to identify where expedited action is needed;
6. Support interpretations and serious actions; and
7. Facilitate fast, efficient research, both soon and long after the occurrence of an event.

Too often, job meeting agendas and minutes are left to be generated by persons without either a clear concept of their importance or the ability to use them properly. Then, job meetings and their minutes can consume large amounts of time and effort, but wind up providing few or none of the foregoing benefits.

Because meetings can be so critically important to the favorable and timely resolution of almost every issue that can affect a project, it follows that the contractor should capitalize on every opportunity to assume the function, either literally or as a practical matter, of preparing the agenda and maintaining the minutes. The manner in which such important documentation is kept and distributed is too important to be left to chance. The extra work and expense involved in producing and distributing the meeting minutes is minuscule when compared with the advantages gained through control over the information content and presentation. Remember the old attorney's adage: "It is not what happened at the meeting that is important, it is what is recorded in the minutes."

Become familiar with Section 10.7.2, "Job Meeting and Minutes Guidelines." Even if the contractor is not formally responsible for production of the minutes, make the most of every opportunity to impose the desired format and procedure on the persons responsible for the minutes. Keep the pressure on the person preparing the minutes to ensure that the minutes are distributed. Since official job records serve the contractor the best, the contractor must be certain that the minutes accurately portray the events which occurred at the meeting. If there is a discrepancy in the minutes, prepare and distribute an objection to the minutes to the persons on the distribution list. The contractor must develop the habit of reviewing meeting minutes, completing item identification and correcting cross-references as soon as the minutes are received and while the information is freshest in your mind.

10.7.2 Job Meeting and Minutes Guidelines

For the job meeting itself:

1. *Schedule morning meetings if at all possible.* A mid-morning start will allow the work to get started and will leave a few precious minutes to complete last-minute preparations. It will also give meeting attendees time to get to the meeting without having to fight morning or noon-hour traffic. Most people are more energetic and effective in the morning. Also, the remainder of the day is left available to get the jump on new problems and to expedite solutions for old ones. In contrast, meetings held after lunch are not as well attended, move more sluggishly, and are more constrained by the lack of time.

2. *Always start job meetings on time.* It is at the very least considerate to those who have taken the trouble to arrange their schedules to meet their commitments. Common courtesy aside, meeting attendees will get the message that the schedule is serious. A few times of having to "sneak" into an ongoing meeting may be embarrassing enough to end the problem. Chronic offenders should be confronted with their lack of attention at the meeting. Finally, let the people who get to meetings on time know that their efforts are appreciated.

3. *Make meeting attendance **mandatory**.* All major subcontractors as well as all subcontractors doing or about to perform *any* work on the site *must* attend all job meetings for the period. Subcontractors must be there to participate in all coordinating activities and discussions. Lack of attendance by any subcontractor affected by a discussion creates problems starting with the extra efforts necessary to coordinate information. It must also be made clear that all subcontractors are responsible for all information contained in the job meeting minutes. Section 10.7.3, "Sample Letter to Subcontractors Regarding Mandatory Job Meeting Attendance," will help to make this requirement clear.

4. *Confirm attendance prior to the meeting.* The day before the meeting, call or have an assistant call all subcontractors who are important to the meeting. Reaffirm the expectation of their attendance. If a specific issue is critical and the attendance of a particular person is required to resolve the issue, call again early on the morning of the meeting. Be clear that attendance is absolutely necessary.

5. *Notify absent parties **immediately after the meeting**—or even during the meeting, if necessary.* Despite the contractor's best efforts, some parties may consistently fail to attend job meetings—even after they have "confirmed" that they would be there. If the lack of attendance precludes the resolution of a particular issue (or moving an issue closer toward resolution), telephone that person immediately after the meeting. In some situations, it may be particularly effective to call him or her *during* the meeting. Be sure that the absentee (and the meeting attendees) know just how conspicuous his or her absence is and the impact of his or her absence on the effectiveness of the meeting. Be clear about the inconvenience that the absentee has caused the attendees who are at the meeting. In the meeting minutes, for each item affected, note the reason for lack of resolution to be the subcontractor's failure to attend. Document the case, and bring such documentation to the absentee's attention. If an item that affects an absent subcontractor has been decided in a certain way, advise that party and note the phone conversation in the meeting minutes. For all expected parties who failed to attend the meeting, send the Section 10.7.4 "Sample Letter to Subcontractors Regarding Lack of Job Meeting Attendance." Be certain to express dissatisfaction with their lack of attention and concern.

In the meeting minutes themselves:

1. *Use outline format.* Keep the minutes in outline form to maintain clear, to-the-point presentations. Number the meetings. Begin the minutes with the separation of "Old Business" (prior discussions) and "New Business." Number each successive item in each section. This will make all references to any job meeting discussion fast and accurate (such as "Reference Job Meeting 4, Item A.6") and will eliminate any question in subsequent discussion or correspondence.

2. *Use a title for each item.* A summary description clarifies a paragraph's subject. They make research fast and the correlation of topics easy. Use exactly the same wording in each meeting that the item is discussed. Issues that continue through multiple job meetings will be clearly tied together through the use of consistent titles.

3. *Include all appropriate references in an item title.* If it involves a change order, include the change order number in the description. If it involves a change estimate, architect's bulletin, or something else, note it as such. The inclusion of these numbers will relate the discussion to all the affected files. Research will become faster and more complete.

4. *Name names.* Avoid remarks such as "the architect stated ...." Instead, use "Mr. _____ stated ...."

5. *Use short but specific statements.* Be as concise but complete as possible. Read back the exact language at the meeting. Get agreement that the representation is entirely accurate and that everyone understands the implications, as well as the obvious.

6. *Require definite action.* Include the names and the precise dates for required actions. Ask frank questions at the meetings. Firmly and persis-

tently narrow complex or difficult issues down to the next step required in the resolution process. Confirm who is responsible for the next step and write it down.

7. *Include safety, cleanup, and construction scheduling as prominent items.* Place them high on the agenda of *every* meeting. Give these common but typically neglected issues the critical importance they deserve, and they will be handled more effectively.

8. *Notify all recipients of the meeting minutes to advise the writer of any errors or omissions in the presentations.* Include such a request on the meeting minute form itself.

9. *Request acknowledgment of the accuracy of the "Old Business" at each meeting.* As part of the discussion of "Old Business" and before proceeding to "New Business," confirm any acknowledgments, objections and corrections to the prior meeting's minutes. Include any discussion of such acknowledgment, objection or correction as the first item of the "New Business."

Section 10.7.5, "Sample Job Meeting Minutes Form," follows through with the recommendations in this section.

10.7.3 Sample Letter to Subcontractors Regarding Mandatory Job Meeting Attendance

The Sample Letter to Subcontractors Regarding Mandatory Job Meeting Attendance is designed to accommodate the Section 10.7.2 job meeting guideline 3. Specifically, it notifies all subcontractors that:

1. Anyone performing or about to perform *any* work on the site is absolutely required to participate in all job meetings during the period. This is a mandatory requirement that will not be compromised.

2. Meetings will be held on the dates scheduled and will start *on time*. Attention is expected to be given to these requirements.

3. It is each subcontractor's responsibility to be aware of all information as it relates to his or her work and to make all necessary efforts to ensure proper coordination.

4. Each subcontractor is absolutely responsible for all information contained in the job meeting minutes. This includes completeness, accuracy of description, noted commitments, and timetables.

SAMPLE LETTER TO SUBCONTRACTORS REGARDING MANDATORY JOB MEETING ATTENDANCE

LETTERHEAD

(Date)

To: (List all project subcontractors)

RE: (Project no.)
(Project title)

SUBJ: Mandatory job meeting attendance

Your subcontracts require your participation in regular job meetings. These meetings have been scheduled to begin on (insert date) and will be held on alternating (insert day of the week). During critical or problem periods, the meetings may be held weekly, as determined by our company. It is your responsibility to be aware of the current job meeting schedule.

(Name major mechanical and electrical subcontractors) are required to attend all meetings. All other subcontractors performing or about to perform work on the site are required to attend *all* meetings throughout the period of their work.

Please note that this is not a request. Your attendance at these meetings is *mandatory*. Your failure to attend job meetings will result in extra efforts by others to coordinate their work with your work. Because these extra efforts disrupt the progress of the work and inconvenience other persons on this project, absences will not be tolerated.

Please be advised that you will be held responsible for all information contained in the meeting minutes, including timetables, commitments, and determinations of responsibility as set forth in the minutes.

Thank you for your cooperation.

Very truly yours,

Project Manager

10.7.4 Sample Letter to Subcontractors Regarding Lack of Job Meeting Attendance

The Sample Letter to Subcontractors Regarding Lack of Job Meeting Attendance is designed to follow the Section 10.7.2 job meeting guideline 5. Specifically, it confirms a conversation with the subcontractor who, despite the best coordination efforts, has failed to attend an important job meeting. The letter confirms that:

1. The subcontractor's lack of attention is creating unnecessary interferences and inconveniences;

2. Interferences, delays, and additional costs resulting from the subcontractor's lack of attention will be the subcontractor's responsibility;

3. It is also the subcontractor's responsibility to be aware of all project requirements included in the meeting minutes, and to comply with said requirements in every respect.

4. "Finally, the letter reminds" the subcontractor of the next job meeting. By a copy of the letter to the subcontractor's home office, the letter makes certain that the field representative's supervisor is aware of the absence and the inconvenience and dissatisfaction created by the absence. If the copy to the home office does not elicit some reaction, then the contractor probably has other problems with the subcontractor.

SAMPLE LETTER TO SUBCONTRACTORS REGARDING LACK OF JOB MEETING ATTENDANCE

LETTERHEAD

(Date)

To: (Subcontractor failing to attend a specific meeting)

RE: (Project no.)
(Project title)

SUBJ: Lack of job meeting attendance

Mr. (Ms.)   :

As we discussed earlier today, your failure to attend today's job meeting as required is interfering with job coordination and completion. As you know, it continues to be your responsibility to be aware of all project requirements and to accommodate them completely and in a timely manner. Please be advised that you will be held responsible for all interferences, delays, and added costs resulting from your lack of attention to these requirements.

The next job meeting will be held on (insert day and date) promptly at (insert time).

Very truly yours,

Project Manager

cc: Owner
    Architect
    Subcontractor Field Representative's
       Immediate Superior

10.7.5 Sample Job Meeting Minutes Form

The Sample Job Meeting Minutes Form is arranged to comply with requirements that must be a part of all job meeting minutes. In addition, it is designed to accommodate Section 10.7.2, "Job Meeting and Minutes Guidelines" by:

1. Providing appropriate areas to include all relevant job meeting identification and distribution information;
2. Encouraging documentation in accordance with the "outline" recommendations;
3. Providing a convenient area for highlighting important action items, including the person responsible and required action date; and
4. Including the important notification to correct any errors or omissions in the noted discussions.

SAMPLE JOB MEETING MINUTES FORM

Job Meeting No. _____
Date:
Location:

Project No.:

Page 1 of ____

PRESENT		DISTRIBUTION	
Name	Company	Name	Company
		Attendees:	

NOTICE to attendees and minutes recipients:
 If any of the following items are incomplete or incorrect in any way, please notify the writer. Failure to advise of such corrections by the next job meeting constitutes acceptance of all information contained herein.

SUBJECT	ACTION REQUIRED	
	By	Date
A. <u>OLD BUSINESS</u>		

Job Meeting No. _____		Page ___ of ___		
Date:	Project No.:			
Location:				

| SUBJECT | ACTION REQUIRED ||
	By	Date
B. <u>NEW BUSINESS</u>		

10.8 SHOP DRAWINGS AND APPROVAL SUBMITTALS

Shop drawings are detailed, large-scale representations that describe how a contractor intends to fabricate and/or install the various components that will make up the project. They are prepared by subcontractors and suppliers (and the prime contractor for work that the prime contractor intends to performs itself). Shop drawings are intended to describe how the work being provided meets the design criteria established by the architect and engineers via the plans and specifications.

10.8.1 Approval Responsibility

In most contract documents, the general contractor (or construction manager) and the architect share the responsibility for the review of shop drawings. Further, most contracts contain a time schedule for the preparation, submission and review of shop drawings and submittals. On most projects, the general contractor receives the shop drawings from its subcontractors, reviews them, stamps them with its approval, and forwards them to the architect and engineers for their review and approval. The design professionals then review and comment on the shop drawings. These comments typically include some variation of the following:

1. Approved;
2. Make corrections noted;
3. Revise and resubmit; or
4. Rejected.

Remember that the design team may use the review of shop drawings and submittals to make changes to the plans and specifications. Be sure to scrutinize every drawing marked "make corrections noted" or "revise and resubmit."

After the design team's review is complete, the shop drawings are returned to the general contractor for distribution to the subcontractor originating the drawings. If the design professional's review of the shop drawings is favorable, the contractor is authorized to proceed with the work in the manner detailed in these submissions.

The interface between the general contractor, subcontractors and the design professional may be the cause of a variety of problems in the early stages of a project. These problems interfere with the submittal routine and may have an undesirable effect on the outcome of the process.

If the responsibilities of the contractor and designer are not absolutely clear both in the contract documents and in the actions of the parties, questions arise. These questions must be removed fast or the shop drawing approval process will be delayed. If the parties are left to decide for themselves how to shoulder their responsibilities, the contractor will include less information than the designers want to see and the architect is more likely to return the submittals to the contractor without action, pending the receipt of additional information.

At the other extreme, many designers have become increasingly aware and concerned with their own professional liabilities. Although attitudes and practices vary, many design professionals now attempt to narrowly define the scope of their shop drawing reviews. These "limitations" have often led to increased confusion, which has resulted in the opposite effect, that is, the assumption of even more liability. It cannot be argued that the roles of the designer and the contractor are in an evolution created by changing contract structures. The concern here, however, is that certain responses from the design team can compromise thoroughness and attention to detail. Similarly, contractors often try to place too much of the burden on the designers. The combination of these two circumstances has contributed to substandard shop drawings and shop drawing review. This situation ultimately creates dissatisfaction at every step of the process, increases the confusion and friction surrounding shop drawing review responsibilities, and may increase liabilities for the designer, the contractor, or both.

The architect is normally responsible for approving shop drawings, reviewing them to make certain that the drawings "conform with the original design concept" or some similarly obscure process. Unfortunately, these terms and many other contract terms do not clearly define the architect's responsibility in the shop drawing review process. To complicate matters further, there is not a consistent standard in the industry. Problems arise due to the vagaries of the phrase "conformity with design concept." Does this phrase mean that connection details, performance calculations and/or sizing determinations will not be checked if they "conform to the design concept"?

To make matters worse, certain design offices treat the shop drawing review and approval responsibility with less importance than it truly deserves. While it may be true that many designers understand the significance of the information incorporated into the shop drawings and treat the shop drawing review with respect and proper attention, some do not. Time and expense pressures create a temptation to assign the shop drawing review task to a junior person in the office. Lack of proper attention, time and expense pressures and improper delegation, among other things, all increase the likelihood of errors sliding through undetected.

The lack of clear definition and assumption of approval responsibilities has afforded many design professionals (and sometimes their professional liability insurance carriers) the opportunity to determine the professional's responsibility in the shop drawing review process. In addition, many architects and engineers have attempted to limit their liabilities by avoiding the use of the word "approved" in their shop drawing remarks. Phrases such as "No Exceptions Taken," "Furnish as Submitted," or "Examined," have now become the rule rather than the exception. Moreover, the shop drawing stamps have been supplemented with elaborate language explaining what is and is not being done, in an effort to define the review process and minimize legal exposure.

There is legal opinion, however, that supports the idea that the designers' stamps may actually increase their liability, if their contract obligations for shop drawing review exceed the limits of the language included on the stamp. The owner and contractors rely on the designer's approval responsibilities as defined in the contract. If a designer operates in a narrower capacity based on the loose language on a stamp, it is a clear admission that the designer is performing less than his or her obligations under the design contract. This can be a powerful argument against the architect during the shop drawing review process or after.

10.8.2 Approval Response Time

Beyond basic approval responsibility, another major concern of the contractor involves the designer's response *time*. Common language in construction agreements notes that a designer will "review and approve shop drawings with reasonable promptness so as not to cause a delay in the work" or something similar. "Reasonable response" is another term that lacks precision in its determination. "Reasonable" becomes defined by trade practice in the project's geographic location. If nothing better is available in the existing contract language, establish the definition of "reasonable time" at the very first job meeting. Ten working days (except in unusual circumstances) is a good starting point on many projects. Additionally, a priority system should be established so that the submissions that involve items required in the early stages of the project are reviewed first. To emphasize these requirements, indicate on each submission transmittal the required response date. Maintain a log or summary schedule to track the status of submissions and to spot trends in the review process.

10.8.3 Treatment of Differing Conditions

Regardless of whether or not conditions are a result of a design error or any other effect described in Chapter 7, Where and How to Find Potential Change Orders, it is the contractor's responsibility to highlight *all* differences from the requirements included in the contract in its approval submission. Even if the condition has not been addressed with specific contract language, the contractor will be held responsible for changes (even those bearing an unwary designer's approval) if the differences from the contract requirements are not brought to the designer's attention for his or her direct consideration. In contrast, by properly highlighting the differences and detailing all justifications for each change, the contractor will speed up acceptance of the changes. Moreover, the ultimate approval *will* carry with it the designer's responsibility that might have been lost if the change was not properly highlighted.

10.8.4 Absolute Contractor Responsibility

One of the keys to fast action and the correct treatment of contractor's responsibilities is the proper application of the pass-through clause (refer to Section 3.4.3). It is important for the contractor to recognize and to make all suppliers and trade contractors understand that the obligations and responsibilities related to the preparation of shop drawings are part of the obligations of each supplier and subcontractor. For example, if the "contractor" is "responsible for dimensions …," application of the pass-through clause clarifies that "the steel contractor is responsible for steel dimensions," "the concrete contractor is responsible for concrete dimensions," and so on. As a prime contractor, do not assume more responsibility than required by the contract documents. If the architect is responsible for checking the work as it relates to "conformance with the design concept" and each subcontractor is responsible for the preparation of shop drawings specific to the subcontractor's portion of the project, the role of the prime contractor (or construction manager) should be limited to:

1. Policing the activities of both the subcontractors and the designers to ensure that their contract responsibilities are being met completely and on time; and
2. Coordinating all information and instructions in a timely manner so that the information will pass smoothly in both directions.

Refer to Chapter 12, Section 12.17, "General Contractor As a Conduit," for more on this idea.

10.9 TIME AND MATERIAL TICKETS

10.9.1 Introduction

Section 9.4 contained an introduction to the time and material (T & M) concept and procedures, as related to managing subcontractors. This section is concerned with the use of T & M to substantiate change order pricing and positions to the owner.

An important consideration in using time and material tickets with the owner occurs in those instances where the responsibility for certain work is not clear, or where the contractor has been directed to proceed with work that you believed to be a definite "extra." If circumstances require that the contractor proceed with the work pending a final decision, or in anticipation of a later claim, the contractor should use T & M tickets to keep accurate, indisputable records. The tickets should be prepared in the same detail as if the owner is clearly paying for the work. Arrange with the owner's site representative to have the daily tickets signed only to acknowledge that certain work was performed that day by specific workers. Clarify that the objective of the daily acknowledgment is to document the facts surrounding the work and that the acknowledgments do not alter the fact that the owner has not agreed to any payments for the work at this time. Section 10.9.2, "Sample Letter to the Owner Regarding Acknowledgment of Actual Work Performed," will help to accomplish this result. These records, acknowledged to be correct by the owner as the work progresses, will become invaluable during later negotiations and dispute resolution proceedings. The discussions will be easily narrowed to the issue of entitlement, with the acknowledged costs being reduced to arithmetic calculation.

10.9.2 Sample Letter to the Owner Regarding Acknowledgment of Actual Work Performed

The Sample Letter to the Owner Regarding Acknowledgment of Actual Work Performed is an example of a confirmation to the owner that the contractor expects daily verification of actual work performed. It is used when proceeding with the construction of work which the contractor believes is outside the scope of work described in the contract documents. Whether the owner's decision regarding responsibility is pending, or if the contractor has already been directed to proceed with the questioned work, the letter advises the owner that:

1. The contractor is proceeding with the questionable work under protest, per the owner's direction, and that there is not any agreement with regard to the responsibility for the added costs;

2. The contractor will be preparing daily T & M tickets to detail the actual work performed; and

3. The owner's on-site representatives will be expected to sign the daily tickets, only to acknowledge the actual work performed and the actual resources consumed, regardless of which party will be ultimately responsible for paying for the work.

If the owner refuses to sign the tickets, prepare the tickets anyway and present them to the owner's field representative for signature on a daily basis. Each time that the field representative refuses to sign, note the occurrence on the respective ticket. These kinds of records will go a long way in later displaying the good, or bad, faith nature of the owner's actions throughout the problem period. It will also demonstrate that the contractor has made every effort to ensure the accuracy of the records on a daily basis. The owner's failure to acknowledge the fact that the work was performed (without accepting responsibility to pay for the work) may actually help in any dispute resolution proceedings.

SAMPLE LETTER TO THE OWNER
REGARDING ACKNOWLEDGMENT OF ACTUAL WORK PERFORMED

LETTERHEAD

(Date)

To: (The owner)

RE: (Project no.)
(Project title)

SUBJ: (Change file no.)
Acknowledgment of actual work performed

Ms. (Ms.)   :

Per your direction of (date), we are proceeding with the subject work under protest, in the interest of job progress.

We will be preparing Time and Material tickets on a daily basis to document the actual work performed, along with all resources used. They will be presented to your on-site representative for signature at the end of each day.

We are writing this letter to advise you of our position with regard to this work and to confirm that your representative's signature on the T & M tickets will only acknowledge the accuracy of information contained in the respective tickets, and will not indicate your acceptance of responsibility for the work at this time.

Very truly yours,

Project Manager

cc: Owner Field Representative
    Architect
    File: (Change File)

10.9.3 Sample T & M Form

The sample Time and Material Form is one way to accomplish all requirements noted in Section 9.4, "Time and Material," and Section 10.9, "Time and Material Tickets." A supply can be given to subcontractors to use in submissions on the project. This will create uniformity in T & M reporting procedures and make the review of the T & M reports more routine.

The form can also be used when performing T & M work for the owner and when documenting actual work performed on disputed items (see Section 10.9.2). When utilizing these tickets, be certain that the tickets are prepared daily, that *all* relevant information is included, and that the signatures are secured at the end of each day.

SAMPLE TIME AND MATERIAL FORM

Bill to: _____ Date: _____
         _____ Project _____
         _____ Project No. _____

Description of Work _____ Location _____
_____ Authorized by _____
_____ Approved by _____

LABOR

Date	Employee	Work Performed	Hours REG	Hours OT	Rate	Amount

MATERIALS

Date	Description	Quant.	Unit Price	Amount

EQUIPMENT

Date	Quant.	Type/Size	Work Performed	Hours	Rate	Amount

TOTAL _____

Prepared by: _____ Accepted by: _____ Date _____

Part Five
Change Order and File Administration

11

Keeping Change Orders Under Control: How to Save Time and Improve Records with Administrative Housekeeping

11.1	Introduction	282
11.2	Establishing Easy-to-Research Change Order Files	282
11.3	File Content	284
11.4	Correspondence File	287
11.5	Tracking Change Order Trends	288
	11.5.1 Introduction	288
	11.5.2 Evaluating the Change Order Summary Sheet	289
	11.5.3 The Change Order Summary Sheet Procedure	290
	11.5.4 Sample Change Order Summary Sheet FORM and Sample Completed FORM	291
11.6	Approval Submissions	294
	11.6.1 Introduction	294
	11.6.2 Shop Drawing Review and Coordination	294
	11.6.3 Shop Drawing Submission Requirements	295
	11.6.4 Sample FORM LETTER TO SUBCONTRACTORS Regarding Shop Drawing Submission Requirements	296
	11.6.5 Submittal Review, Distribution, and Follow-Up	299
	11.6.6 Sample FORM LETTER TO SUBCONTRACTORS Regarding Shop Drawing Resubmission Requirements	301
	11.6.7 Shop Drawing Submittal Summary Record Procedure	304
	11.6.8 Shop Drawing Submittal Summary Record FORM	305
11.7	Sample Letter of Transmittal	309
	11.7.1 Sample FORM LETTER OF TRANSMITTAL	309

11.1 INTRODUCTION

Chapter 10 dealt with the various types of documents and records that will support change order prices, both in principle and in dollar amount. This chapter will deal with the practical management of change orders and their documentation. Multiple changes occur at varying frequencies and intensities throughout the life of a project. Some are easily approved and settled quickly. Others can drag on at an agonizing pace, in terms of real work and from the standpoint of paperwork. Each change may present the contractor with a unique or different situation, requiring varied negotiating strategies and timing. At any point in time, several (or many) changes may be in process or about to begin process. Each change will require different types and amounts of attention. All these effects have a critical impact on the way in which change order records are maintained. A filing system must be established and maintained throughout the organization that will encourage comprehensive monitoring and treatment of all change orders and related documentation. It is only when the in-process and later research requirements are understood and accommodated that all project change orders will be kept under control, each being managed through to its optimal resolution.

11.2 ESTABLISHING EASY-TO-RESEARCH CHANGE ORDER FILES

The underlying reasons for a particular filing system begin with the nature and size of the contractor's organization. A subcontractor's requirements are different from the requirements of a general contractor. Both are different from the needs of a construction manager. In the following discussion, the filing system requirements of a general contractor will be explored. A construction manager's requirements will be similar, but a subcontractor's requirements will normally be less elaborate. This is because subcontractors typically deal with much fewer subcontractors and suppliers than does a general contractor. After reviewing the entire chapter, the contractor should be in a good position to evaluate the needs of his or her organization. The contractor will then be able to select those provisions that will best suit his or her own operation.

An efficient filing system begins with the idea that when *any* item is researched, *all* required pieces of information are either physically present in the file or clearly referenced in terms of file location and subject matter. There should be no need to mentally recall circumstances in an effort to identify related issues and files. If the filing system is dependent upon an individual's specific recollection of a topic, a relationship or a file location, the system will be unmanageable. The difficulty increases exponentially, even for those intimately involved with an issue, if the research is required after any appreciable lapse of time.

The first problem associated with a full and complete filing system is the increased amount of paper to be filed. This extra effort must be weighed against the potential losses associated with missed opportunities resulting from overlooked significant documents that were filed in not-so-obviously related locations. The other difficulty encountered when developing a proper filing system is that a large number of contracting professionals are concerned that if they do not actually file important documents *themselves*, they will not be able to find the documents when they need them. This is a legitimate concern if filing is handled inefficiently and without regard for filing and retrieval systems. This concern is not restricted to the construction industry and until everyone in the contractor's organization understands the importance of maintaining proper files, the concern is valid. Section 11.4, "Correspondence File," will help you improve the documentation and filing process. If adopt the process, the filing function can be given back to the secretarial or administrative staff.

As a result of the electronic explosion in recent years, many contractors are maintaining project files electronically or on compact discs. The use of electronic media for the storage of project records intensifies the need for a proper coding and cross-referencing system. Although the use of compact discs may save space, eliminate the need for multiple sheets of paper and give the impression of being highly organized, electronic records will be of no use to the contractor if they cannot be accessed easily. The content and style of letters written by the contractor will have a great impact on their understandability and the ability of the contractor's staff to file the letters properly. Letters that cover multiple topics must be filed in several files. This particular practice results in unnecessary and unrelated information being included in a variety of files. It can be confusing for the recipient of the letter as well as the person responsible for the filing. A few simple rules will contribute to authoring effective correspondence and maintaining good records. These rules include:

1. *Confine the subject of each letter to a single issue, or a small group of closely related issues.* If there are two or more unrelated items which must be addressed with another party on the project, keep them in separate letters. Picture the attention span of the busy individuals who must review and act upon these letters. Separate items will allow these people to quickly understand the problem, perform the necessary research and respond to the issue. The net result will be a faster response. In contrast, a letter with several unrelated items may confuse the reviewer or cause the reviewer to lose focus on the key issue. Each subject must be individually researched, and the separate reactions must be consolidated into an equally confusing response. Both letters, each containing multiple unrelated items, must then be filed in the various issue files which have been established for each topic mentioned in the letters. Further, if the person preparing the response elects to respond to each of the issues in a single letter, the response will be delayed until the last issue is fully researched.

   In addition to causing confusion at various stages and extending the time required to respond, there is one additional problem created by letters involving multiple issues. This problem arises in the dispute resolution process when it becomes necessary to introduce the letter discussing a variety of issues to prove a single point. At the very least, valuable (and expensive) time will have to be spent explaining why all the unrelated items do not apply to the issue at hand. In some situations, the contractor might be required to introduce information that the contractor would have preferred to leave out of the particular debate.

2. *Confine letters to a **single page.*** Except in the most rare circumstances, all correspondence should be confined to a single page. Again, consider the attention spans of the busy individuals who must review the letter. Long, laborious letters are disfavored. They require a significant amount of time just to understand the main point. Make the point(s) in all letters as clear as possible and get to them fast. If the issue is a complicated one, make the main point(s) in a cover letter and put the details and supporting information in an attachment. The immediate result will be a quick understanding of the letter by the recipient. The recipient also has the supporting information in the attachment if further review is required. With the points clearly articulated and the supporting information provided, all the ingredients for a prompt response are provided.

3. *Use outline form, if possible.* If a long, complicated letter is required by the situation, try to break down the issues into manageable (understandable) parts. This applies equally well to short letters that contain multiple items. Clearly identify each individual item. If the letter is being sent to confirm several facts, separate each particular fact in the body of the letter and number them. If the letter contains several conclusions, requires two or more specific actions, or makes any statement combining more than one issue, consider the form of the letter closely. If the issues are unrelated, consider using separate letters. If, on the other hand, the matters are related, try to break the topics into separate components, list each item on its own line, and number them. This procedure will clarify the thought process for the author and will assist the author with the structure of the letter. If properly structured, the presentation of the facts will be more understandable to everyone who must review and act upon the information. Finally, future research will be facilitated because attention will be quickly directed to the main points.

When filing incoming correspondence, note on the document the date of the response and where the response has been filed. Be sure to include all appropriate file references and codes. A note such as "see (company name) response dated ( )" will do. This effort will quickly confirm that the item has, in fact, been responded to. Additionally, it will reduce the research time associated with locating important documents.

11.3 FILE CONTENT

Immediately after award of the contract, the entire filing system for the project, both on- and off-site, must be established. A policy should be in place to determine the location of certain key documents, such as contracts, estimates, and personnel files. Once these issues have been decided, the filing system for the project staff can be implemented. The entire general file must be maintained in duplicate so that the project staff at the job, as well as the staff at the central office, have a complete file. The items that are not required in both locations are the submittal documents that are not to be used for construction and the file books discussed below. With these general considerations, the files will be assembled according to the following guidelines.

1. Correspondence and submittal file separation

For each subcontract and major purchase order, the in-process files will be divided between the folders containing the contracts and correspondence and the folders containing the bulky submittal files. This separation will make the constant searches into the relevant areas much more convenient. Bulky shop drawings will not have to be handled when looking for a letter, and vice versa.

2. General file folders

Depending upon the general contract structure and the method used for assembling subcontracts and purchase orders, the general working file folders will be headed with either the specification section number or bid package number. Thereafter, add the subcontractor or supplier name to the file tab. In any case, be sure to have a separate file for each subcontractor and supplier. Under no circumstances should files for different trades or bid items be combined. All filing systems are different. However, many businesses use hanging files that will allow the insertion of manila folder subfiles.

Besides establishing a file folder for each subcontractor and supplier and arranging the folders by either specification section or bid package number, each project file drawer should contain general folders with the following headings:

- Backcharges
- Building Permit/Certificate of Occupancy
- Change Orders
- Construction Schedule—Baseline
- Construction Schedule—Updates
- Contractor—Notifications
- Contractor—Requisitions for Payment
- Contractor—Schedule of Values
- Correspondence—Architect/Engineer
- Correspondence—Owner
- Guarantees/Warranties
- Job Meetings
- Maintenance and Operating Manuals
- Progress Photographs
- Progress Reports

- Project Narrative Report (if required)
- Punch Lists
- Subcontractor Approvals (if required)
- Subcontractor Bonds
- Subcontractor Certificates of Insurance
- Subcontractor Schedules of Values
- Substantial/Final Completion
- Others as appropriate

3. Subfiles

Within a general file folder, any specific issue generating at least two pieces of correspondence, or with the potential for generating more, will justify the preparation of a subfile. This is a manila folder appropriately headed and placed into the general hanging folder. Each subfile folder must be headed with the same designation as the specific general file folder. A description of the issue is also placed on the file tab. For example:

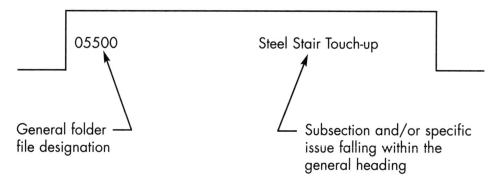

4. Correspondence and submittal books

There should be a duplicate of *everything*, with the exception of the actual submittal drawings themselves (only because of their bulk). In most cases, the original file should be maintained at the central office and the duplicate file should be kept at the jobsite.

At the same time that the general file drawer is being prepared, two large-capacity three-ring binders should be prepared. The binders should be entitled "Correspondence" and "Submittals," respectively. The correspondence file is customarily referred to as the "chronological" file or the "reading" file. Each book should be identified with:

1. Project name
2. Project numbers—assigned by the owner as well as the contractor
3. Book number (1, 2, etc.)
4. Beginning date
5. Reserved area to insert the end date for each book

Have all the project binders (correspondence, submittals, notes, daily field reports, etc.) in the same color, and in a color distinct from all other projects falling under the responsibility of any individual manager or executive. This simple procedure will assist with locating the books and keeping them filed where they belong.

A copy of all letters to and from your company, and to other parties (copies of which are received by your company), are to be placed in the correspondence book in reverse chronological order (most recent up front). This arranges the project file as it occurs. It also organizes the duplicate file in a manner that is distinct from the drawer file. The result is a duplicate file that can be researched by date instead of by file reference. This can prove to be an extremely valuable time-saver in many cases.

Stamp all incoming letters with the date of receipt, or mark the date of receipt in the upper right-hand corner and initial it.

Copies of all transmittals (in and out) regarding everything but shop drawings (approval submissions) are to be placed in the correspondence book in the manner described above.

Copies of all transmittals regarding shop drawings and other approval submissions are to be placed in the submittal book in the manner described for the correspondence book (Refer to Section 11.7, "Sample Letter of Transmittal," for a related discussion). As a matter of good practice, the contractor should prepare a transmittal directed to the appropriate subcontractor for each submission returned by the architect or engineer. The contractor's transmittal should be attached the transmittal from the architect or engineer and should reference the same item. A copy of the entire package (cover transmittal, architect/engineer transmittal and the submission) should then be filed in the submittal notebook based on the date set forth on the contractor's transmittal.

11.4 CORRESPONDENCE FILE

Every document should be noted with the precise file folder location in which it is to be filed. If a copy is to be placed in other locations, each one should be clearly marked in the document file reference. If only the cover correspondence (such as a letter of transmittal) is to be filed in the respective area, only the file designation is to be noted. If the entire package, including all attachments, is to be filed in a particular location, "with attachment" (w/att., w/l, etc.) must be indicated on the first page of the document. These filing instructions are a critical part of the exchange of information and the filing system because the instruction advises the person who is making the distribution copies and performing the filing exactly what he or she is to do with the document. In addition, the instructions advise whoever is reviewing or researching the document where the file copies are as well as their status. The location of the file instruction on the contractor's own correspondence is a matter of personal preference. One convenient location for the filing instruction in the contractor's own correspondence is in the "subject" portion of the letter. The section can be used to identify the issue as well as the proper file location. For example:

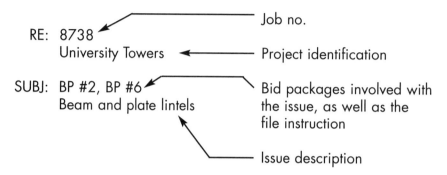

Another location that works well for every type of file instruction is the last item noted in the distribution listing at the letter's conclusion. For example:

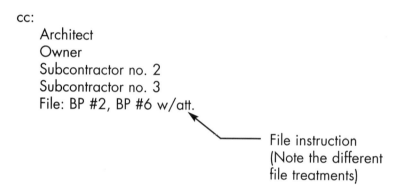

On incoming correspondence, add the same information on the document before the document is forwarded to the filing system. A handwritten instruction in the upper right-hand corner of the document is a fast, accurate way to accomplish the important instruction. For example:

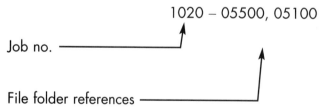

11.5 TRACKING CHANGE ORDER TRENDS

11.5.1 Introduction

Because change orders vary widely due to their timing, dollar value, scope, the people involved, and so on, it is important to develop a mechanism to monitor the overall change order situation on the entire project. This mechanism must be easy to update and must be kept updated. The system must also allow for the instantaneous identification and analysis of a particular change, as well as all changes on the project at any point in time. This might appear to be a formidable task at first glance. However, these objectives may be accomplished through the use of a convenient, relevant change order summary. This summary should:

- Identify all project changes.
- Record all relevant impact information.
- Summarize the submission status.
- Cross-reference the status with appropriate files.
- Indicate owner action or inaction.
- Compare the final effects of a change to the projected effects.
- Present all this information in summary form so that the data can be assimilated quickly and favorable or unfavorable trends in the overall change order process can be identified.

The Change Order Summary Sheet is discussed in the next sections, 11.5.2 and 11.5.3. A similar log should be developed to monitor and track trends with the review and return of submittals, as well as responses to Requests for Information ("RFI's") or Requests for Clarifications ("RFC's").

11.5.2 Evaluating the Change Order Summary Sheet

The Sample Change Order Summary Sheet is a log that accomplishes the objectives set forth in the previous section. It is designed to summarize all relevant change order information for each change that has been identified, is in process, or has been completed. The purpose of this information is to allow fast but accurate analysis of the change order situation from the standpoint of the total project. Finally, it provides a convenient checklist for the status of each change. This format also exposes incomplete items that need attention.

At a glance, the Summary Sheet provides:

1. The total amount of change orders in relation to the total contract, along with their effects.
2. Negotiation trends:
   a. Number of proposal submissions; and
   b. Magnitude of price reductions by the owner.
3. Times between change order issue identification and key owner actions (i.e., proposal requests, directives to the architect to provide revised designs).
4. Response times on the part of the owner for the approval of change order proposals. (Remember, slow owner responses may cause additional interference to the construction schedule.)
5. Whether the final proposal amounts are close to the initial estimates.

6. Visual reminder of approved change orders that are to be included on the requisitions for payment to the owner.

7. Clear references to related files that provide the detailed explanations.

8. Significant remarks that highlight actual or potentially serious project effects and other key information.

9. Identification of all outstanding items.

11.5.3 The Change Order Summary Sheet Procedure

The form should be used as soon as a potential change is identified. It is important to recognize this event before the documentation has an opportunity to proceed without proper change identification. After the change is identified and entered onto the summary, the summary is constantly referred to at every step throughout the life of the change, until its eventual resolution and payment. The procedure amounts to this:

1. First, refer to the form to determine the next available change estimate number, beginning with 001. That number is permanently assigned to the potential change. From that point, all correspondence regarding the change will have a subject identified with the same change estimate number. (Refer to Section 11.4, "Correspondence File.") In addition, *all* documents involving or potentially affecting the change must reference the appropriate file (also described in Section 11.4). These two activities combine to ensure that all related documents important to accurate change assessment and advantageous resolution are cleanly tied together.

2. Immediately identify the change item. Establish a summary description that clearly represents the issue. This description will be included in exactly the same language in the subject of each letter and other records so that there will not be any confusion over letters, meeting minutes, and other items that should be linked to each other.

3. Insert the date discovered (i.e., use the earliest date on which the item was realized to be a change in the work).

4. The "Date Design Finalized" column should contain the date on which all design information necessary to allow responsible pricing is *complete* and in the contractor's *possession*.

5. "Date Proposal Requested" refers to the date when the contractor receives an owner acknowledgment of the change. This most often comes in the form of a change order proposal request.

6. Include the relevant information as it applies to "1st" and "2nd" Proposal Submission columns. Only two columns are provided on the form to emphasize that all efforts are expected to be directed toward resolution of the change by the second submission. If a proposal requires a third submission, it will require a new line on the form. In this case, include an appropriate cross-reference in the "Remarks" column.

7. The "Approved" column indicates the receipt date for owner approval.

8. The "Project Impact/Remarks" column is used to note items such as:

a. Significant references to other files and correspondence;
b. Backcharges;
c. Job delays;
d. Reasons for proposal rejections or price reductions;
e. Correlations with other changes that contribute to compounding effects;
f. Probable claim actions; and
g. Reasons for delays in any step in the process.

9. If any category does not apply for the particular change (such as the "2nd Submission" column for a proposal that has been approved on the first round), clearly cross it off with a diagonal line through the box as soon as the status is confirmed. This will accomplish two things:
   a. Having the box crossed off will save valuable time and trouble by indicating that the item has been considered and that further consideration is not requited.
   b. It is visually apparent that each box *not* crossed off represents an outstanding item with the potential to affect the job. This draws your attention to incomplete items and continues to keep them on the front burner until they have been resolved.

11.5.4 Sample Change Order Summary Sheet Form and Sample Completed Form

The Sample Change Order Summary Sheet described in Sections 11.5.2 and 11.5.3 is included for your reproduction and use. Following that is a sample form, filled out to illustrate the manner in which the form is used in practice. A review of the completed form will clarify the instructions in the previous section.

CHANGE ORDER SUMMARY

File	V	C.O.	Description	Date Discovered	Date Design Finalized	Date Proposal Requested	1st Proposal Submission Date	1st Proposal Submission Amount	2nd Proposal Submission Date	2nd Proposal Submission Amount	Approved Date	Approved Amount	Projected Impact/Remarks

CHANGE ORDER SUMMARY

File	∨	∧	C.O.	Description	Date Discovered	Date Design Finalized	Date Proposal Requested	1st Proposal Submission Date	1st Proposal Submission Amount	2nd Proposal Submission Date	2nd Proposal Submission Amount	Approved Date	Approved Amount	Projected Impact/Remarks
001	6		7	MISSING SPEC. PAGES - DIV 15 BALANCING (AUTOMATE CHEM TREATMENT)	6/5/86 JOB MTG.	6/12/86	8/7/86	11/2/87	$2,557.50	1/2/87	$2,536.61	3/13/87	$2,512.08	1ST PROPOSAL RETURNED FOR ① SUBCONTRACTOR BREAKDOWN
002	8		4	FOOTING CHANGES AT COLS F-1 & F.6-1	6/12/86 JOB MTG	6/24/86	8/6/86		871.83			11/7/86	871.83	WORK PERFORMED AS NECESSARY, ALONG WITH CONTRACT WORK
003				CHANGES TO TRANSFORMER LOCATION	6/12/86 JOB MTG.	9/18/86 JOB MTG				9/18/86 JOB MTG		11/7/86	0.00	SEE RAC 9/29/86 LETTER. CONTRIBUTES TO DELAY IN TRANSFORMER DELIVERY
004	4		1	CHANGES TO SLABS & FNDTNS. AT UNIT "A" NORTH	6/27/86 8/21/86	7/10/86	7/27/86		6,098.82	8/22/86	$7,687.76	10/8/86	7,687.76	REDESIGN DELAYS ENTIRE UNIT "A" - NORTH DELAYED TO 6/15/86 MASONRY START OF 10/18/86
005	2		3	CHANGES TO FIRE & DOMESTIC WATER	6/12/86 JOB MTG	7/10/86	8/6/86	8/30/86 9/13/86 10/13/86	19,078.44	8/30/86 9/13/86 10/13/86	12,277.60 10,526.57 20,526.57	11/9/86	19,978.68	DELAYS BACKFILL/SOG. @ 1-LINE DELAYS WATER & FIRE SERVICE ADD'L COST TO PLACE SLAB OUT OF SEQUENCE
006	3		8	CHANGES TO 4, 5, 6, 21 TO COORDINATE W/ DIV.15 PIPING	6/24/86 JOB MTG. 8/15/86	7/25/86	1/16/87	2/20/87	2,547.26	2/20/87	1,319.50 ①	3/4/87	1,237.50 ① ②	① SUBCONTRACTOR DRAFTING CHARGES SEPARATED ② RAC "PRINTING" CHARGES DEDUCTED
007	1		6	VENTILATION FOR ELEVATOR MACHINE ROOM	6/12/86 REC D 7/8/86	7/9/86 7/18/86	12/19/86		3,937.73			3/4/87	3,180.21	
008		CHANGE		2½" DISCREPANCY BETWEEN STRUCTURAL & ARCH. DRAWINGS	7/30/86									INCORPORATE W/009
009		WITHDRAWN		EXCESSIVE COORDINATION & INTERFERENCE	SUSTAINING									DELAYS PLUMBING UNDERGROUND. STRUCTURAL STEEL
010	5		2	REIMBURSE ELECT. SERVICE CHARGES	8/17/86		8/17/86		2,795.10			11/7/86	2,795.10	DELAYS ORDERING TRANSFORMERS

11.6 APPROVAL SUBMISSIONS

11.6.1 Introduction

Section 10.8, "Shop Drawings and Approved Submittals," described the shop drawing review responsibilities ("shop drawing" is a term that includes all approval submissions). It reviewed the important interface among subcontractors, the prime contractor (or construction manager), and the design professionals. This section will consider the process itself and the procedure for:

1. Receiving shop drawings;
2. Reviewing submittals for form and content;
3. Submitting documents for architect/engineer approval;
4. Evaluating designer action;
5. Following up on required resubmissions; and
6. Distributing (coordinating) the information to be used in construction.

More important, this effort will be accomplished while keeping each individual subcontractor and supplier, as well as the design professionals, practically and administratively under control. Utilizing the procedures discussed in the following sections will ensure that each step will be priced so it is performed completely and on time, and that each submittal processed by the contractor's office will be complete and correct.

11.6.2 Shop Drawing Review and Coordination

The shop drawing preparation and submittal procedure is normally very clearly detailed in the contract. The items to be submitted for approval, size and number of copies, information required, and distribution are among the requirements specified. The Pass-Through Clause (see Section 3.4.3) included in the contract and the opening statements in the technical specification section that reference and incorporate the general conditions obligate each "per plans and specs" subcontractor and supplier to comply with all technical submittal requirements.

Submittal documents are one key to effective project control, and should be taken very seriously. Shop drawings and samples should be submitted in strict accordance with the contract requirements. They should conform to all specified parameters and be delivered in the requisite number of copies for proper distribution. By fully complying with the requirements, the contractor will eliminate the risk of shop drawings being returned without action because of a paperwork problem. The correct number of copies submitted the first time makes the contractor's own coordination of the work, now and later, easier. Proper form and content means the inclusion of the information necessary to allow the work to proceed without interruption, interferences, or delays.

In this regard, it is to project management's advantage to offer a certain amount of assistance to subcontractors and suppliers so that simple misunderstandings can be minimized and the entire process can be expedited. Any assistance, however, should be informational only. Do not prepare submissions for a subcontractor unless such preparation has been made a specific condition of the agreement with the subcontractor. If any material or equipment is being submitted as "equal" or as a "substitution," do not take these categories and the contractual meanings of these words for granted. The semantics may mean a world of difference in the submittal process. It can also affect the owner's approval procedure. The contractor must know how the contract defines these terms and whether the contract differentiates between "equal" and "substitute" products in terms of submission procedures. After the procedures are identified, the contractor must comply with the respective procedures to the letter.

Using Section 11.6.4, "Sample Form Letter to Subcontractors Regarding Shop Drawing Submission Requirements," will force the contractor's management team to become familiar with the technical requirements of the submittal process and will highlight the requirements for the subcontractors. It will make it easier for the subcontractors to understand their contractual requirements and that the requirements will be taken seriously. The letter will also serve as a convenient checklist for follow-up.

11.6.3 Shop Drawing Submission Requirements

It is project management's responsibility to be certain that all subcontractors and suppliers meet their submittal obligations squarely. Project management must be certain that all submissions submitted for approval contain the following:

1. Project title and job number;
2. The contract identification;
3. The date of submission, as well as the dates of all previous submissions and revisions;
4. The names of the contractor, supplier, and/or manufacturer;
5. Identification of the product(s) by specification identification numbers;
6. Field dimensions, clearly identified as such;
7. Relationship to adjacent and/or critical features of the work or materials;
8. References to applicable standard specifications;
9. Clear identification of deviations from the contract documents;
10. All other pertinent information as may be required by the specifications, such as:
    a. Model numbers;
    b. Performance characteristics and capabilities;
    c. Dimensions and clearances; or
    d. Wiring or piping diagrams and controls;
11. In the case of a manufacturer's standard or schematic drawings or diagrams:

a. Modifications or notations deleting information that is not applicable; and
b. Specific inclusion of any supplemental information which is applicable.

Check the project specification for any additional or special requirements.

11.6.4 Sample Form Letter to Subcontractors Regarding Shop Drawing Submission Requirements

The Sample Form Letter to Subcontractors Regarding Shop Drawing Submission Requirements is designed to accommodate the Section 11.6.2 and 11.6.3 shop drawing coordination considerations and submission requirements. It is a form letter that can be filled out quickly by project management. It communicates very clearly that the submittal procedure is a routine operation and that the recipient is expected to complete all of the requirements that are included in the contract. The letter is to be filled out and sent to each subcontractor and supplier concurrently with, or immediately following, the transmittal of the respective subcontract or purchase order. The letter:

1. Calls specific attention to the submittal requirements to lessen the probability that any submissions will be overlooked.
2. Serves as the first written notification that the recipient is definitely expected to comply with all of the various specification requirements and that proper attention must be given to the subject contract.
3. Serves as a definite written notification that the recipient is responsible for all information contained in the plans and specifications, that the contractor will supply all necessary documents if the recipient is missing any documents, and that the recipient is responsible for being aware of all information necessary to properly coordinate the work.
4. Serves as written notification that all approved materials required to coordinate the subcontractor's work with the work of other trades are open to inspection at the jobsite.
5. Requires the subcontractor to include the job number and other relevant information on all correspondence. Filing, requisition preparation, accounts payable, accounts receivable, and so on will then proceed smoothly. In a polite and professional tone, the letter emphasizes that ignorance is not an excuse and that it will not be tolerated.

Following the simple procedure for implementing the letter will also provide the contractor's management team with a complete understanding of the specific requirements of the particular subcontract or purchase order. If the project team reviews the individual specification sections, contracts, and miscellaneous requirements in detail, at the start of the contractual relationship, they will wind up on top of the situation fast. It will send a powerful message to the subcontractors, suppliers and other recipients that the contractor is in control of the project. Before using the letter, the contractor must:

1. Review the owner's specification requirements that are included in the respective subcontract or purchase order and reference them in the letter.
2. Review the subcontract or purchase order and its standard and special conditions so that these specific requirements may be indicated in the space provided in the letter.
3. Send the letter to the addressee, with copies to the superintendent and other relevant project people, so that everyone who will be involved in the management of the subcontract or purchase order is aware of all considerations and existing documentation.

SAMPLE FORM LETTER TO SUBCONTRACTORS REGARDING SHOP DRAWING SUBMISSION REQUIREMENTS

LETTERHEAD

To: _____ Date: _____ 19____

_____ Project: _____

_____ No.: _____

Attn: _____ Subcontract/P.O. no.: _____

SUBJ: Submittal requirements sections(s) _____

_____

The contract plans and specifications require you to submit the items listed below. Please note that your failure to submit any item required by the contract document will not relieve you of the requirement. You will be required to comply with your specifications in every respect. This list is for your convenience only.

PLEASE SUBMIT:

__ CERTIFICATE OF INSURANCE   __ INSTALLATION INSTRUCTIONS
__ PERFORMANCE BOND   __ DELIVERY TIME AFTER APPROVAL
__ LAB. & MAT. PAYMENT BOND   __ ERECTION/INSTALLATION TIME
__ CERTIFIED PAYROLL REPORTS   __ TEST/TEST RESULTS
__ SHOP DRAWINGS; __ COPIES   __ GUARANTIES/WARRANTIES
__ ERECTION DRAWINGS; __ COPIES   _____
__ PRODUCT SPECIFICATIONS   _____
__ EMPLOYER EEO REPORTS

If you are not in possession of the necessary plans and specifications, please contact me immediately. Copies of contract plans, specifications, all approved shop drawings of all subcontractors, and the project construction schedule are available at the jobsite for inspection.

Please include our job number and your contract number as indicated above on all project correspondence relating to this project.

Thank you for your cooperation.

Very truly yours,

Project Manager

cc: Project Superintendent

11.6.5 Submittal Review, Distribution, and Follow-Up

Review

It is the contractor's (or the construction manager's) responsibility to review shop drawings, product data, and samples to confirm their compliance with the criteria set forth in Section 11.6.3, "Shop Drawing Submission Requirements." In addition, project management must:

1. Ensure that all subcontractors and suppliers submit their materials with promptness and in accordance with the submittal and construction schedule so that there will not be any delay or interference in delivery or to the work.
2. Determine and verify:
   a. That the subcontractor has incorporated or will guarantee all field dimensions;
   b. That all field conditions and construction criteria have been accommodated;
   c. That the product either complies with the specification requirements in every respect or that any deviations have been properly identified and explained.
3. Coordinate each submittal with both the field and contract document requirements.
4. Research and confirm all "justifications" for any deviations from the contract requirements *before* submitting the documents to the architect for approval.
5. Determine if a credit or addition to the contract is in order, based on any changes incorporated into the submission. (Refer to Chapter 7, Where and How to Find Potential Change Orders, for further assistance in this regard.)
6. Determine if any backcharges to any subcontractors are in order as a result of changes necessitated by such subcontractor's acts or omissions.
7. Determine that the submission is timely and that the material delivery schedule conforms to the confirmed delivery times following the receipt of approved shop drawings. (This information is based on the terms set forth in the respective subcontract agreements and the current construction schedule.)
8. Positively identify the responsibility for all "Not by Subcontractor" or "By Others" notations, and correct as necessary *before* the submission of the submittal to the architect for approval. (Often, for example, the "furnish and install" subcontractor may be submitting its supplier's shop drawings for approval. Those drawings may be covered with "Not by Supplier" notations, particularly as they relate to the installation of the item and miscellaneous attachments. As appropriate, simply cross out the notations and correct them all to read "By (Subcontractor's Name)." Another way to accommodate this situation is to clarify with the subcontractor, in writing, that the "Not by Supplier" notations do not alter the

subcontractor's obligation to perform the work or provide the product and that the item is still the responsibility of the subcontractor. In any event, be sure that *all* by "By Others" notes have been definitely assigned to the responsible party or have been corrected to be included in the subcontractor's responsibilities.

9. Compare all *resubmissions* with the file copy of the previous submission to confirm that all required corrections have been made. (Note that the Section 11.6.6 Sample Form Letter to Subcontractors Regarding Shop Drawing Resubmission Requirements is an effective method for expediting the return and resubmission of the submittals that were nonconforming in the first round.)

A helpful practice in the shop drawing review process is to use a marker in a different color from the color being used by the architect and engineers (typically red). The contractor should also request that the design professionals differentiate between the colors used in modifying submittals. This makes it very easy at a later time to decipher the office that originated a particular remark.

Distribution

Upon receipt of submittals bearing the architect's approval ("Approved," "Approved as Noted," "No Exceptions Taken," etc.—refer to Section 10.8, "Shop Drawings and Approved Submittals"), distribute copies of submittals bearing such stamps to:

1. The jobsite file.
2. The record documents file.
3. The subcontractor or supplier making the submittal.
4. The other affected subcontractors and suppliers.
5. Anyone else who may need the information to properly coordinate their work.

Except under special circumstances, do not forward rejected submittals to the jobsite; only the final approved documents which are to be used for construction should be maintained at the jobsite. Maintaining unapproved submittals at the jobsite creates an unnecessarily bulky site file and increases the chance that an inappropriate shop drawing will be used in the performance of the work.

Follow-Up

Project management is responsible to:

1. Monitor the time it takes for the approval process to be sure that the architect is giving proper, timely attention to the submissions.
2. Be certain that the design professionals:
    a. Include all information required of them by way of questions in the submittals;
    b. Do not overstep their authority;

c. Do not overstep their professional capacities or licenses;

d. Do not add work without regard to the established change order procedure;

e. Include only meaningful comments on the submission, which comments facilitate the completion of the work or the proper completion of the submittal;

f. Affix the accepted stamp and initial/sign it; and

g. Clearly indicate any requirements for resubmittal or approval of the submittal.

3. Upon distribution of the submittal back to the subcontractor or supplier who originated it, project management should:

a. Reconfirm the delivery schedule(s);

b. Confirm that the submission is being returned in adequate time for the subcontractor or supplier to meet its own requirements;

c. Note any significant information for the next construction schedule update; and

d. Begin any actions that may be necessary to resolve problems that have been exposed by the review process.

As mentioned earlier, the contractor must be certain to note all delays and inappropriate actions by the design professionals in correspondence. Additionally, the contractor must include all delays in the updates to the construction schedule. The use of a submittal log (which may be very similar to the Change Order Summary Sheet mentioned in Sections 11.5.2 and 11.5.3) will facilitate the contractor's management of the submittal process. The submittal log will also show trends in the overall process. The record is discussed more fully in Section 11.6.7, "Shop Drawing Submittal Summary Record Procedure." Further, a sample Summary Record Form is included in Section 11.6.8. Finally, Section 10.8, "Shop Drawings and Approved Submittals," also contains a related discussion.

11.6.6 Sample Form Letter to Subcontractors Regarding Shop Drawing Resubmission Requirements

The Sample Form Letter to Subcontractors Regarding Shop Drawing Resubmission Requirements is designed to follow up on submittals that have been returned to a subcontractor or supplier for correction and resubmission. The letter has been arranged to compel an otherwise unresponsive subcontractor or supplier to devote more attention to *timely* compliance with the contractual paperwork requirements. Specifically, it:

1. Serves as a reminder to the subcontractor or supplier that he or she must devote more attention to the project.

2. Serves as important documentation of the contractor's efforts to coordinate the exchange of information and the work as well as provide additional support for backcharges and other more serious claims against the recipient.

3. Advises the recipient in writing of potential backcharges that may result from his or her inattention to the project.

The letter is based in part on the "squeaky wheel" theory of motivation to action. It reemphasizes that the contractor will not relax until the submittal process has been completed. If the letter is to be used effectively, the contractor must be keenly aware of the status of each and every submission. To use the letter effectively, periodically review the shop drawing submissions in accordance with the recommendations in Section 11.6.7, "Shop Drawing Submittal Summary Record Procedure." The Resubmission Requirements Letter should be sent to each subcontractor and supplier who is delinquent in his or her response to a required resubmission. Keep sending additional letters until the desired result is achieved or the lack of response prompts more drastic action. As mentioned earlier, these records will become extremely important in the event the subcontractor's untimely action results in delays or interferences to the overall project.

SAMPLE FORM LETTER TO SUBCONTRACTORS
REGARDING SHOP DRAWING RESUBMISSION REQUIREMENTS

LETTERHEAD

To: _____   Date: _____ 19_____

    _____   Project: _____

    _____   No.: _____

Attn: _____   Subcontract/P.O. no.: _____
                   SUBJ: Submittal resubmission
                         requirements
                         sections(s) _____

    Copies of the shop drawings *requiring immediate correction and resubmission for approval* were returned to you on the dates listed below. By contract, you were required to resubmit these drawings promptly. Your failure to resubmit these drawings in a timely manner has caused delay and additional costs. Immediate resubmission of these drawings is required if additional delays and extra costs are to be avoided.

Drawing No.	Date Returned to Your Office	Description

    Thank you for your cooperation.

Very truly yours,

Project Manager

11.6.7 Shop Drawing Submittal Summary Record Procedure

This section describes the procedure for the implementation and use of the Shop Drawing Submittal Summary Record Form discussed in the next section. The form is to be used for all submissions for approval. It records the characteristics of each submittal along with resulting action and distribution information. The form has been designed to systematically identify and record submissions that must be approved by the architect and/or the owner, as part of the contract. It provides the following information in an understandable and easy-to-read format:

1. A listing of all required submittals, which will serve as an important checklist;
2. The chronological order of submissions;
3. The due date required for submissions;
4. The required response date for architect action;
5. The owner/architect action;
6. The required and actual distribution.

In addition to providing an important checklist, the form will be an invaluable record that can be used in nearly all project research and control efforts.

Procedure

1. The forms should be placed in a 1" three-ring loose-leaf binder and arranged by specification section, bid package number, subcontract number, or subcontractor name. (This decision is a matter of preference by persons who must work with the form continually. Many contractors prefer to have the forms arranged and identified in the same manner as the submittal files. Refer to Section 11.3, "File Content.")
2. Research the individual specification sections included in the scope of responsibility for each subcontractor or supplier and list all required submissions on the form.
3. Determine any other submittal requirements that may be necessary for the proper coordination of the work of the subcontractor or supplier.
4. Add to the list as the shop drawing approval process proceeds so that information and additional submittal requirements disclosed during the process are identified and tracked.
5. Under the "Copy" headings, insert the names of the subcontractors who require copies of the approved submittal for coordination. (For example, if you know that the concrete subcontractor will need the steel subcontractor's anchor bolt layout, insert the concrete subcontractor's name in one of the columns in the steel subcontractor's Summary Record Form. Then place a checkmark in the corner of the concrete contractor's distribution box adjacent to the anchor bolt submittal line. This will serve as

a valuable reminder of the distribution requirement. Upon receipt of the approved anchor bolt drawing from the architect, note in the same box the date that the drawing was sent to the concrete subcontractor).

6. As information requests or distribution requirements are received during the overall shop drawing process, include those names in the proper distribution sections. (For example, if early correspondence from the concrete subcontractor requests the anchor bolt locations for the light poles that are being provided by the electrical subcontractor, immediately turn to the electrical contractor's Summary Form, insert the name of the concrete subcontractor at the top of a distribution column, and place a check next to the submittal item for the light poles. Once again, this check will provide an important reminder to send the submittal to the concrete subcontractor when the submittal goes through the system.)

7. Insert the dates for the submission of the respective submittals. Get the information from the construction schedule and use "Early Finish Dates" if the schedule will identify such information.

8. Insert the due dates for the architect's response and, as set forth above, use the "Early Finish Dates" from the construction schedule.

9. Have the Summary Form for the respective subcontractor available when writing the transmittal to the architect or back to the subcontractor. After completing the transmittal, *immediately* transfer all relevant information to the Summary Form so that the summary is always accurate and up to date. (Remember your mother's admonition: "Do it now or chances are it will not get done.") Maintaining this summary in an electronic file should facilitate the prompt updating of the summary.

10. Process the transmittal as described in Section 11.7, "Sample Letter of Transmittal."

11.6.8 Shop Drawing Submittal Summary Record Form

The procedure for use of the Shop Drawing Submittal Summary Record Form is outlined in Section 11.6.7. Set forth below is a sample form, along with the precise definition of all included terms.

Definition of Terms

1. COPIES

   The number of copies required to be submitted for approval that will allow complete subsequent distribution.

2. DESCRIPTION

   Title of the drawing or item being submitted.

3. ____ DRAWING NO.

   If required by the specification or by project management for any reason, submissions may be stamped to correlate with the applicable specifications section or bid package. Its chronological order of submission may also be identified. For example:

If no particular numbering system is being used, leave the box blank.

4. SUB DRAWING NO.

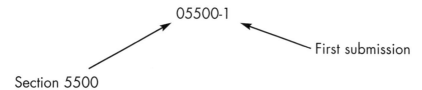

Insert the drawing number indicated in the title box or elsewhere on the subcontractor's submittal that will positively identify the document.

5. REQ'D BY

Insert the required submission date as per the construction schedule. Use the "Early Finish Date" if the scheduling program allows such identification.

6. APPR REQ'D BY

Insert the date for the architect's approval, as dictated by the construction schedule. As above, use the "Early Finish Date" if the date is available.

7. FIRST SUBMISSION

Insert the date that the submission was initially transmitted to the architect for approval or other action. The date should correspond with the date on the transmittal prepared for the submission. (Remember the discussion earlier regarding the content of letters. Most letters should be limited to a single topic. This rule also applies to the transmittals. Submissions from different specification sections should be transmitted under the cover of separate transmittals.

8. RETURNED FROM ARCHITECT

Insert the date that each submission is *received* back from the architect.

9. SECOND SUBMISSION

Insert the date of resubmission of each item requiring correction or resubmission. Refer back to FIRST SUBMISSION.

10. ACTION/ACTION CODE

Identify the action taken by the architect on each submittal, for example, A = Approved, AN = Approved as Noted, RR = Revise and Resubmit, and so on. Add additional actions and codes as required by the particular project. Note the Action Codes on the form for easy reference.

11. FILE COPY

Insert the date that a copy of the submission is filed into the system (refer to Section 11.2, "Establishing Easy-to-Research Change Order Files," and Section 11.3, "File Content"). Every submission with any action or inaction by the architect or owner must be kept on file. Remember that rejected submissions should not be filed at the jobsite office.

12. JOB COPY

    Insert the date that a copy of the respective submission is sent to the project superintendent for use in construction of the work. As mentioned previously, send only submissions approved for construction to the jobsite.

13. COPY

    Insert the date that copies of the submittals are transmitted to the subcontractors and suppliers that need the submittal for the proper coordination of their own work. Note the name of each subcontractor at the Copy heading.

SAMPLE SUBMITTAL SUMMARY RECORD FORM

Project _____ Sub. _____ Sect _____

Owner No. _____ Phone _____

No. _____ Persons _____

ACTION CODE: A   Approved   ANR App. As Ntd; Rev. & Resub.   FU For Your Use   page _____ of _____
              AN Approved As Noted NA   Not Approved

Description	Company	Sub. Dwg. No.	Dwg. No.	First Submission	Returned From Arch.	Action	Second Submission	Returned from Arch.	Action	File Copy	Job Copy	Copy	Copy	Copy	Copy

Copies

11.7 SAMPLE LETTER OF TRANSMITTAL

The Sample Letter of Transmittal is one of the most common correspondence types used on every construction project. Its purpose is to document the movement of information or materials. Each time that a shop drawing or sample is transferred from the custody of one party to another, it should be accomplished by way of the transmittal.

Despite the fact that the transmittal is such a common and important communication, too many individuals and companies do not use the transmittal properly. The use of form transmittals or letters containing self-explanatory categories should facilitate the prompt and efficient transmission of information. However, they are seldom put to their proper use. In the rush to fill out the form and get it out, mistakes occur. Some of the most common mistakes include:

1. Individual names are neglected;
2. File references are omitted;
3. Communication is assumed, and not necessarily completely indicated;
4. The method of delivery is left out;
5. The items being transmitted are inadequately described;
6. The items being transmitted are incomplete or omitted;
7. Requested action is left unclear;
8. Important remarks and qualifications are abbreviated or are left out altogether; or
9. Proper distribution is not accommodated.

Do not let these kinds of mistakes transform a potentially powerful project record into a worthless, time-wasting exercise. Instead, afford the Letter of Transmittal the proper amount of attention and use it properly. The contractor's organization must understand the significance of the communication and treat the categories included in the letter with respect. Exercise the discipline to be sure that the information in the letter is correct and complete. It should be a routine application that will leave the contractor with complete records where and when required.

11.7.1 Sample Form Letter of Transmittal

Use the Sample Form Letter of Transmittal when sending all shop drawings, a third-party letter, samples, or almost anything else to another party. In each case, be sure to:

1. *Send it to a specific person's attention.* If the person responsible for responding to the transmittal is unknown, find it out before sending the transmittal. This step will facilitate follow-up if the response is not forthcoming.
2. *Always note the complete file reference on the transmittal.* Refer to Section 11.4, "Correspondence File," and apply the guidelines to the transmittal. As

with any other letter, two convenient locations for the file instruction are in the project reference itself or below with the noted "cc:" instruction.

3. *Include the required action or the purpose of the letter.* The boxes on the form are there to be checked. Use them. If a document is attached, note it. If the document is to be sent under separate cover, check that off.

4. *Indicate the method of delivery.* This can be an important piece of information, particularly when time or a delay in response becomes an issue. It might make all the difference in the world, for example, if a document has been sent by mail or express delivery, or has been hand-delivered.

5. *Use document identification numbers if there are any.* Describe the item(s) being sent concisely, and in a manner that positively identifies them and distinguishes them from other submittals. A description such as "Duct Drawings," for example, means nothing. Use the same description that is in the title block of the drawing, such as "Unit A-North Third Floor Ductwork." Although it may take more time to write at this stage, it will save much more time in the future.

6. *Request definite action.* Again, the boxes are there to be checked off. There is no excuse for assuming that the architect "knows" that the submission needs to be approved. The box is there—check it off. If a response has not been conveniently categorized and preprinted, take the time to positively indicate the requested action.

7. *Use the "Remarks" section for the important qualifications and notifications demanded by the particular situation.* This is a legitimate project record document. The recipients are responsible for all information contained in it. Do not overlook repeated opportunities to request action, require responses, and include technical notifications.

8. *Indicate the distribution of the letter and its attachments.* Refer to Section 11.4, "Correspondence File," for additional related instructions.

CHANGE ORDER AND FILE ADMINISTRATION

FORM LETTER OF TRANSMITTAL

To: _____ Date: _____ 19 ___
    _____ Project: _____
    _____ Project No.: _____
    _____

Attn: _____

TRANSMITTED: ☐ Attached ☐ Under Separate Cover via _____ are the following:
☐ Shop Drawings   ☐ Contract Drawings   ☐ Brochures   ☐ Specifications
☐ Samples   ☐ Letter (Copy)   ☐ Change Order   ☐ Coordination Dwgs
☐ _____

Quant	Date	I.D. No.		Description

(If items received are not as listed, please notify us immediately.)

☐ Approval   ☐ Approved   ☐ Submit _____ Sepias & _____ Prints for:
☐ For your use   ☐ Approved as Noted   (Approval) Distribution)
☐ As Requested   ☐ Rejected
☐ For Fabrication ☐ Revise & Resubmit   ☐ _____

Remarks _____
_____
_____
_____
_____
_____
_____
_____
_____

cc: _____
    _____
    _____ By _____
File: _____

# Part Six
Dispute Resolution

12

Winning in Change Order Negotiation

12.1	Introduction	317
12.2	Acceptance Time	317
12.3	Agenda	317
12.4	Gentlemen's Agreements	318
12.5	Agreement vs. Understanding	318
12.6	Allowances	318
12.7	Alternatives	319
12.8	Arbitration and Mediation	320
12.9	Aspiration Level	321
12.10	Assumptions	321
12.11	Authority	321
12.12	Averages	322
12.13	Boilerplate	322
12.14	Catch-22	323
12.15	Change Clauses	323
12.16	Change the Negotiator	323
12.17	General Contractor As a Conduit	324
12.18	Contingency	325
12.19	"Convenience" Specifications	325
12.20	Concessions	325
12.21	Constructive Changes	326
12.22	Correlation of Contract Documents	327
12.23	Cost Perceptions	328
12.24	Credits—Turning Them Around	328
12.25	Deadlines	329
12.26	Deadlock	329

12.27	Deliberate Errors	329
12.28	Level of Detail	330
12.29	Discipline	330
12.30	The Eighty–Twenty Rule	331
12.31	Elaboration	331
12.32	Empathy	331
12.33	Designer's Estimates	331
12.34	Equitable Adjustment	332
12.35	Exceptions	333
12.36	Excusable Delays	333
12.37	Use of Experts	334
12.38	Face-Saving	335
12.39	Job Meetings	335
12.40	The Power of Legitimacy	335
12.41	Letter Wars	336
12.42	Lost Notes	337
12.43	"Nonnegotiable" Demands	337
12.44	Objections	338
12.45	Off-the-Record Discussions	338
12.46	Patience	339
12.47	Personal Attack	339
12.48	Presentations	339
12.49	Proceed Orders	340
12.50	Promises	341
12.51	Questions	341
12.52	Quick Deals	342
12.53	Reasonable Review	342
12.54	Reopening Change Proposals	342
12.55	Split the Difference	343
12.56	Statistics	343
12.57	Telephone Negotiations	343
12.58	Plain Hard Work	344
12.59	Unit Prices	344
12.60	Value of Work Performed	345
12.61	Conclusion	345

12.1 INTRODUCTION

A great deal has been written on negotiation in the general sense, covering in detail just about everything that could possibly come up in nearly every theoretical situation. This chapter is not intended to reiterate those exhaustive treatments. Rather, this chapter identifies the ideas, strategies, tactics, and countermeasures that are usually employed or encountered in the negotiation of change orders. Many of the ideas presented have been tested and proven by contractors throughout the country. The strategies and tactics discussed in this chapter have been used to get the most out of each change order situation. They are included in this book because of their relevance to the subject matter and their consistent record of success.

The concepts and considerations can be easily applied when formulating the answers to the questions contained in Section 6.3.3, "Change Order Research Checklist." After the checklist has been completed for a particular change, return to this chapter and review all the ideas that will help in the accurate assessment and ultimate determination of the proper course of action.

The information and ideas are presented in a "bullet point" format. This format was selected to allow busy construction professionals to get right to the point. Information can be assimilated quickly and put to immediate use with a minimum of time and effort.

12.2 ACCEPTANCE TIME

People need time to accept new ideas. Resistance to change is universal. People need time to rationalize and justify a change to themselves. This applies to *any* change, be it a multimillion-dollar problem or deciding to accept a spouse's choice of a vacation spot. More importantly, people need time to justify a change to their own organization.

Allowing time for new ideas to sink in gives the other person time to rationalize, to think of his or her own reasons why the new deal is not so bad for his or her side. It is often easier for the other side to collect reasons to justify a bad deal than it is to work harder to obtain a better deal.

An unresolved issue creates tension. The tension will grow into stress and will only be relieved when the change is resolved. Leave room in all negotiations for acceptance time.

12.3 AGENDA

Agenda are often underestimated and overlooked in terms of their importance in negotiations. The party that formulates and controls the agenda will formulate, control, and time the negotiations. By controlling the agenda, the contractor can leave a good amount time for the discussion of the topics that the contractor wants to discuss and can arrange to have items which the contractor wishes to avoid scheduled for late in the day.

The agenda is the first test of purpose, and sets the stage for what will follow. The agenda can:

1. Clarify or hide motives.
2. Establish fair or biased rules.
3. Keep talks on track or permit digressions.
4. Force quick decisions or permit delays.

If the contractor can control the agenda, the contractor will be able to control what will or (perhaps more important) will not be said.

12.4 GENTLEMEN'S AGREEMENTS

Gentlemen's agreements are very dangerous and should be avoided in the present business environment. Even if the other party to the agreement has absolute integrity, there is always the possibility that he or she may be removed from the deal before it has a chance to be consummated. People get shifted to other areas of their organizations; they get promoted, they get fired, or they just plain go back on a deal. If there is an agreement on a particular issue, get it in writing.

12.5 AGREEMENT VS. UNDERSTANDING

It is not enough merely to reach an agreement on some issues. Even when two parties have the best of intentions, agreements break down for many reasons. These reasons include:

1. The persons responsible for implementing the agreement do not understand the common viewpoints, attitudes, and backgrounds that brought about the agreement;
2. Neither party knows how to make the agreement work; or
3. Neither party knows how to prove that the agreement is or is not working.

A good agreement should describe the understanding behind the written words in addition to spelling out the specific terms (scope, money and time). It should also include a procedure for measuring costs in the event that additions or deletions become necessary.

It is not enough to reach an agreement on terms. The responsible contractor should evaluate whether there are any special understandings or procedures that need to be laid out as part of the agreement.

12.6 ALLOWANCES

Allowances are items that are identified in the contract documents as part of the contractor's scope of work, but for some reason, cannot be priced with any specificity when the bid or estimate is submitted. They are there to remove a portion of the uncertainty surrounding the item. This is accomplished by the inclusion of a budget amount for the item that keeps the contract at a realistic level, and a qualification that the contract sum will be adjusted upward or downward, based on the difference between the allowance and the actual final price.

The inclusion of an allowance in the contract price creates uncertainty for the owner. The owner is relying upon the accuracy of the allowance estimate as a reliable measure of the ultimate price. The contractor or construction manager, on the other hand, should welcome the inclusion of allowances because it eliminates the risk associated with the contractor's inability to definitively price certain portions of the work. While it is generally true that allowances preclude the opportunity for large savings through effective purchasing of an item, they also reduce or eliminate the risk associated with purchasing the item. The contractor's purchasing efforts can proceed with the confidence that the item will definitely meet the budget, because the budget will be adjusted to meet the item.

When negotiating any of several types of construction agreements (including change orders), make the most of every opportunity to incorporate allowances. Today's construction industry already presents the contractor with a wide variety of risks. Take all available steps to stabilize the project environment and reduce that risk.

12.7 ALTERNATIVES

Presenting an individual with a choice of alternatives is a time-tested method of steering that person toward the desired resolution of an issue. People do not like to be put into a box. Like many dangerous animals, people can be backed up only so far and when they are finally against the wall, they stop the retreat and attack. In contrast, if several doors have been strategically placed in the wall, the animal still has some feeling of security through options. The key is to design and provide the options. Real but less desirable options must be strategically removed from consideration. Simply not including them in the list of options is one way to avoid them. The one door in the wall leading to the outside must be locked or at least camouflaged to look like the rest of the wall. Attention will then be diverted away from the disfavored option and focused on the desired decision. If designed properly, all options will lead the decision maker to the desired decision.

If the tactic has been effective, your opponent will feel some consolation in the fact that there is *some* choice. Just having a choice makes a person feel that he or she has some control over the situation, even though exactly the opposite may be the case. In the best scenario, your opponent will feel good about making the desired decision and will believe that it was his or her own decision.

12.8 ARBITRATION AND MEDIATION

The use of arbitration or mediation as a dispute resolution option is so important that a few short statements here will not do these procedures justice. For this reason, these procedures are discussed in more detail in Chapter 13. There, specific considerations and recommendations are outlined for evaluation and implementation.

In general, the right to request arbitration or mediation is governed by the contract. In an arbitration, the parties select an arbitrator (or arbitrators) and present oral and documentary evidence. At the end of the arbitration hearing, the arbitrator(s) makes a decision in favor of one party or the other. With a mediation, the parties select a mediator, but there is no presentation of "evidence." In most cases, the mediator does not make any type of decision. After hearing statements from each side, the mediator tries to negotiate a compromise agreement between the parties through the use of "private caucuses" with the parties and "shuttle diplomacy." In very rare cases, the parties empower the mediator to make a decision if the parties reach an impasse. However, the mediator's power is usually severely curtailed (i.e., select the position presented by one party or the other, which position was developed in confidence and sealed prior to the start of the mediation sessions).

Arbitrations are usually faster and less expensive than court actions, although there are exceptions. Mediations can take place very quickly once the parties and the mediator are ready. In many cases, arbitrators are professionals with construction backgrounds. For this reason, arbitrators are able to grasp and understand the issues presented on a complex construction project. Mediators, on the other hand, are professional settlement officers. While some mediators specialize in the resolution of construction disputes, many mediators simply focus on the resolution of disputes. Both of these procedures present a sharp contrast to a theory-minded judge and a jury with no construction background and who cannot appreciate that delay costs are real.

Arbitration generally is more informal than court proceedings. Mediations offer the parties even more flexibility. In arbitration, the rules of evidence are less stringent and different types of evidence may be easier to introduce. Since the mediator is not required to make a decision in most proceedings, the rules of evidence do not play a part in the proceedings. Both arbitrations and mediations may be scheduled to fit more conveniently into everyone's busy day. Finally, the structure of the arbitration hearings or the mediation sessions varies widely, from very formal and legalistic to conversational, depending on the personality, style and background of the arbitrator or the mediator.

There are pitfalls with both arbitration and mediation. For instance, arbitration awards are "final" and there are only a few grounds on which to appeal or set aside an award. In mediation, if one party takes a very hard line or exercises power over the other party, the mediation process will not be successful. In arbitration, the more informal style and relaxed rules may benefit one party more than the other party in certain cases. Legal and technical defenses to particular claims may not be considered by the arbitrator. With mediation, there is not a final resolution of the dispute unless the parties are able to reach an agreement. If an agreement is beyond the reach of the parties, the parties will be required to proceed to arbitration or litigation.

12.9 ASPIRATION LEVEL

It is too easy to say "Aim higher and you will come out better." The fact of the matter, however, is that it is true. If the contractor consistently sets higher targets in his or her negotiations, and truly commits to those targets, the contractor will do better in the long run. As a result of this process, there is an increased risk that the negotiations will end in an impasse. However, if the contractor starts high and leaves room to make minor concessions, the risk of an impasse is minimized.

A good understanding of the change order components and their structures, as outlined in Chapter 9, Substantiating Change Order Prices; and the records that are required to substantiate a change order request, as detailed in Chapter 10, Using Project Records to Discover, Define, Support, and Track Change Orders and Claims, will facilitate the presentation and recovery of higher prices. It is important always to have very good, believable reasons for starting high. As mentioned previously, building and maintaining credibility is very important to successful negotiations.

Higher demands on the part of the contractor will tend to lower the aspirations and convictions of the other party. By making a higher demand and by offering only small concessions during the negotiation process, the contractor conveys to the other parties that it is going to be a lot harder to get what they want.

12.10 ASSUMPTIONS

Never assume *anything*. Assumptions are dangerous. They are more likely to be wrong than right. Thoughts as to what constitutes "reasonable" are going to be diametrically opposed to the "reasonable" thoughts of the other side. This is not being overly pessimistic; it's merely expressing a fact of life. A party's impressions of a situation as well as its ramifications are based upon the experience, training, and disposition of the party. Motives are unique, and organizations are distinct. For these reasons, among others, perspectives will be different.

12.11 AUTHORITY

Try to negotiate only with those people who have enough authority to make the deal. It is important to understand that there are at least two types of authority. There is formal authority as indicated on the organization chart, as well as informal authority. Persons with informal authority have the power in their own organizations to advise their superiors on their courses of action. Site inspectors, for example, can have this kind of informal authority, by virtue of the way in which the "facts" are reported to their superiors. Be observant and identify the real decision makers as early as possible on the project. Do not be constrained by the structure set forth on the organizational diagram.

Too much effort is spent with front-line personnel who have been placed on the project to delay and divert the contractor's endeavors. Recognize these situations for what they are, and avoid them if at all possible.

Even if the contract documents empower a consultant with the "authority" to make modifications to the contract "not involving a change in cost or time," get the authorization in writing from the owner. This "extra" step will end many arguments when unanticipated components of what was previously a "job condition" make the change expensive and time-consuming.

12.12 AVERAGES

Averages are always negotiable. This is because they are both true and untrue at the same time. It's been said that the average American family has 2.4 children. It is true and untrue at the same time.

A price on any manufacturer's standard list reflects the average product, shipped to the average location, to an average buyer, for average use. Do not be hypnotized by a meaningless number that looks right. Be skeptical. Question every "standard" rate and statistical representation. The contractor, project and change order are all unique. None of these things is average.

12.13 BOILERPLATE

Specifications would easily be one-third their size if it were not for boilerplate. Boilerplate is the collection of those long-winded, cryptic, and confusing "exculpatory" phrases, paragraphs, and entire sections that use the shotgun approach to problem resolution. Just as with averages, boilerplate is designed as a net to catch everything that falls through the cracks.

Boilerplate is only effective for the owner if the contractor accepts it. The boilerplate language is put in the contract to justify actions by architect and owner in instances where the rest of the contract failed to accommodate a particular oversight. In analyzing and attacking boilerplate, the term "oversight" is significant. The designers know how to include specific language and contract requirements in the contract documents. These terms allow the responsible contractor to understand the requirements and to price them accordingly. Boilerplate, in contrast, is placed in the contract in an effort to cover the errors and omissions of the architect or the owner. It does not address specific situations.

When faced with boilerplate as a justification for the denial of a change order, protest loudly. Work to identify all the reasons why the situation surrounding the change order request is the exception to the intention of the catch-all phrase. Be firm. Do not stand for the inappropriate application of these often meaningless terms.

12.14 CATCH-22

It may not always be advisable to be smart, decisive, knowledgeable, or rational. This is sometimes a difficult posture to comprehend, because people prefer to think of themselves as reasonable. They like to appear intelligent and authoritative and in control. The truth is, however, that the contractor may be a lot better off and get more if the contractor is slower to understand, less decisive, and slightly irrational. The phrase "you can't argue with a crazy man" has some application in the negotiation of change orders. When the other party gets frustrated with the slow or indecisive responses, the other party will typically make more statements in an attempt to increase understanding and prompt a firm decision. Along with all the extra discussion comes an increased probability that they are going to slip and offer an important concession. Smart may be stupid, and stupid may be smart after all.

12.15 CHANGE CLAUSES

Section 3.4.2 introduced the idea that every good construction contract should incorporate some kind of Change Clause. The clause gives the owner the right to alter the work within the scope of the original contract. If the owner requires work to be done, and the work is in excess of the work required by the original contract but still within the general scope of the contract, the contractor must:

1. Perform the extra work;
2. Request and receive extra compensation under the formal change order procedure described in the contract or under the constructive change doctrine (see Section 12.21).

The Change Clause can provide the vehicle for the incorporation of the contractor's suggestions into the work. The clause also includes the procedure for the contractor to claim additional compensation to cover extra work done at the direction of the owner.

12.16 CHANGE THE NEGOTIATOR

Once an organization becomes accustomed to dealing with a specific individual, it becomes uncomfortable, at least, to start all over again with someone new. The new negotiator can:

1. Have a perspective diametrically opposed to the old one;
2. Retreat from previous concessions;
3. Introduce new arguments;
4. Question previous "givens" and assumptions;
5. Delay agreements; or
6. Change the subject or tone of the discussions.

A change in the negotiator puts pressure on the other party to bring the new person up to date concerning past arguments and agreements.

If your *opponent* introduces a new negotiator:

1. Do not exhaust yourself by repeating old arguments;
2. Be patient if the new person goes back on a previous agreement or understanding; but at the same time, demand that he or she get the story straight;
3. Find a good reason for breaking off the discussion until the original person becomes available; or
4. Feel free to change a prior position if the other side changes its position.

12.17 GENERAL CONTRACTOR AS A CONDUIT

Section 3.4.3 introduced the pass-through principle. It is important to understand this concept clearly. The primary responsibility of every general contractor, prime contractor, or construction manager is to *coordinate* the work. Coordination does not mean inventing the information. It means:

1. Securing relevant information from the parties responsible for generating it;
2. Fitting that specific information into the project requirements;
3. Confirming the appropriateness of the information;
4. Forwarding the information to the persons ultimately responsible for approving its incorporation into the project; and
5. Distributing the information to all parties requiring the information to complete their work properly.

If a subcontractor disputes a condition, pass that dissatisfaction on to the owner for resolution of the problem by the owner with the owner's own documents. If the owner rejects a subcontractor's claim for additional compensation, be sure to advise the subcontractor that it is the *owner* who has denied the change. Keep moving information back and forth in both directions. Try to be a pane of glass or a simple conduit.

12.18 CONTINGENCY

A contingency is a way to reduce the uncertainty associated with a particular item. In a budget, it is included to reduce the risk of overshooting costs. If a change proposal is complex and riddled with uncertainties, contingencies can be used to reduce the risk to the contractor. Even if it is a clear agreement that an unused contingency will revert back to the owner, its inclusion will improve the contractor's position relative to the questionable item.

In other cases, it may not be to the contractor's advantage to be too clear about a contingency. For example, if the construction schedule shows the "extra" time that has been built into an activity as a result of a contingency, the affected subcontractors and suppliers will develop a false sense of security. The activity will not be afforded priority attention. Accordingly, the activity will not receive the proper attention until it is too late. The result will be the rapid transformation of an activity that should have been completed with a comfortable margin, to one that starts late, requires extra attention, and ultimately delays the project.

12.19 "CONVENIENCE" SPECIFICATIONS

What is good for the goose is good for the gander. The contractor is held to dot the i's and cross the t's. Do not let the owner, the owner's representatives, and the design professionals operate under a set of rules that is more relaxed. The contractor should be intimately familiar with the precise requirements of the specifications and should hold all contracting parties to these requirements.

If the terms of a contract have been skewed in one direction, the opposite consideration may also apply. For example, the contract language may provide something like "The contractor is responsible for disclosing previously unknown conditions to the owner," and then stop. In this situation, the contractor may have every right to expect that the owner has a similar obligation to disclose previously unknown conditions to the contractor. In other words, contracts work in both directions. If the contract language itself does not include these kinds of considerations, they may be implied by custom, usage, or the law. Consult with an attorney if presented with this type of specification.

12.20 CONCESSIONS

Concessions may be real or not real. An easy concession may be listening to the other party. Remember what has been called the "call girl principle"; that is, the value of goods and services is greater before they are rendered than after.

Some do's and don'ts for concessions are:

1. Allow room to make some concessions. Have a good reason for starting with the amounts in the initial demand, whether requesting money or time or both. Be able to substantiate the request with proper supporting information so that credibility can be maintained. Start high and concede small amounts slowly (see Section 12.9, "Aspiration Level").
2. Make the other parties work hard for everything they get. People do not appreciate something for nothing. The harder they work, the more they will appreciate it.
3. Conserve concessions. Later is better than now. Make the other party work.
4. Avoid "swapping" concessions or "tit for tat" negotiations. If the other party says, "Let's split the difference," consider saying, "I can't afford to."
5. Train yourself to get something for every concession.
6. Do not be afraid to say no.
7. Do not lose track of concessions. Keep a list.
8. Do not be afraid to back away from previous concessions. The agreement that counts is the one at the bottom line.

12.21 CONSTRUCTIVE CHANGES

A constructive change order can arise not through the stated change procedure, but through the conduct of the parties. Actions or inactions by an owner's representative may change the scope of the work and increase the cost of performance. If so, the contractor should be entitled to an "equitable adjustment" of the contract amount (see Section 3.4.10). Any action by the owner that requires a response from the contractor that is different from the response required by the terms of the original contract may constitute a constructive change order. A constructive change may be said to have occurred in cases where:

1. Extra work is required by some job condition and even though the owner is aware of the requirement, the owner does not authorize the change or object to the performance of the work.
2. Extra work is ordered by the owner or its representatives either orally or in writing outside of the change clause.
3. An approved change results in additional work which was not anticipated by the original change order.
4. The parties consistently disregard the procedures set forth in the change clause and the requirement for written change orders in the regular course of doing business.

The key to resolving and getting paid for the constructive change order involves the immediate recognition of the fact that a change has occurred or is about to occur. The contractor who simply proceeds with the work which is the subject of a constructive change, without prior confirmation of the condition, does so at great risk. The contractor must be aware of the scope of the work contained in the contract documents which formed the basis of his or her bid. If the action of the owner or the owner's representative alters this scope of work, a change has occurred. The contractor must be vigilant to protect his or her contractual rights and preserve his or her profit.

The Change Clause (see Section 3.4.2) entitles the owner to order changes in the work. As such, there is a risk to the contractor if the contractor refuses to perform the work of a constructive change. If written confirmation of the change cannot be secured from the owner by the time the work must be accomplished, the contractor should notify the owner that he or she is proceeding with the constructive change under protest.

Be sure to follow the procedures contained in the disputed work provision and to maintain the required documentation. If the item is significant, a number of other actions may be appropriate. Further, it may be smart to get an attorney on the scene to assist with the identification and evaluation of options, as well as the implementation of any procedures required to preserve the contractor's rights.

12.22 CORRELATION OF CONTRACT DOCUMENTS

The mechanism to decide between certain types of conflicts in the contract documents may or may not be completely detailed in the specifications. If the specifications address the issue of conflicts and precedence, the treatment of these issues varies significantly. Some specifications will simply state that in instances of conflicting details or descriptions, "the more expensive will apply." Other specifications are more responsible, outlining the precise rules of precedence. For example:

- Addenda take precedence over the specifications.
- Specifications take precedence over the plans.
- Schedules take precedence over other information in the plans.
- Large details take precedence over small details.
- Stated dimensions take precedence over scaled dimensions.

While it is true that the foregoing list is little more than a restatement of the rules of common sense, the guidelines reflected in this list are very helpful in dealing with otherwise senseless directives. The contractor must determine if the contract documents includes a precedence clause, and then the contractor must use the rules embodied in this clause to resolve any perceived conflict in the contract documents.

12.23 COST PERCEPTIONS

There are always two value systems at work in any negotiation. The value of extra work to the contractor reflects the cost of the work and an allowance for overhead and profit. The value of the work to the owner may extend far beyond the addition of the contractor's overhead and profit to the actual cost of the work. The work in question may have a bearing on the owner's ability to continue with its own affairs. For example, if the owner and the contractor are unable to reach an agreement on a change, the cost to the contractor may be the cost of the work. The cost of the change to the owner, however, may be the cost plus contractor overhead and profit, *plus* the lost opportunity costs arising from the owner's inability to use the space by a certain point in time. The contractor must be observant of the reasons for the change and its impact on the owner's intended use for the project when analyzing the total cost (real and potential) for a change. If the contractor understands the import of a particular change to the owner, the contractor will be in a better position to accurately assess the strength of his or her hand in the negotiations.

12.24 CREDITS—TURNING THEM AROUND

The moment any portion of the work is deleted from the contract, the word "credit" pops into the minds of the owner and the architect. Suddenly, despite the fact that the contractor has been coordinating the item all along, submitted shop drawings, placed material orders, and so on, the owner requests a big price break. Consider the owner's request and each item of coordination carefully. Specifically, the contractor should:

1. Analyze the original value of the work.
2. Catalog all the activities that have been performed in relation to the work, along with all the effects of the deletion on all other components of the project. Take all the time that you need to be certain that the list is complete.
3. Assign a value or price to each of the items that have been completed or which will be impacted by the deleted work (including all cancellation and restocking charges). Refer to Section 8.2.3, "Assembling Component Prices," for a detailed discussion of the relevant considerations.
4. Deduct the value of the completed and affected items from the original value of the deleted work to establish the amount of credit.

As a result of this analysis, the owner and the contractor will come to realize that it might actually cost the owner more money to delete work with a small original cost from the contract in the latter stages of the contract. Additionally, larger items will not be the windfall credits that the owner might otherwise expect as a result of a reduction in scope. The contractor must "know the details" with regard to deleted work and must present a credible explanation of the costs associated with coordinating the deleted work through the date of the owner's request, if the contractor wants to avoid "giving the store away."

12.25 DEADLINES

Always give specific dates for the acceptance of a change order proposal. Make the deadlines real and enforce them. People typically do not believe in deadlines until one has been uncomfortably imposed upon them. When the date for the acceptance of the proposal has run out, immediately submit a revised proposal reflecting all appropriate cost increases. If the deadline and request for additional compensation are real, properly substantiated and enforced, people will begin to take the deadlines seriously. In setting the deadline, the contractor should understand that the deadline will very likely be the earliest date of the requested response. In contrast, if a deadline is not clearly specified in the proposal, the stage will be set for procrastination.

12.26 DEADLOCK

There is nothing wrong with a deadlock in and of itself. If the contractor is not apprehensive about deadlocks or the problems that they can create, the contractor will be able to consistently demonstrate a firm attitude to the other party. The concern for the responsible contractor, of course, is how to break the impasse that the contractor does not want. In these situations, the contractor should consider:

1. Changing the negotiator.
2. Changing the time, place or shape of the negotiations by postponing a difficult part of the agreement for renegotiation at a later time.
3. Changing the shape of some future aspect of the subject, such as offering additional guaranties.

Deadlocks put severe pressure on both organizations. If the other party does not make the first move to break the deadlock, they might very well welcome an attempt by the contractor. Be creative. Think outside of the box. Reinvolve the other party in the discussion and try to create a climate in which new alternatives can be developed and evaluated.

12.27 DELIBERATE ERRORS

Deliberate errors serve a purpose. Deliberate errors are often used to misdirect and deceive. In today's fast-paced world, people:

1. Add wrong;
2. Multiply wrong;

3. Leave out words;
4. Move decimal points and commas;
5. Make incorrect statements; and
6. Change clauses after agreements have been made.

A very subtle word change can alter the entire meaning of a statement.

If an error is discovered after an agreement is made, someone in the organization must have the courage to explain the error to management and explain why the error was not caught in the first place. From the error-maker's point of view, the penalty is low, but the potential rewards are high. There is nothing to lose. "Anybody can make a mistake."

The responsible contractor must be on the lookout for such errors. Everyone in the contractor's organization must be forced to check the arithmetic, read the fine print, and confirm all understandings. When presented with deliberate error, get angry. Never assume everything is right. Be skeptical and check.

12.28 LEVEL OF DETAIL

Section 8.3.5, "Presenting Change Order Components," introduced the idea that a large number of small items is better than a small number of large ones. Large items must be broken out in some manner to be evaluated and understood by the other party. The contractor should not make the evaluation of a requested change more difficult than it needs to be. Individual small prices (and a lot of them) are easy to discuss, are readily demonstrated, and do not create much friction. It is easier to give in to persistent requests for smaller dollar amounts than requests for large-ticket items. People do not like to appear pretty (even if they are). Break apart big-ticket changes into as many small activities and/or components as possible. As mentioned previously, this exercise will allow the contractor to recover higher prices than originally contemplated.

12.29 DISCIPLINE

Exercise the discipline to:

1. Perform the necessary research;
2. Check the calculations;
3. Follow through on procedures;
4. Follow through on commitments;
5. Complete the documentation; and
6. Promote professionalism.

The effectiveness of the procedures and techniques discussed in this book will seriously deteriorate if gaps are left in the change order file or in the contractor's administration of the file.

The opinions of people who are known to work hard will carry greater weight than the opinions and/or decisions of people who are not. The contractor's statements and decisions will be more credible and less susceptible to criticism if the contractor has completed the action items identified above. Furthermore, criticisms can be met with facts, dates, and quantified illustrations when the contractor has done his or her homework on an issue.

12.30 THE EIGHTY–TWENTY RULE

The Eighty–Twenty rule states that 80% of the items are worth 20% of the price, and 20% of the items are worth 80% of the price. This should tell the contractor something about how the contractor should be spending his or her time. Very often, people have a tendency to spend nearly an equal amount of time for each individual item rather than prioritizing their time and devoting more time and effort to the items with the largest financial impact.

12.31 ELABORATION

Keep presentations and explanations short. The more that a contractor rambles on about his or her position, idea, or anything else, the more likely the contractor is to slip, let out information that should remain private or contradict a previous statement. Keep statements brief and to the point. Try to get the other parties to babble and explain their positions or ideas. Look for precious indicators of weakness, motivation and contradictory positions. If the other party lets something slip, try to use the slip to gain an advantage.

12.32 EMPATHY

Before formulating any negotiating strategy, try to understand the other party's position. See the situation through the other party's eyes. Be sure to know:

1. The decision that the other party needs to make;
2. Why the other party has not already made the decision; and
3. The actions that can be taken to make it easier for the other party to make the desired decision.

12.33 DESIGNER'S ESTIMATES

In most instances, the owner, its agents, and its advisors will have their own ideas concerning the proper amount for a particular change order proposal. It should not be any surprise to the contractor that the owner's estimated price will be lower than the price that the contractor and his or her subcontractor's developed. The contractor must be prepared to respond to the assertion that the proposal, with all its components, terms, and conditions, is much more costly than everyone on the owner's team was "hoping" to see.

The "engineer's estimate" has been defined as the cost of construction in heaven. In preparing their estimates, the engineer or other design professional probably overlooked the costs associated with the indirect and consequential cost portions of a change (refer to Section 8.1, "Change Order Components"). In addition, the engineer or architect probably did not consider things such as:

1. Travel time.
2. Worker/machine idle time.
3. Opportunity costs.
4. Minimum rental periods.
5. Minimum employment periods.
6. Interest charges.
7. Design costs.
8. Inspection expenses.
9. Consulting costs.
10. Waste.

12.34 EQUITABLE ADJUSTMENT

The concept of an "equitable adjustment" is one of the most important factors in any settlement. To equitably adjust a contract, the contractor must be left in the position that he or she would have been in had the change not occurred. Equitable adjustments should include:

1. All actual costs;
2. Home and site overhead;
3. Legal costs;
4. Interest;
5. The cost of special meetings;
6. Travel;
7. Special studies; and
8. Similar costs attributable to the change.

The firm application of the three-cost approach detailed in Chapter 8, Designing and Constructing Effective Change Order Proposals, will guarantee complete consideration of all possible cost categories.

Most federal contracts, or state contracts containing federal money, include an Equitable Adjustment Clause that describes a contractor's rights in a change situation. Even if an individual contract does not include a specific clause, the principle may still apply. Check with a construction law attorney.

DISPUTE RESOLUTION

12.35 EXCEPTIONS

For every rule there are exceptions. See the trees instead of the forest. Look beyond the obvious at everything. This type of an analysis or review requires mental training, but it can be learned and developed.

Exceptions are usually the focus of the creative thinker's inquiry. When faced with a troublesome rule, the creative thinker asks:

1. What is being taken for granted in this situation?
2. What is being presupposed?
3. What assumptions have been made, and are the assumptions valid?
4. What are the exceptions and can the exceptions be applied here?
5. Is there a better way?

Look beyond the obvious. Thoroughly investigate the surrounding circumstances to find the constraints that make the situation unique.

12.36 EXCUSABLE DELAYS

Excusable delays generally permit an extension of the contract time where:

1. The delays are the fault of neither party to the contract; or
2. The delays are the fault of the owner or its agents.

The general effects of an excusable delay clause are that:

1. The owner may not declare the contractor to be in default for finishing late;
2. The owner cannot assess liquidated damages; and
3. The owner cannot demand that the contractor accelerate construction to finish on time.

To obtain an extension of the contract time for an excusable delay, there must be a clause in the contract that permits an extension of time. Without an excusable delay clause, the contractor may be bound to perform the contract on time, unless it is impossible to complete the work or inequitable to enforce the time provision. The courts have recognized the following circumstances as acceptable justifications for implying an excusable delay provision (i.e., extending the contract time without imposing liability on the contractor):

1. Acts of God.
2. The law.
3. The fault of the other contracting party.

Closely related to the concept of excusable delays and the excusable delay provision is the "no damages for delay" provision. This provision typically provides that the owner shall not have any liability to the contractor for excusable delays and that the extension of time is the contractor's sole remedy. Some provisions go so far as to provide that the owner shall not have any liability to the contractor even if the owner caused the delay. The latter clause will not be enforced in every state or circumstance. (Remember the discussion of exceptions in Section 12.35). Further, some states have declared that this type of provision violates public policy. For instance, California Public Contract Code Section 7102 provides that:

> Contract provisions in construction contracts of public agencies and subcontracts thereunder which limit the contractee's liability to an extension of time for delay for which the contractee is responsible and which delay is unreasonable under the circumstances involved, and not within the contemplation of the parties, shall not be construed to preclude the recovery of damages by the contractor or subcontractor.
>
> No public agency may require the waiver, alteration, or limitation of the applicability of this section. Any such waiver, alteration, or limitation is void . . . .

If confronted with this type of clause, consult with an attorney familiar with construction law in your area.

12.37 USE OF EXPERTS

Deadlocks often occur in a "your word against mine" situation or under conditions that rely heavily on the "construction experience" of the participants. The types of circumstances may not lend themselves to a reasoned analysis or discussion between the parties.

In these difficult situations, the contractor should consider bringing in an expert to give authoritative guidance in the discussions and/or negotiations. Some categories in which experts may help include:

1. Law.
2. Claims analysis.
3. Accounting.
4. Project management procedures.
5. Estimating.
6. Architectural design.
7. Civil, mechanical, or electrical engineering.
8. Delay analysis.
9. Scheduling.
10. Soils.
11. Concrete.
12. Roofing or waterproofing.
13. Any other highly specialized or very technical matter.

With a little effort, the contractor should be able to find a recognized expert in almost any field, however esoteric the subject matter. If the expert's qualifications are verified, his or her opinions, remarks, and statements may cast a new light on the situation and facilitate the resolution of the deadlock.

If the other party to the contract engages an expert to support his or her position on a particular issue, the contractor should consider:

1. Retaining a better expert; and
2. Digging into the expert's analysis to try to uncover flaws in the expert's analysis.

In the event the position advocated by the other party's expert contains errors or omissions, the errors or omissions can be used to chip away at the expert's credibility and opinions.

12.38 FACE-SAVING

A bad deal for a person's organization may not necessarily be a bad deal for the person. If the contractor gives the owner's representative, the architect or the construction manager a good explanation of his or her position, that person will be able to justify the deal to others with confidence and without damage to the person's status in his or her organization.

Conversely, a person who has no way out will fight with conviction, and with the support of his or her people.

In all negotiations, try to maintain avenues that will provide the other party with the reasons and justifications that will allow them to present all proposals to their organizations without the risk of having their judgment questioned.

12.39 JOB MEETINGS

Section 10.7 introduced the importance of job meeting minutes in the project record. The value of proper job meeting minutes cannot be overstated. If managed properly, minutes can be used to trace the discussion surrounding a particular issue or tie together a complicated project history. Refer to Section 10.7 and to Section 12.3, "Agenda," for important guidance concerning this critical project record.

12.40 THE POWER OF LEGITIMACY

Do not underestimate the power of standardized forms. People are hypnotized by forms. They assume that the terms have been worked out over a period of time and are there because they have a track record of working. People hesitate to mark up neat, expensive-looking forms. They have the mistaken belief that a standardized form cannot or will not be altered. This effect is most pronounced when the form is presented by a large bureaucratic-appearing organization. After all, standard procedure is standard procedure.

The standardized form is a trap for the unwary. Remember the motto of every teenager: "Question authority." Look past the form to see the language and meaning. Do not hesitate to rewrite the entire form if required by the situation.

On the other hand, the contractor should use the power of the standardized form to his or her own advantage. Always print contracts in standard form, even if the contract represents a one-time arrangement. The contractor should make his or her contracts, subcontracts, change orders and other regularly utilized documents into standardized forms. Take full advantage of the "business as usual" and "standard procedures" phenomena. Also, put requests for information in a standard form. People tend to fill out *all* the information on a form, even if it is irrelevant.

12.41 LETTER WARS

The greater the distance between a person and the day-to-day operations at the job site, the easier it becomes to escalate a problem into a letter war. It is like driving a car. The minute some people get behind the wheel, they become different. If they know that they will not have to look the other driver in the eye, it is easy to cut him off at an intersection. The same occurs in writing a letter. It becomes easy to make abrupt statements and determine absolute positions if the author is not required to look the recipient in the eye.

After concluding a heated discussion, always think twice before sending the letter which contains all of the passions of the moment. Write it down. Get the issue "off your chest." But hold the letter. Do *not* mail it. Put it down and let it cool. If it truly is a rational response to the situation, it will be equally rational in the morning. If the proposed letter withstands the light of day, then by all means, let it fly. If upon further reflection (with a cooler head) the letter is determined to be a more emotional rather than rational response, patience will have saved an embarrassment.

The foregoing is especially true in today's world of electronic mail and instant messaging. Letters and e-mail messages represent permanent project records. While these records are very important to the responsible administration of a construction project, they can be very harmful to the contractor if not used properly. A letter or e-mail message sent in the heat of the moment may engender ill will and move the parties farther apart. Additionally, the contractor may be required to recant the positions taken in the passionately prepared letter. This is not to say that passionately prepared letters are bad. Passion is an effective tool of the successful advocate. However, passion must be coupled with rational arguments.

Writing an impassioned letter in the heat of the moment may create an important record even if the letter is not sent. If researched properly, it can be a useful index or summary in the future.

12.42 LOST NOTES

One way to reveal a position without taking responsibility for it is to "lose" your unsigned notes in the corridor, in the field office, or out on the site somewhere. Note that:

1. Information obtained from indirect sources is more credible than that provided openly;
2. Lost notes are studied line by line by your opponent; and
3. The same information passed freely might not even be looked at.

Do not always trust information that is received freely and without responsibility for the accuracy of the information. Some information is deliberately planted or designed to mislead.

12.43 "NONNEGOTIABLE" DEMANDS

Nonnegotiable demands are so extreme that compromise seems impossible at first. The introduction of such demands creates hostility because they threaten important professional, economic, or even personal beliefs of the other party. Nonnegotiable demands may also inflame the other party so that deadlock becomes inevitable.

The contractor should not make a nonnegotiable demand until the contractor has fully evaluated:

1. The risk of a deadlock;
2. The likelihood of obtaining mutual acceptance;
3. Whether there is an opportunity or need for face-saving; and
4. The contractor's ability to get internal support for the position in the demand.

Some advantages commonly associated with the use of nonnegotiable demands are:

1. Extreme demands demonstrate conviction.
2. Nonnegotiable demands lower the expectations of the other party (see Section 12.9, "Aspiration Level").
3. The demand may make the other party more willing to compromise rather than risk a serious confrontation.
4. The nonnegotiable demand may be used as a rallying point for people in the contractor's own organization, while defusing the opposition.

If there are people in the other party's organization who believe that the basis for the demand has at least partial merit, the decision maker must evaluate the risk of a deadlock as well as his or her ability to gain support for a position, before rejecting a nonnegotiable demand.

12.44 OBJECTIONS

Some change order proposals will be met with objections. The objections will involve either some or all of the proposal. The contractor's ability to effectively respond to the other party's objections will determine whether the proposal will be accepted in the manner presented by the contractor.

Set forth below are several steps which the contractor can utilize to increase his or her success rate:

1. Before submitting, or even preparing, the proposal, write down every objection that might be raised against the change order proposal or its requirement;
2. Know all the answers to the anticipated objections;
3. Go through a practice session or "dry run" with other persons in the organization raising objections and practice answering the objections as though they are being raised by the other party;
4. Be sure to understand an objection completely before answering;
5. Rephrase the objection into a question which requires a "yes" response from the other party to confirm the correct understanding of the objection;
6. Do not reinforce the objection by agreeing with it before answering ("Yes, but....").

12.45 OFF-THE-RECORD DISCUSSIONS

Formal discussions serve a propaganda role. Individuals have to cover their actions and show others in their organization that they are fighting hard.

Off-the-record discussions can be used to get at the real issues and problems. It is a legitimate form of communication that allows both parties to discuss person-to-person ideas and to complain about their own organizational constraints. Informal, off-the-record discussions permit:

1. Problems to be resolved;
2. Assumptions to be tested;
3. Alternatives or other options to be discusses or explored;
4. Assessments of integrity to be made;
5. Informal and unofficial leaders of the negotiation to act without upsetting the official relationships and procedures.

Some of the dangers associated with off-the-record discussions include:

1. People sometimes make a record or try to utilize the off-the-record discussion at some point in the future; and
2. Some "confessional" sessions may be designed to mislead (see Section 12.42, "Lost Notes") or to plant false information.

12.46 PATIENCE

Do not give in too quickly. Take time to evaluate the proposed deal and other options. Hasty evaluations and decisions often result in bad judgments and unfavorable deals. Hold out for the terms requested in the change order proposal. In many instances, patience will:

1. Divide the other party's organization;
2. Lower the other party's expectations and aspirations;
3. Get other people involved; and
4. Force a reevaluation of priorities.

Patience allows the contractor to collect and evaluate all viewpoints before settling. It also gives the other party time to get used to new ideas (see Section 12.2, "Acceptance Time").

12.47 PERSONAL ATTACK

Be careful. People who are attacked on a personal level try to get even. They have a greater need to defend, and will always put up more resistance. Personal attacks seldom, if ever, help the contractor's cause. Calling a man a liar will not make him honest, but it may give him the impetus he needs to justify himself to his own organization.

When faced with a personal attack, walk out. Protest as loudly as possible to everybody who will listen. The amount of personal abuse that a man or woman is subjected to is equal to the amount that he or she is willing to take.

Do not tolerate personal attacks or abusive tactics. Most construction projects present the parties with enough challenges without the problems created by personal attacks. The contractor and the members of his or her organization should try to conduct themselves with professionalism at all times. "Keep an eye on the ball." Will the personal attack move the ball forward? In most instances, it will not.

12.48 PRESENTATIONS

Pictures can be worth more than thousands of words. They can also be worth thousands of dollars. Do not underestimate the power of a professional presentation. The people who criticize the time and expense of putting on a rehearsed and effective road show are usually the first ones to be the most dramatically swayed by it.

Effective presentations arrange complicated issues into logical and easier-to-understand format. Well thought-out, clean, bright graphics that facilitate understanding are truly appreciated and thankfully received by persons who are asked to understand and make decisions involving complex interrelationships. In many situations, easy-to-understand graphics become the focal point of future discussions. Whether they realize it or not, each time that the other party refers to the contractor's presentation materials when making points, he or she psychologically buys into the contractor's position. It is like the door-to-door sales rep whose own presentation is a series of questions designed to get repeated "Yes" answers.

Do not sell this technique short. The largest corporations understand the power of good presentations as an effective tool to sell products and raise money. Savvy contractors incorporate this technique into their business procedures. With the reduction in the size and cost of laptop computers, the PowerPoint presentation is becoming a part of many contractor presentations. If a contractor does not have the ability to prepare presentation graphics in-house, the contractor may want to find a consultant who can provide the needed assistance.

12.49 PROCEED ORDERS

The contractor needs to know only three things about Proceed Orders and Construction Change Directives:

1. They are risky;
2. They are Risky; and
3. They are RISKY.

A Proceed Order is a change order that the owner just does not want to recognize, finalize or accept. In some situations, the Proceed Order may reflect that a change order is warranted, but that the owner does not accept the contractor's price or schedule analysis. In other instances, the Proceed Order is issued because the owner wants the work to proceed regardless of any dispute regarding entitlement, price or impact. In essence, the Proceed Order is a directive to proceed with the work without an agreement that any time extension is in order or that the contract price will be adjusted upon the completion of the work. If the owner will allow the contractor to proceed with the work on a time and material basis, the risk to the contractor is minimized. However, even under the time and material arrangement, the precise method for determining the final price and time should be established. After the basis for adjusting the contract time and price is determined, the only remaining item is to actually go out, do the work, and add it up.

In contrast, if a complete change order cannot be executed before the commencement of the additional work, and the Proceed Order is open-ended, the contractor may be assuming a significant amount of risk. If there are unresolved issues surrounding the changed work when the work is begun, the contractor will lose leverage with the owner for the resolution of these issues as the work is completed. Stated simply, the contractor will be proceeding with the work as if he or she were shooting craps in a casino.

12.50 PROMISES

A promise is a concession with a discount rate attached. Some promises are not worth anything, and some are worth only a portion of their face value. But a promise is better than nothing.

If the contractor is unable to get a signature, the contractor should get a promise. Promises must be credible. Be sure to ask, "If the promise is so good, what is the problem with securing a signature?" If the explanation is not plausible, do not proceed. Oral promises are difficult to enforce. When relying on any promise, be sure to implement a monitoring mechanism to follow up on the promise or to be certain that the other party is complying with the terms of the promise.

Thoroughly document all promises and confirm all significant promises and understandings in writing. Notes should include names, dates, places, times, and so on.

12.51 QUESTIONS

The most direct route to understanding the other party's position is a good question. Unfortunately, most people usually think of the really good questions in the car on the way back to the office. Set forth below are some helpful guidelines with regard to question-and-answer sessions:

1. Prepare the questions in advance.
2. Ask questions that pry into the other person's affairs.
3. Ask questions innocently and encourage good answers.
4. Vary the type of question (i.e., open-ended questions when eliciting information or trying to obtain an explanation and pointed, specific inquiries when trying to pin down a specific issue).
5. Ask questions even if the questions may be evaded.
6. Ask a few questions to which the answers are already known to calibrate the accuracy and credibility of the other party and the other answers.
7. Do not ask antagonistic questions unless one of the goals of the discussion is to start a fight.
8. Do not cancel a teammate's question by asking another question before his or her question has been answered.
9. Do not pick any time to ask a question. Try to pick the right time.
10. Do not ask questions that question the honesty of the other person.
11. Do not stop listening in anticipation of asking a question. Write the question down and wait for the next break.

If the other party provides an evasive or incomplete response to a direct question, the response may be a useful indicator of the status of the other party's position. The contractor should make arrangements to follow up on all incomplete responses. Be certain to understand all responses before ending the discussion.

12.52 QUICK DEALS

Quick deals are generally foolish. They may be interpreted as a sign of bad judgment. Quick deals can, however, be advantageous if the contractor is fully and properly prepared and the other person is not. When the contractor's team is fully prepared and familiar with each component in the change order proposal, a short negotiation will be a "quick deal" for the other person, but not for the contractor.

12.53 REASONABLE REVIEW

The word "reasonable" is the epitaph on too many contractors' tombstones. Contract specifications are riddled with the imprecision attached to the term. When discussing any issue that incorporates the word "reasonable," make every effort to define the issue in precise terms as soon as possible. If, for example, the specifications indicate that the architect will review and approve shop drawings "with reasonable promptness," use the first job meeting to define "reasonable promptness" as "ten working days." Apply the same idea in every situation in which the word is encountered. Nail down this vague reference with a quantitative description whenever possible.

12.54 REOPENING CHANGE PROPOSALS

Change orders can be reopened if either:

1. Some condition of the proposal or its acceptance has been violated; or
2. Conditions arise that were either nonexistent or not apparent at the time of the change.

Beyond these circumstances, the ability to reopen a change order is the primary objective of the qualifications, terms and conditions of the proposal. Chapter 8, Designing and Constructing Effective Change Order Proposals, treats this subject in detail.

The fact that a change order has been signed should not in and of itself preclude a contractor from requesting that the change order be reopened when confronted with an unanticipated condition. The contractor must be keenly aware of the facts and circumstances surrounding the terms of each change order. The contractor may have a variety of options if unanticipated issues arise.

12.55 SPLIT THE DIFFERENCE

Splitting the difference can be a fast way to reach an agreement. People are accustomed to getting and giving equal shares in their personal lives, so splitting in the middle is simple.

In the case of change order proposals, splitting the difference may be equal, but it may not be equitable. The next time the other party says "Let's split the difference," try saying no. In many instances, the contractor will end up with more than half.

12.56 STATISTICS

Numbers are a subtle source of power. People who assemble statistics can control decisions. Even with the best of intentions, errors can occur. A good estimator or cost analyst can come up with almost any number that he or she wants, and even be able to back it up to some extent.

A major problem with statistics is that people have difficulty separating facts from assumptions. There is also a significant problem legitimizing the sources behind the "facts." Statistics hypnotize. Below the surface are incorrect "facts," correct facts, interpretations, assumptions, value judgments, and a few mistakes.

Be skeptical of statistics. Dig in and get as much information as possible about how the statistics were prepared.

12.57 TELEPHONE NEGOTIATIONS

Do not negotiate over the phone if at all possible. If a telephone negotiation is necessary, be entirely prepared.

Some of the most common problems with telephone negotiations are:

1. The caller has the advantage of surprise.
2. The person called is disorganized and not prepared for the negotiation.
3. Phone calls usually come when the recipient is busy or has other things on his or her mind.
4. Proof cannot be given or checked.
5. You cannot see the other person's reactions.
6. It is easier to misunderstand the other person's position.
7. Presentation materials cannot be used.
8. There may not be not enough time to think.
9. It is easier for the other party to say "no."

Set forth below are several helpful rules with regard to telephone negotiations:

1. When called, get the full story. Do not make a decision. Call back after the issue has been reviewed.
2. The less that you talk, the more the other person will talk.
3. Take detailed notes.
4. Restate any agreements promptly in your own words.
5. Confirm all agreements and significant statements promptly with a written letter or memorandum.
6. Have an excuse handy to end the call.
7. Make a checklist to avoid omissions.
8. Do not conclude a telephone negotiation unless the issue is completely understood and an acceptable position has already been prepared.
9. If after thinking about it the telephone deal begins to look bad or an error is discovered, immediately call back.

12.58 PLAIN HARD WORK

The entire change order negotiation process, from recognition of the change through the negotiation and execution of the final contract change, takes a lot of hard work. The person who is willing to work the hardest will usually control the outcome. The opinions of the person who has put in the work necessary to thoroughly understand the issues will be given greater weight. When properly prepared for a negotiation, it is easier to make effective decisions. The contractor who has done his or her homework will exude confidence because he or she knows that all loose ends have been tied up. Even in the face of resistance from the owner or the owner's representatives, the properly prepared contractor's confidence will not be shaken.

12.59 UNIT PRICES

This idea was first introduced in Section 8.3.5, "Presenting Change Order Components." When constructing a change order proposal, break each activity down into its component activities and parts. Work out a separate price for each activity. The more detailed the breakdown, the greater the chances of maximizing the bottom line of the change proposal. The greater the number of items, and the smaller the amount for each individual item, the less argument will be generated.

After the individual component amounts are accepted, add overhead and profit. Be sure to consider all direct, indirect, and consequential costs.

12.60 VALUE OF WORK PERFORMED

Courts have rendered a variety of decisions regarding the value of work performed. In certain instances, the courts have held that the contractor's costs will be used to determine the value of the work, even if another contractor could have performed the work for less money. In other situations, the cost of work is not the only measure of the value of the work. If a contractor realizes a special benefit, he or she may be entitled to retain it.

12.61 CONCLUSION

Do not hesitate to use novel, detailed, and unique approaches to negotiate change orders. Contractors must strive to get everything that they are entitled to on each and every change. Get inside the minds of the owners. Make them feel the difficulties associated with a particular change. Demand to be heard. Everyone admires a fighter, especially when his or her cause is right. Look to find the person on the other side who will consider the position set forth in the change order proposal and who will act to do what he or she can to correct an unjust situation. Persist.

13

Preparing for Winning When Changes Become Claims

13.1	Introduction	347
13.2	Turning Around Change Order Rejections	348
	13.2.1 Introduction	348
	13.2.2 "Good" or "Bad" Faith Rejections	349
	13.2.3 The "Nothing to Lose" Attitude	350
	13.2.4 Change Amount vs. Litigation Expense	351
	13.2.5 Meetings at the Highest Levels	351
	13.2.6 CHECKLIST for Meetings at the Highest Levels	352
	13.2.7 Sample LETTER TO THE OWNER Confirming a Special Meeting	353
13.3	Arbitration/Litigation/Mediation—What Is the Difference?	355
	13.3.1 Introduction	355
	13.3.2 Arbitration	355
	13.3.3 Litigation	358
	13.3.4 Mediation	360
	13.3.5 Conclusion	362
13.4	Finding an Attorney	362
	13.4.1 Introduction	362
	13.4.2 Characteristics of the Lion	363
	13.4.3 Characteristics of the Pussycat	363
	13.4.4 How to Find Your Lion	363
13.5	Selecting Consultants	364
13.6	Construction Claims CHECKLIST	366

13.1 INTRODUCTION

Reducing the number of change order rejections is another process that is subject to systematic analysis. The specific reason (or reasons) for a high change order rejection rate will vary between contractors, but the fundamental procedure for turning around this situation and improving the acceptance rate will be very similar for most contractors. This procedure may be summarized as follows:

1. Draft complete construction contracts, clear of legal pitfalls.
2. Maintain complete documentation.
3. Establish a mechanism for organizing and presenting the documentation.
4. Be completely aware of all available dispute resolution options and procedures.
5. Take appropriate dispute resolution action *in time*.
6. Be prepared to move decisively to the next step in the dispute resolution process.

Construction Contracts

Minimizing the number of change order rejections begins with the anticipation of difficulties and/or problems before the contract is drafted and executed. Good contracts anticipate as many difficulties as possible. They go on to define dispute resolution options and procedures with absolute precision. The contract language is clear and devoid of cryptic legal pitfalls. This is probably the most important step in either pursuing a contractual remedy or avoiding the need to pursue one.

Complete Documentation

At this point, the true value of complete records should be very clear. The contractor must have profound respect for the importance of comprehensive documentation. The contractor must educate each member of the project team on the importance of maintaining proper project records. Any project manager, superintendent and/or foreman that fails to maintain proper records does his or her company a serious disservice. The saying, "You don't have to be right, you just have to have good records," rings with more truth than it should.

Mechanism for Organizing and Presenting Documentation

The procedures and practices regarding the organization and presentation of project documentation were discussed extensively in Chapter 10, Using Project Records to Discover, Define, Support, and Track Change Orders and Claims, and Chapter 11, Keeping Change Orders Under Control. If given the attention that they deserve, the contractor will be able to proceed with the confidence that his or her records are complete and well-substantiated. The effective presentation of the relevant project records as part of the contractor's change order proposal will facilitate a complete understanding in the minds of everyone who must consider the proposal. This is the first real step toward the favorable resolution of the proposal.

Awareness of Resolution Options

Determine all options available, along with the probabilities for success and degrees of acceptability of each option. Thereafter, the contractor can weigh each option against the other and assess the amount of time, trouble, and expense that will accompany each option. After completing this analysis, the contractor will be in a better position to present the owner with the available alternatives, and the owner's approval of the desired option will become more probable (see Chapter 12, Winning in Change Order Negotiation).

Take Appropriate Action in Time

Knowledge of the contract terms, effective records, and an accurate response schedule or log will allow the fast, correct determination of all required response dates. The response schedule or log should identify the contractor's notice and response requirements, as well as the response requirements for the architect and the owner. The contractor's awareness of all applicable notice and response requirements, as well as his or her control over the pressures that time can create, will force the owner, architect and engineer to make decisions in a timely manner and stop procrastination.

Move Decisively

Proper, efficient documentation combined with a complete understanding of change order fundamentals and contract relationships will allow precise notifications. Statements requiring action and the response dates will be substantiated and well-documented. Follow-up on all issues will become a matter of strict procedure and, as the contractor moves early on all follow-up actions, the contractor's initial statements will be taken seriously. The contractor will be the tail that wags the dog.

13.2 TURNING AROUND CHANGE ORDER REJECTIONS

13.2.1 Introduction

Chapter 12 outlined many strategies and tactics to consider in specific change negotiations. Much of the chapter was devoted to presenting change order proposals and reversing change order rejections. This section offers additional insight into some of the specific reasons why, despite all obvious logic and support, change proposals are returned in a piecemeal fashion, or with a total or partial rejection.

The process of handling change order rejections can be summarized as follows:

1. *Try to develop a complete and correct understanding of the personal and professional motivations of the other party.* As soon as this can be accomplished, the contractor will start moving in the correct direction.

2. *Determine the actions that the contractor can take to satisfy the motivations identified in step 1.* Pay particular attention to the *personal* motivations. The probability that personal motivations take precedence over professional motivations increases with each higher level in a bureaucracy.

3. *Strategize the scenario.* Weigh each possible action identified in step 2 against the actual benefit to be realized *if* the intended objective is attained. Assess and evaluate the time and cost associated with each action as well as the probability of success.

4. *Be prepared to move forward quickly into the next step in the dispute resolution process if the rejection is not resolved favorably.* Know the next step in advance. Demonstrate commitment and resolve in all communications and actions. The contractor should be clear about his or her intentions to follow through, and should follow through when necessary.

The balance of this section will be devoted to exploring the attitudes and motivations introduced in step 1. From there, refer back to Chapter 12, Winning in Change Order Negotiation, for a discussion of the various tactics and strategies to be employed to move issues toward favorable results.

13.2.2 "Good" or "Bad" Faith Rejections

Just because an owner or its agents are tough and unyielding does not mean they are unreasonable and/or unprofessional. Construction contracts, with all their interdependent relationships, are extremely complex. This is true not only with regard to the many explicit terms contained in the contract, but also with regard to the rights and obligations which are implied in the contract.

The first question to consider upon receipt of a change order rejection is simply whether or not the rejection was made in good or bad faith. In other words, was the rejection squarely based on a provision in the contract or was the rejection baseless? If the rejection originated from at least an attempt on the owner's part to responsibly evaluate the contract before issuing the response, the rejection may be deemed to have been issued in "good faith" and the contractor's response can likewise be based upon a clear application of those portions of the contract that support the contractor's position. Thereafter, the contractor should focus his or her attention on arguments that will convince the owner that the contractor's interpretation of the contract is the proper one.

If, on the other hand, the evaluation of the owner's rejection does not reveal a reasonable contractual basis for the rejection, it may really be the result of any number of other factors. These factors include, but are not necessarily limited to, the following:

1. Incompetent or incomplete review of the proposal;
2. Budgetary or other financial constraints; or
3. Personal agendas.

The contractor must carefully review any response that explains the rejection and must try to ascertain and understand the motivations of the other party before crafting a response to the rejection.

13.2.3 The "Nothing to Lose" Attitude

Picture a budget-bound bureaucrat who is faced with a situation in which the operating costs for a project come out of his or her budget, but litigation expenses come out of the attorney general's office or some other department. A contractor presents this budget-minded bureaucrat with a large, complicated change order which is supported by very bulky documentation.

Assume for the purposes of this example that the contractor:

1. Is right;
2. Is entitled to at least a substantial portion of the proposal amount;
3. Is efficient and that the actual costs are difficult to criticize;
4. Has complete documentation regarding the change.

In this situation, the budget-minded bureaucrat has three options:

1. Pay the change order now;
2. Pay the easy-to-justify portions of the change order and reject the more marginal or questionable portions; or
3. Flatly reject the entire change order proposal.

If the bureaucrat selects the first option and accepts the change order now, all of the money comes out of the bureaucrat's project budget now.

In the second option, the bulk of the change order is accepted, and the approved amount is deducted from the bureaucrat's budget. If the rejected portion of the change roughly approximates the cost of fighting the bureaucrat's action, the contractor is presented with a very difficult decision. It is a no-win situation for the contractor in terms of costs. However, there is a matter of principle in not letting the bureaucrat get away with an improper rejection. In this situation, the bureaucrat knows that there is a very good chance that he or she will wind up with a bargain. Even if this test of the contractor's resolve proves that the contractor has decided to play his or her hand, a move toward a compromise settlement will again, in all probability, leave the bureaucrat paying less than the face value of the change order.

The third situation may not appear to be a viable option if the bureaucrat is convinced that the contractor is entitled to payment. However, if the bureaucrat does not find acting in bad faith distasteful and the bureaucrat's organizational structure does not monitor his or her activities closely, the bureaucrat has got nothing to lose. If the bureaucrat can fabricate any reason at all for rejecting the proposal, the bureaucrat will kick into gear the dispute resolution procedure included in the contract. When that process fails to give the contractor satisfaction, the contractor will be compelled to begin arbitration or litigation proceedings. If the bureaucrat has learned to work well within the bureaucracy, the bureaucrat may be able to pull this off while convincing his or her own organization that the bureaucrat is really fighting hard to "prevent an injustice." The bottom line is that the bureaucrat has preserved his or her budget.

What makes this gross injustice even worse is the idea that every case that goes to arbitration or litigation, no matter how well founded, carries with it some risk that the contractor will lose (or at least lose a portion of the claim). If this should happen, not only will the contractor be out the legal expenses, but the bureaucrat who made it happen will become a hero. The next time a similar situation comes up, the bureaucrat will be able to pull it off with the complete support of the entire bureaucracy.

Admittedly, the foregoing example may be extreme. However, as discussed in the next section, an unfounded rejection may be a tactic to secure some concessions before correcting the improper action. Continually assess the other party's situation and disposition, to allow the prompt identification of these kinds of potential circumstances.

13.2.4 Change Amount vs. Litigation Expense

This idea is closely related to the "nothing to lose" attitude described in Section 13.2.3. If a change is relatively small, or if a part of a change that is difficult to resolve is proportionately small, a flat, unfounded rejection may be utilized by the owner as a negotiating tool. Upon analysis, it may become apparent that the cost of fighting the unwarranted action will meet or exceed the amount of the unfounded rejection. This situation is more common than most people in the construction industry would like to admit. In deciding this issue, the contractor must weigh the cost of the change against the cost associated with the dispute resolution procedure. Additionally, the contractor must evaluate the long-term implications of the decision on the relationship with the owner.

13.2.5 Meetings at the Highest Levels

As disputes escalate, or as the amounts in question increase, the contractor (or the owner) may ask for a meeting between persons at higher levels in the two organizations. Just because a meeting is requested and scheduled with a high-level executive, director or commissioner, the contractor should not assume that the other party's position will change or that the contractor will be treated with consideration.

These types of meetings must be taken seriously. They are important because they afford the contractor the opportunity to make his or her presentation to a new audience. The presentation is also made directly to the decision maker, rather than through "the filters" in the decision maker's organization.

Notwithstanding the foregoing, some of these meetings are simply necessary procedural steps, without substance. Recommendations by the decision-maker's staff may cause the decision maker to be predisposed to pursue a particular course, regardless of the contractor's presentation.

13.2.6 Checklist for Meetings at the Highest Levels

As set forth in the previous section, meetings with decision makers at the highest level in the other party's organization may be required by the contract and can be very useful in the dispute resolution process. Sometimes, these types of meetings are the last available means to resolve a dispute before arbitration or mediation.

In addition to being properly prepared for the meeting, there are several items which the contractor should try to confirm as part of the proposed meeting. A checklist of these items is set forth below:

1. *Meet at the site.* In most instances, the decision maker will not be amenable to coming to the contractor's office. The contractor must endeavor to get the decision maker out of his or her office. Find as many reasons as possible to support a request that the meeting *must* be held on the site. After all, the site is where all the pieces to the problem are out in the open and available for inspection.

2. *Confirm all meeting parameters.* Immediately upon the agreement to meet, send a confirming message, documenting date, time, place, and attendees. It will become an important document if the decision maker does not show up.

3. *Establish a complete agenda.* As part of step 2, confirm the major items to be included on the agenda. This will remove an excuse on the other party's part that they do not have the people available or information prepared to discuss an issue. If that kind of problem comes up in the meeting, the contractor will have every right to call foul.

4. *Confirm expected attendance.* Establish those individuals who are required to be present if an agreement is to be reached. Do not give the other party the opportunity to turn the meeting into one more instance of going through the motions. If the meeting is to address an engineering problem, get the engineer there. Confirm the expected attendance, and send a copy of the confirmation to all required attendees. In some instances, the contractor may want to call each of the attendees before the meeting to briefly discuss the meeting, as well.

After the meeting, the contractor should be sure to *issue meeting minutes.* The contractor should issue these minutes to confirm his or her understanding of the results of the meeting. The minutes should be issued by the contractor even if the contractor's issuance of the minutes is outside the procedure for the issuance of normal job meeting minutes. These meetings are special and specific attention to the minutes is justified.

The contractor should note that the rules governing meetings at which legal counsel are present may be slightly different. If the parties schedule the

DISPUTE RESOLUTION

meeting in an attempt to resolve a dispute prior to litigation, counsel may require that the meeting be held under the rules governing settlement discussions. Discuss this issue with your attorney. Follow the guidelines in Section 10.7, "Using Job Meetings to Establish Dates, Scopes, and Responsibilities," to be certain that this is accomplished as effectively as possible.

Section 13.2.7, "Sample Letter to the Owner Confirming a Special Meeting," incorporates the foregoing suggestions.

13.2.7 Sample Letter to the Owner Confirming a Special Meeting

The Sample Letter to the Owner Confirming a Special Meeting is an example of a confirming letter designed to follow through on the recommendations in Section 13.2.6. Specifically, the letter:

1. Confirms the meeting parameters of date, place, and time.
2. Establishes a complete meeting agenda.
3. Lists the names of all the persons who are expected to attend.
4. Notifies all expected attendees by copy of the letter.

It communicates that the meeting will not be an impromptu discussion, but a serious effort staged to resolve important conflicts.

SAMPLE LETTER TO THE OWNER
CONFIRMING A SPECIAL MEETING

LETTERHEAD

(Date)

To: (Owner) RE: (Project Number)
 (Project Title)

 SUBJ: (Change file no.)
  (Change description)

Mr. (Ms.)   :

    Per our conversation this date, a special meeting will be held at the jobsite on (date) at (time) to resolve the subject change. Each issue on the following agenda will be reviewed:

1. (Description)
2. (Description)
3. (Description)

    The following individuals (insert the names of each person required) are required to be present to resolve these items.

    Thank you for your consideration.

Very truly yours,

Project Manager

cc: (List all those named in the letter)

13.3 ARBITRATION/LITIGATION/MEDIATION—WHAT IS THE DIFFERENCE?

13.3.1 Introduction

Construction disputes that are not resolved in the field or immediately following the completion of construction usually end up in one of three different dispute resolution procedures: arbitration, litigation, or mediation. There are important differences between the three procedures, and the contractor must be acutely aware of the advantages and disadvantages of each option. These advantages and disadvantages must be considered in light of the facts and circumstances surrounding a particular dispute, before selecting an option. In many cases, the contractor will not have an option because the dispute resolution procedure is specified in the contract documents. When the contractor is required to make a choice, the following general guidelines will help the contractor determine which option will best suit his or her situation. In any event, the contractor is strongly urged to consult with a competent construction attorney before making a final selection of the dispute resolution procedure.

13.3.2 Arbitration

Agreement to Arbitrate

Arbitration will not be a dispute resolution option unless there is a specific agreement between the parties to arbitrate. This agreement can be in the form of a provision in the contract documents or a separate agreement between the parties prepared after the dispute has matured to a point where the parties need to implement a more formal dispute resolution procedure. On balance, the second situation is very difficult to accomplish if one party determines that the cause lends itself best to arbitration in that the other party will likely see the disadvantage from his or her perspective. Since the rules of evidence do not need to be applied in arbitration proceedings, arbitration may favor one party over the other.

Informal

Arbitration proceedings are much more informal than is taking the dispute to court. Arbitrations are generally held in a conference room at any of a number of locations. Room arrangements, accommodations, and the proceedings themselves may be arranged and administered through the American Arbitration Association. However, many procedural matters are left to the arbitrator's discretion. In many instances, the meeting considerations such as location, time, agenda, and so on can be worked out at the convenience and availability of the parties and the arbitrator(s). Depending upon the arbitrator's disposition and background, the proceedings may be very structured and formal (i.e., a retired judge as arbitrator) or very informal and conversational (i.e., an architect, engineer, or contractor as arbitrator).

Arbitrators

There will be an arbitration panel of either one or three arbitrators. These people may have various construction-related backgrounds. They might be retired judges, attorneys, contractors, architects, engineers, or anyone else with a familiarity with at least a portion of the construction business. The arbitrators may be less familiar with the technical points of law, but their construction industry savvy can help to keep issues on relevant tracks. The background of the arbitrator may influence the structure and level of formality in the arbitration hearings. The arbitrator's background will also have an influence on the type of information presented by the parties as well as the presentation method.

Expense

Arbitration expenses can be substantial, although not in every case. On small, relatively simple, cases requiring only one or two days in arbitration, the arbitrator's fee may be included in the filing fee. If the arbitrator's fees are not included in the filing fee, the fees will be established by the arbitrators themselves, and will generally be set at levels roughly on a par with the local fee structures of other competent professionals. The fees will begin to accrue at the first prehearing conference with the arbitrator(s), and they can add up rather quickly.

Most organizations, such as the American Arbitration Association, charge a filing fee or an administration fee in addition to the arbitrator's hourly or daily fees. In some instances, the filing fee is based upon the amount in controversy. On large disputes, this fee can be significant.

Many contracts provide that the cost of the arbitrator and the fees should be awarded to the prevailing party.

If there is no allocation in the contract, the arbitrator(s) may award costs at his or her discretion. Please be aware that some contracts provide that the arbitrator's fee shall be split equally between the parties, regardless of the outcome of the hearing. Finally, if the arbitrator is swayed very strongly for or against the merits of either position, it may affect the manner in which the arbitration expenses get charged.

Time

Arbitration is generally faster than a court action. The filing procedure, arbitrator selection process, and calendar arrangements can proceed in as little as a month, although two to three months to complete the process is more common. This is dramatically faster than the court system, which may require years before a case can find its way onto a court calendar. Most arbitrations can be completed within six months of the filing date. However, the complexity of the issues presented, as well the calendars of the persons involved, may prolong the process. In any event, the time span between the filing date and the hearing date is usually significantly shorter than the time required for a court proceeding.

As a businessperson with the usual concerns over cash flow and carrying costs, the contractor should realize that anything that the contractor can do

DISPUTE RESOLUTION

to speed up the decision/award will in all but the most unusual cases be well worth the trouble. The speed associated with arbitration is, therefore, a primary consideration in any dispute resolution decision.

Privacy

Arbitration proceedings are not open to the public. The privacy of the parties is protected, and none of the documents are made a part of the court (public) record. If for any reason the contractor is concerned over public disclosure of private information, this may be an important factor to consider.

No Discovery

There is no "discovery" in arbitration unless the arbitration agreement allows for discovery. Discovery is a court process by which the parties to a proceeding have the right to:

1. Inspect and photocopy any and all business records of the other party, which records have some bearing on the issues presented in the proceeding;
2. Ask written questions (i.e., interrogatories and requests for admission) concerning assertions, contentions and facts at issue in the proceeding; and
3. Ask oral questions of persons involved with the project or knowledgeable about the contentions asserted in the proceeding (i.e., depositions).

The discovery process is designed around the principle of no surprises. It can be very useful when trying to evaluate the relative strengths and weaknesses of a particular case. It is an absolute necessity if events were not documented or records were not exchanged during construction, or if the other party's records contain information which is vital to a particular claim or defense. On projects where everything has been fully documented and documents have been exchanged during the course of the construction, the need for discovery is reduced.

Discovery is very expensive and time-consuming. Imagine the time required for each attorney to sift through each file, and multiply that time by his or her hourly rate, and it will be easy to recognize the amount of time and expense associated with the discovery process.

Inasmuch as many owners and contractors maintain elaborate paper trails surrounding all significant issues, the lack of the long, tedious, expensive discovery process will not leave either party at a real disadvantage. A complete review of the documentation that is already spread around will clarify almost all positions that your opponent is going to take in the proceeding. If surprises come up during the proceedings, your attorney should know the procedure for properly handling them on an item-by-item basis.

Evidence

The rules of evidence are much more relaxed in arbitration than in a court proceeding. The requirements for the introduction of evidence are much less stringent than the court's requirements, and the admissibility of evidence is again a matter of the discretion of the arbitrator(s). The admission of evidence by arbitrators is a sensitive issue, because the failure of an arbitrator to consider an important item can be a basis for appeal. Therefore, even if an item is clearly inadmissible under the rules of evidence, the arbitrator will allow the evidence to be admitted, but will give the evidence "its appropriate weight" when balancing the scales of justice and making his or her decision. The important point is that the suspect evidence (evidence that would not be admitted in a court) will be brought to the arbitrator's attention for his or her consideration.

Arbitration Award

After the conclusion of the arbitration hearings (and in some cases, the exchange of closing briefs), the arbitrator(s) makes an award. Once the award is rendered, it may be confirmed and enforced through the courts. The courts may also correct and/or vacate an award. However, there is a strong public policy in favor of arbitration and the making of awards final and conclusive. As such, there are only a few available grounds for vacating an award. These grounds include:

1. Corruption or fraud;
2. Prejudicial misconduct;
3. The arbitrator exceeded his or her powers; or
4. The arbitrator making the award was subject to disqualification.

13.3.3 Litigation

Requirement

If there is no agreement to arbitrate, either established in the contract or subsequently arranged between the parties, there is no option. The contractor must use the conventional legal system.

Formal Proceedings

Every court action is formal and rigidly structured. There is an elaborate set of rules that permits only technically correct claims to be filed. The technical aspects of a particular case may play a disproportionate part in the proceedings, and may significantly increase the time necessary to present the case and/or defense. Additionally, technical defenses and procedural tactics may eliminate and/or narrow certain claims before the claims ever get to the trier of fact. Like every other consideration in this section, this can work for or against one party or the other depending upon the circumstances of the particular case.

Trier of Fact

Judges and juries will be deciding the merits of the case. It will be rare if either one has any construction experience, but the contractor can count on the fact that both will be long on common sense and that they will be able to evaluate the credibility of witnesses. A serious disadvantage may be in their ability to grasp and believe important complicated concepts, such as off-site overhead expenses in a delay situation. However, their ability to see through the phony or the fake may outweigh this disadvantage.

Expense

Although the court filing fees are generally much less than the arbitration filing and arbitrators' fees, the other attributes can increase total costs substantially. For instance, court mandated conferences and procedures, congested court calendars and delays attributable to the technical rules of evidence will increase the cost of this dispute resolution procedure. Only an experienced attorney will be able to lay out the details of the projected costs of litigation.

Time

Court calendars are usually very congested. In many courts, cases are called for trial within twelve to twenty-four months after the filing of the complaint. However, some courts are more backlogged and the time differential is greater. In any event, the time period is many times longer than with arbitration, and this can lead to many times the total expense.

Open to the Public

Court complaints and all other filings are made a part of the court record and this record is open for inspection by the public. Further, the proceedings themselves are open to anyone who cares to drop in. This aspect of litigation may present a problem for certain litigants. For instance, private financial or proprietary information may need to be filed and discussed during the course of the litigation. This issue must be thoroughly reviewed with counsel throughout the course of the dispute resolution process. There are certain procedural devices that can be put into place to protect certain information.

Discovery

The discovery process was described under "Arbitration." As set forth in that section, discovery is designed to prevent surprises for either party. It is a process that is part of every litigation matter. The process can increase the costs attendant to litigation substantially. As mentioned previously, discovery may be necessary if the contractor's case is complicated and dependent upon information that is not already included in the contractor's own documentation.

Evidence

The strict rules of evidence in court are there (in theory) to protect a naive jury from considering items that are not credible or are not relevant to the case. The net effect of the evidentiary rules is that it is very time-consuming to introduce each new piece of evidence. The most basic assumptions and construction business idiosyncrasies must be described in fundamental detail. Each piece of evidence must be evaluated against its admissibility criteria and debated by the attorneys. Technically questionable items may never be presented to the trier of fact. If these materials are important to the contractor's case, the exclusion of these materials under the rules of evidence will have a serious detrimental effect on the case. As with each other consideration, however, the effect can work for the contractor if it is working against the other party.

Judgment

At the conclusion of the court hearings, the trier of fact will render a decision. If the case is submitted to a jury, the jury's decision is called a verdict. After the decision is rendered, either party can try to attack the decision. These attacks include:

1. A request to enter judgment notwithstanding the verdict;
2. A request to reduce the amount of the verdict; and
3. A request for a new trial.

If the decision is converted to a judgment, either party can file an appeal. On appeal, the reviewing court will confirm that the trial court made the proper legal rulings throughout the course of the trial and applied the law properly. The appellate court will not admit any further evidence. The appellate court may:

1. Deny the appeal and affirm the trial court's decision;
2. Remand the case to the trial court for further proceedings; or
3. Reverse the trial court's decision.

As the contractor can imagine, all of these procedures are expensive.

13.3.4 Mediation

As mentioned previously, the important distinction between mediation, on the one hand, and arbitration and litigation, on the other, is the fact that the mediator does not render a decision. The mediator tries to facilitate the resolution of the dispute through a series of discussions with the litigants.

Agreement to Mediate

Many construction contracts require the parties to participate in mediation as a condition precedent to the filing of an arbitration or a court action. Additionally, since mediation is a creature of contract, the parties may elect to proceed to mediation at any time during the arbitration or litigation process.

Proceedings

The American Arbitration Association, as well as a variety of public and private organizations, conducts mediation proceedings. The agreement between the parties will determine the structure of the mediation. In almost all cases, the mediation process consists of four basic steps:

1. The execution of the mediation agreement;
2. The selection of the mediator;

3. The mediation sessions; and
4. The documentation of any settlement reached during the mediation.

As with arbitration, the mediation may be scheduled to accommodate the schedules of the parties and their counsel.

Mediators

A large number of retired judges, practicing attorneys, and other construction professionals serve as mediators. The mediator should be a disinterested third party who has experience in the construction industry and/or the resolution of disputes.

Expense

The fees charged by mediators vary widely. In most instances, mediators set their own fees. When evaluating the mediator's fee, the contractor should consider, among other things, the mediator's success rate, availability, and experience in the construction industry. A substantial mediator's fee may be a wise investment if the mediator is able to resolve the dispute prior to the expenditure of significant litigation costs.

Time

If the time parameters for the submission of a dispute to mediation are not defined by contract, mediation may proceed as soon as the parties and the mediator are ready. In almost every instance, mediation may be scheduled and completed much faster than the other dispute resolution procedures.

Privacy

As in arbitration, the materials submitted to the mediator are kept confidential. They are not part of the public record and are not subject to public review.

Discovery

The availability of discovery will depend on the circumstances surrounding the submission of a dispute to mediation. If mediation is a condition precedent to arbitration or litigation, the parties will not have conducted any discovery. In the event mediation is not required by contract, but is being proposed as an alternative to arbitration or litigation, one party may request certain discovery as a condition to agreeing to mediation. Finally, if the mediation is conducted during the course of a court action, the parties may have, or may require, certain discovery before the commencement of the mediation sessions.

Evidence

Since the mediator does not have the power to make an award or to compel one side to adopt the views of the other, the rules of evidence do not apply to the mediation sessions. However, the mediator may use the rules of evidence to point out certain problems associated with presenting a particular claim or argument in court.

In most instances, all materials submitted to the mediator are deemed to be confidential and protected settlement communications. In other words, the materials prepared for the mediation may not be used against the party preparing the materials in any future proceeding if the mediation is unsuccessful. Most mediators will require the execution of a confidentiality agreement as part of the mediation process.

Award

Once again, the mediator does not render an award. In its most basic form, mediation consists of the use of a disinterested third party to achieve a mutually acceptable resolution of a particular dispute. If the parties reach an impasse during the course of the mediation sessions, the mediator may adjourn the mediation or continue the mediation sessions to another day. If the mediation sessions are unsuccessful, the parties will be free to pursue any other available dispute resolution procedure.

13.3.5 Conclusion

These guidelines are intended to provide a general awareness of some major considerations when evaluating the different dispute resolution options. The contractor is strongly urged to discuss these and other considerations with an *experienced* construction attorney before making any final decision in a particular case.

13.4 FINDING AN ATTORNEY

13.4.1 Introduction

Anyone who tries to get into a game in which he or she does not know the rules and cannot understand the procedures may be playing with less than a full deck. All commonsense understandings of the issues aside, the contract upon which the contractor has based his or her actions is riddled with both obvious and very subtle language. While the contractor's experiences and familiarities may make the implications of certain provisions very clear to the contractor, the hidden meanings place the interpretation and enforcement of most construction contracts through a formal dispute resolution procedure outside the realm of the attorney with a general business or civil practice. Beyond the contract language and its interpretations, there are very definite rules and formalities that must be observed with absolute precision to even bring a case to its first step. Only a competent, experienced legal professional with a construction law practice will be suited to complete the arrangements and proceed with the action.

In general, attorneys fall into several categories. The first distinction is litigation versus transactional or corporate attorneys. While transactional attorneys fulfill extremely important functions in our society, including the preparation and negotiation of contract documents and the identification and evaluation of options during construction, it is the litigating attorney who must take the lead in developing strategy and following through on the tough negotiations and ultimate court battles. The balance of this section therefore is devoted to describing and selecting a litigation attorney.

In general, litigation attorneys fall into two categories: Lions and Pussycats. Both have advantages and disadvantages, depending upon the contractor's organization and culture, the circumstances surrounding the dispute and the nature of the proposed legal services.

13.4.2 Characteristics of the Lion

The Lion is generally young (but not too young) and extremely aggressive. He or she enjoys the fight for the fight's sake and will work incessantly to find the new angle, discover the different approach, and find the key (but often overlooked) information that will sway a decision. The Lion figures that he or she is smart enough and mean enough to kill anything in the jungle. The Lion is not overly preoccupied with muscle-bound opponents and can, therefore, confine his or her attention to the tactics necessary to work around even the most formidable opponent.

The Lion will generally be found in a small or midsize law firm or in a small practice group in a large firm. The Lion has drive, initiative, and need for power and control. These traits should be evident during the initial discussion. He or she is a winner.

The Lion has at least a few battle scars, and knows the ropes. He or she is not someone just out of law school. The Lion's successful track record has resulted in a string of good contacts, from powerful clients to other professionals and consultants of similar high caliber.

13.4.3 Characteristics of the Pussycat

The Pussycat is generally a senior partner in a well-known firm who knows everybody. He or she has numerous contacts in the industry and may have the power and the influence to steer the contractor to the right person. If the contractor's organization is big enough, the Pussycat can be a powerful associate who can place a wealth of resources at the contractor's disposal. If, on the other hand, the contractor's organization is small and the dispute does not merit the interest of the Pussycat, the matter will be handed off and the contractor will end up dealing with a junior partner or a senior associate. The work will be done by paralegals and clerks. The Lion hired by the other party will eat the Pussycat's team alive.

13.4.4 How to Find Your Lion

Finding your Lion will not be easy, but the time and effort (and even expense) may very well be the best long-term investment the contractor make for his or her company.

1. *Ask respected professionals and consultants.* If you have been to court or arbitration, you may have seen an impressive witness, expert, or consultant. Those people's experiences along the way have probably exposed them

to competent and incompetent lawyers. Their suggestions may become an important starting point.

2. *Ask transactional attorneys.* Start with the attorney who provides business and/or corporate advice. If this attorney has been honest enough to acknowledge that construction law is a specialty that demands a specialist, the attorney's suggestions may be very valuable.

3. *Find business associates who have had successful experiences.* Their experiences may or may not have relevance to the subject dispute, but a fighter is a fighter. Get some names.

4. *Watch local elections.* If an attorney is running for office, he or she is a Lion. It might pay to give him or her a hand. If he or she loses, the contractor ends up with a good attorney. If he or she wins, the contractor has a powerful ally, plus a source of excellent recommendations for good attorneys.

5. *Observe performances.* If possible, go sit in the courtroom for a while and watch a prospective attorney in action. Observe the manner in which he conducts himself, leads the team, and the overall level of preparedness. See how well her actions are received and responded to by the court.

6. *Do **not** "ask around."* Do not casually ask around and expect to get quality information or recommendations. The recommendations will be for relatives, friends, and their friend's attorneys.

The time that the contractor spends finding the right person to represent his or her interests will almost certainly return benefits. Selecting the right attorney will increase the likelihood of success and may result in some significant cost savings.

13.5 SELECTING CONSULTANTS

Section 13.4, "Finding an Attorney," outlined some important considerations in selecting the individual who will lead the team through the dispute resolution process. Consultants can also play important roles not only in the dispute resolution process, but in each negotiation along the way. Further, the efficient use of consultants in a preventative approach to project management, before or during construction, may facilitate the resolution of potential disputes during the change order process. Effective analysis and convincing presentations of cause, effect, and consequences may provide the necessary catalyst to resolve an issue and prevent costly legal actions.

During legal actions, presentations can clarify and summarize an otherwise complicated and confusing chronology of events detailed in the project record. The stories, arguments, and counterpoints that have been spread out over a period of months can be tied together in a meaningful, clear presentation. If the consultant's presentation is well received, it will be depended upon by the trier of fact or the arbitrator at decision-making time.

Some areas in which consultants may be useful include:

1. Project management.
2. Construction scheduling.
3. Claims/delay analysis.
4. Technical design analysis.
5. Cost controls.
6. Productivity improvement.

Good consultants can uncover the real facts by conducting efficient, detailed reviews of all relevant project documents. They can reconstruct events, and then be able to demonstrate all resulting effects. The consultant should be familiar with client–attorney relationship and should be able to understand his or her role. The consultant can aid trial counsel in understanding the technical aspects of an issue, so that the attorney can prepare the issue effectively. In this capacity, the consultant can also aid the attorney in document review, document and evidence assessment, and witness preparation. Good consultants will also be able to prepare and present a variety of impressive graphic devices to deliver points with precision and power.

On the other side (in a defense or rebuttal situation), it will be the competent consultant who will best be able to refute technical evidence or any presentation by the other side. If your opponent, for example, presents a detailed, complicated construction schedule to show the effects of a delay, find a good schedule consultant who will be able to find all the flaws in it (and fast) so the attorney will be able to chip away quickly at the schedule's legitimacy and destroy the presentation.

Consulting fees are determined by the consultant's caliber and reputation, but also by the scope and extent of the work involved in a particular case. The extent of research necessary, the sessions required to coordinate the work with the attorney, and the hours necessary to prepare demonstrative evidence, presentations, and testimony will vary greatly. Fees, therefore, cannot be established until after the consultant knows more about the dispute and the extent of his or her involvement with these kinds of details.

When evaluating the retention of a consultant, take the time to review the details of the consultant's involvement with the trial attorney. Try to develop a scope of services and a budget for the consultant. In other words, know what services the consultant will be providing in the case, the benefits to be derived from these services and the estimated cost of the services. Ask trial counsel to weigh the value of certain services against the cost of the services, the benefits to be derived from the services and the potential risks associated with not performing the services.

After making the decision to retain the consultant, the contractor should monitor all early operations very closely. Be certain that everyone understands the concern for cost efficiency as related to the consultant's fee, and make them aware that close attention is being paid to the outputs.

13.6 CONSTRUCTION CLAIMS CHECKLIST

Because of the dramatic variety in extent and composition of construction claims, the manner in which such claims may be presented, the disposition of the contractor's own organization and that of the trial attorney, and a variety of other factors, an absolute, comprehensive checklist that will cover all situations is really not practical. However, the preparation of most claims have certain common characteristics. Set forth below is a listing of the most common steps in claim preparation:

1. Interview key project personnel.
2. Review documents and available data sources to determine what transpired on the project. Documentary data sources include such items as daily logs, meeting minutes, job diaries, photographs, correspondence, progress schedules, RFI's, change orders, and RFI, shop drawing, submittal and change order logs.
3. Review cost data sources to determine where the money was lost, during what period of time and for what reason. The cost data sources include monthly labor reports, monthly material cost reports, equipment cost reports and project expense and overhead reports.
4. Review the physical characteristics of the project and any and all constraints on the construction of the project.
5. Compare the final construction drawings used at bid times with the as-built drawings to confirm that all significant changes have been identified and submitted.
6. Compare the as-planned sequences and durations to the as-built sequences and durations to confirm that all major issues (changes and/or problems) which were encountered on the project have been considered and, if appropriate, submitted to the owner.
7. Review the contract documents and the change orders to be certain that the contractor has been fully compensated for all changes.

If the contractor is required to submit a formal claim or a request for an adjustment to the contract to the owner (or the contractor elects to submit such a claim), the claim document should contain the following sections:

1. An introductory section which contains a brief overview of the basis for the claim and the requested relief;
2. A summary of the contract requirements, including a review of the critical contract provisions;
3. A description of the individual claim items, which description should include, among other things, the basic facts involved with each claim, the rationale as to why the contractor should be compensated for the claim item, and a summary of the costs associated with the claim item;
4. A summary of the damages;

DISPUTE RESOLUTION

5. A brief legal analysis; and
6. The project documents which support the claim, as well as any demonstrative exhibits which explain the claim.

Inasmuch as each claim situation is somewhat unique and the circumstances surrounding the particular claims vary widely, the claim preparation procedure and the claim document can only be organized by the construction attorney and the consultant. The suggestions set forth and the Construction Claims Checklist presented in this section are designed to provide the contractor with an introduction to the kinds of documents and information that the attorney will require and of the categories of people who may become necessary along the way. The checklist should be reviewed with trial counsel to determine what and who is or is not appropriate for involvement in each particular claim or dispute. After all of the prearranged categories have been considered, the contractor and the attorney should continue the process during the same meeting to expand the list of remaining items and people that will need to be involved with the preparation of the claim document and the dispute resolution process. After this analysis has been completed, the contractor will have a good understanding of the real level of involvement of all persons named, and the impact of the dispute resolution process on the contractor's future operations. Additionally, the contractor will be in a position to evaluate the costs associated with the various dispute resolution procedures.

Everything listed on the checklist will not be necessary for every claim. However, the inclusion of the items on the form will continually bring them to the attention of the people in the contractor's organization in each new situation. If an item does not apply, simply cross it off the list.

SAMPLE CONSTRUCTION CLAIMS CHECKLIST

People necessary for information and/or testimony:
   Key project people

   Subcontractors involved: _____

   Project executive: _____
   Project manager: _____
   Project engineer: _____
   Project superintendent: _____
   Scheduler: _____
   Architect: _____
   Engineers: _____

   Experts/consultants: _____

   Lawyers: _____
   Accountants: _____
   Other: _____

Key documents (originals)
   Plans, specifications, addenda .... _____
   Related files .... _____
   Notice date establishment .... _____
   Pass-through reference .... _____
   Change clause .... _____
   Dispute clause .... _____
   Arbitration clause (agreement to arbitrate) .. _____
   Other: _____

Project records
   Contract/subcontract/P.O. .... _____
   Files: .... _____
   Subfiles: .... _____
   Change order files .... _____
   Change order summary log .... _____
   Daily field reports .... From: _____ To: _____
   Schedule: Baseline (Target) .... _____
            Updates .... _____
            Projections/analysis .... _____
            Narratives .... _____
   Job cost records .... _____
   Shop drawings and transmittals .... _____
   Shop drawing log .... _____
   Other: _____

Appendix
Sample Contract Change Order

The final contract change order form is the formal amendment to the original agreement. The considerations associated with processing a given change are generally the same whether the amendment is to the prime contract, subcontract, trade contract, or purchase order. Identical principles apply. A change order is a mini contract, and as such, the final change order document must be in writing and should include the following items:

1. A description of the work to be performed (or deleted) as part of the change;
2. The adjustment to the contract price or the methodology for calculating the adjustment (e.g., unit prices, cost plus a fee);
3. The adjustment to the contract schedule or time; and
4. Any special terms (e.g., limitations, assumptions or reservations) which relate to the change.

The following discussion focuses on the Change Order AIA Document G701, 2000 edition. Alternative forms are also included in this appendix to illustrate subtle differences in the manner in which certain objectives are accommodated.

The first portion of the change order document identifies the same parties named in the primary agreement (the original contract). As a matter of protocol, be certain that every detail of each name is duplicated *precisely*. ABC Co., for example, can be an entirely different entity from ABC Associates, ABC Limited Partnership, and so on. Carelessness in these designations may result in inconsistencies that will leave the contractor exposed to technical contract restrictions, should difficulties arise. Other forms may go one step farther by adding a more direct reference to the actual original contract (or subcontract).

The change order number is the formal contract amendment designation in chronological sequence. It can actually represent the consolidation of several distinct change order proposals that were developed and prepared individually. Chapter 11 details an efficient system for tracking multiple changes so that when they are grouped into a larger change to the contract, they can be controlled and properly identified.

The body of the change order document describes the work affected by the change. The key consideration again is the complete and correct identification of the individual items, in language that is identical to the documents referenced in the change or describing the work. It is not necessary to reiterate all the specific elements of the work and their specific conditions. It is, however, critical to accurately define the documents that include all of these components in precise detail. Consistency of the descriptions through all documents that make up a change order's paper trail is crucial. If the specifications, for example, refer to an item as a "rolling counter shutter," the change order documentation should not refer to it as an "overhead door."

The next portion of the form recapitulates all dollar changes to the original agreement. It summarizes the total changes to the contract through the date of the particular change. The tabulation applies whether it is a general contract, a construction management arrangement with a guaranteed maximum price, or any other type of construction agreement.

Beginning with the original contract price, it follows with a line that summarizes all changes that have preceded the particular change. It is a simple but powerful indicator of the current status of the project. Its size and direction (add or deduct) immediately displays the relative performance of the entire project, given all other considerations. This line is followed with a sum total of the original contract amount plus all changes to date. It represents the current contract amount prior to adding the effects of the present change. The next two lines add the amount of the present change order and calculate the new contract price, inclusive of the change.

It is easy to stop here, as many change order forms do. However, it is important to understand that all change orders affect both dollars and *time*. For this reason, there is a category included on the AIA form that addresses the contract time. If appropriate, the change order should identify any adjustment (add or deduct) to the original contract time. The AIA form recalculates the actual revised date of substantial completion that results from the change. In this way, the change order becomes a clean summary that removes all misunderstandings.

If change orders are prepared in this manner, billing for changes becomes a simple matter of adding a line item on the general requisition that references the change order number and corresponding price. Further descriptions, breakdowns, and so on are unnecessary.

In addition to identifying the changes to the scope of work, contract price and completion date, the AIA form includes a notation that the change order does not include any adjustment which is the subject of a Construction Change Directive (Proceed Order). Any adjustment to the contract or contract time for the work described in a Construction Change Directive is subject to some final determination in the future. (Remember the discussion in Section 12.49, Proceed Orders are risky, risky and risky.)

The AIA form stops here. Other forms might go on to include language to the effect that all other terms and conditions to the original agreement remain in effect, and apply to the work of the change. As mentioned above, if the change order is subject to certain special conditions, the conditions must be spelled out on the change order to avoid misunderstandings and preserve important contractual rights.

Whatever the final form, it is a good idea to incorporate the form itself (or at least refer to it) in the change clause (refer to Section 3.4.2) of the original construction agreement. If done this way, there will be no question as to its appropriateness when its use becomes necessary. Extraneous references to prior contracts then become redundant and can be eliminated.

Finally, it is important to understand that the final change order document is a portion of a complete contract. It is intended to represent the full understanding of the parties. As such, the rules discussed in Chapter 3, Proven Strategies for Applying Construction Contracts, apply to the change order as if it were a stand-alone agreement. In other words, all prior negotiations, discussions, (tentative) agreements, and so on are superseded by the final written form. As such, it is incumbent upon the contractor to be certain that the written agreement incorporates all these understandings clearly and completely. The contractor must be certain that the final form, with all its references, does not include ambiguities that are subject to misinterpretation, and that all potential change order effects have been directly addressed.

CHANGE ORDER
AIA DOCUMENT G701

Distribution to:
OWNER ☐
ARCHITECT ☐
CONTRACTOR ☐
FIELD ☐
OTHER ☐

PROJECT:
(name, address)

TO (Contractor):

CHANGE ORDER NUMBER:

INITIATION DATE:

ARCHITECT'S PROJECT NO:

CONTRACT FOR:

CONTRACT DATE:

You are directed to make the following changes in this Contract.

> AIA copyrighted material has been reproduced with the permission of The American Institute of Architects under permission number 86038. Further reproduction is prohibited.

> This document is intended for use as a "consumable" (consumables are further defined by Senate Report No. 94-473 on the Copyright Act of 1976). This document is not intended to be used as model language (language taken from an existing document and incorporated, without attribution, into a newly created document), but is a standard form intended to be modified by separate amendment sheets or by filling in of blank spaces.

> Because AIA Documents are revised from time to time, users should ascertain from the AIA the current edition of the Document reproduced herein.

> Copies of this AIA Document may be purchased from The American Institute of Architects or its local distributors.

SAMPLE

Not valid until signed by both the Owner and Architect.
Signature of the Contractor indicates his agreement herewith, including any adjustment in the Contract Sum or Contract Time.

The original (Contract Sum) (Guaranteed Maximum Cost) was .... $
Net change by previously authorized Change Orders .... $
The (Contract Sum) (Guaranteed Maximum Cost) prior to this Change Order was .... $
The (Contract Sum) (Guaranteed Maximum Cost) will be (increased) (decreased) (unchanged)
   by this Change Order .... $
The new (Contract Sum) (Guaranteed Maximum Cost) including this Change Order will be ... $
The Contract Time will be (increased) (decreased) (unchanged) by   ( ) Days.
The Date of Substantial Completion as of the date of this Change Order therefore is

Authorized:

ARCHITECT	CONTRACTOR	OWNER
Address	Address	Address
BY	BY	BY
DATE	DATE	DATE

AIA DOCUMENT G701 • CHANGE ORDER • APRIL 1978 EDITION • AIA® • © 1978
THE AMERICAN INSTITUTE OF ARCHITECTS, 1735 NEW YORK AVE., N.W., WASHINGTON, D.C. 20006

G701 — 1978

DISPUTE RESOLUTION

CHANGE ORDER
AIA DOCUMENT G701

Distribution to:
OWNER ☐
ARCHITECT ☐
CONTRACTOR ☐
FIELD ☐
OTHER ☐

PROJECT: CIVITELLO TOWERS
(name, address) 888 88TH STREET
HAMDEN, CT. 06510

TO (Contractor):

QUALITY CONTRACTORS, INC.
870 WINTERGREEN AVE.
BETHANY, CT. 06525

CHANGE ORDER NUMBER: 2

INITIATION DATE: 3/24/87

ARCHITECT'S PROJECT NO:

CONTRACT FOR:

CONTRACT DATE: 8/22/86

You are directed to make the following changes in this Contract:

1. Add steel bracing per Quality Contractors, Inc.   ADD $10,000.00
   Proposal No. 890-2 dated 2-22-87

2. Add marble at entrance vestibule per Quality   ADD $19,200.00
   Contractors, Inc. Proposal No. 890-6
   dated 3-20-87, as corrected by the
   owner on 3-22-86

                                                     TOTAL $29,300.00

Not valid until signed by both the Owner and Architect.
Signature of the Contractor indicates his agreement herewith, including any adjustment in the Contract Sum or Contract Time.

The original (Contract Sum) (Guaranteed Maximum Cost) was .... $ 2,620,000.00
Net change by previously authorized Change Orders .... $ 71,600.00
The (Contract Sum) (Guaranteed Maximum Cost) prior to this Change Order was .... $ 2,691,600.00
The (Contract Sum) (Guaranteed Maximum Cost) will be (increased) (decreased) (unchanged)
by this Change Order .... $ 29,300.00
The new (Contract Sum) (Guaranteed Maximum Cost) including this Change Order will be ... $ 2,720,900.00
The Contract Time will be (increased) (decreased) (unchanged) by Sixteen   (16) Days.
The Date of Substantial Completion as of the date of this Change Order therefore is 5-6-86

Daniel, Nicholas & Joseph, Inc.	Quality Contractors, Inc.	Authorized: Civitello Associates, Inc.
ARCHITECT	CONTRACTOR	OWNER
22 22nd St.	870 Wintergreen Ave.	888 88TH St.
Address	Address	Address
Hereandere, CT.	Bethany, CT. 06525	Hamden, CT. 06510
BY_____	BY_____	BY_____
DATE 3-28-86	DATE 3-26-86	DATE 3-28-86

AIA DOCUMENT G701 • CHANGE ORDER • APRIL 1978 EDITION • AIA® • © 1978
THE AMERICAN INSTITUTE OF ARCHITECTS, 1735 NEW YORK AVE., N.W., WASHINGTON, D.C. 20006   G701 — 1978

ALTERNATE CHANGE ORDER FORM OR SUBCONTRACTOR AMENDMENT

[LETTERHEAD]

Date:

To:	Project:

	Change No: _____
	(Or Subcontract Amendment No.)

Gentlemen:

Your proposal dated _____ in the amount of $_____
for _____
_____
_____
_____
has been accepted and incorporated into contract no. _____ dated _____ as Change Order Number _____. The contract is being (increased) (decreased) by the sum of $_____. All other conditions of the contract remain unchanged.

  1. Original Contract Amount $_____
  2. Previous Additions $_____
  3. Previous Deductions $_____
  4. Previous Total $_____
  5. This Change Order $_____
  6. New Contract Amount $_____

The Contract Time will (remain unchanged) (be increased) (be decreased) by _____ (working) (calendar) days, and is to be complete on _____.

ACCEPTED BY:

_____   _____
Date _____   Date _____

INDEX

Acceptance time for negotiation, 317
Access to site as owner's responsibility, 17-18
Accuracy, technical, of documents as architect's responsibility, 23
Adhesion contracts, 42
Adjusted Schedule, 258-259
Administrative control of change orders, 281-311 (*see also* Change order and file administration)
Administrative staff expenses, use of to substantiate change order pricing, 251-252, 255
Administrative Time Sheet, 252, 254
Agency approvals as owner's responsibility, 18
Allowance items, 119
Allowances in negotiations, 318-319
Alternatives in negotiations, presenting, 319
American Institute of Architects (AIA), 9, 20, 46, 48, 49
   The General Conditions of the Contract for Construction, 9, 25-26
   The Standard Form of Agreement Between the Owner and Contractor (AIA No.101), 38
American Arbitration Association (AAA), 5, 355, 356, 360
American Welding Society, 32
Arbitration, 320-321, 355-358
   agreement for, 355
   arbitrators, 356
   award, 358
   evidence, 357-358
   expense, 356
   informal, 355
   no discovery, 357
   privacy, 357
   time, 356-357
Architect, role of in general contracting relationship, 8-9
   code compliance, responsibility for, 25
   for diligence, skill, and good judgment, 28
   evaluation of work, 27-28
   and "final" interpretation of documents, 20-21, 25-26
   owner, responsible to, 21
   plans and specifications, production and coordination of, 22-23
   for specific design (not design criteria), 23-24
   response, prompt, 27
   responsibilities, ten categories of, 22-28
   shop drawing approvals, 272
   for submittal review and approval, 26-27
   technical accuracy of documents, 23
   in traditional owner-contractor-architect associations, responsibilities of, 13-35
   for workability of design, 24-25
As-Built Schedule, 258-259
As-Planned Schedule, 258-259
Associated General Contractors (AGC), 46
Assumptions in negotiations, 321
Attorney, need for, 355, 358-360, 362-364
Authority in negotiations, 321-322
Authorizations and easements, owner's responsibility to secure and pay for, 14, 17-18 (*see also* Owner, categories of contractual responsibilities of)
Averages in negotiations, 322
Award date of contract, extending, 111-113

"Bad faith" rejection, 349-350
Bid documents, 111-121 (*see also* Contract and bid documents)
Bidding system, owner's responsibility to uphold, 14, 15
Boilerplate:
   language, 322
   nonapplicable, as reason for change order, 77
Boring data, 104-105
Building code compliance, 105-106
Building code requirements as reason for change order, 76
Building industry, changing, 4-5

California Coastal Commission, 107
"Call girl principle," 325
Ceiling spaces, conflict, 123-126
Change clauses, 44-46, 323, 327
   description, 44
   elements, basic, 44
   problems, 45
   sample, 45-46

Change order and file administration, 279-311
  approval submissions, 294-308
    introduction, 294
    sample letter to subcontractors regarding submittal requirements, 295, 296-298
    sample letter to subcontractors regarding shop drawing resubmission requirements, 300, 301-303
    shop drawing review and coordination, 294-295
    submission requirements, 295-296
    submittal review, distribution, and follow-up, 299-301
    submittal summary record procedure, 304-308
    terms, definitions, 305-307
  correspondence file, 287-288
  file content, 284-287
    chronological file, 286
    correspondence and submittal file separation, 285, 286-287
    general file folders, 285-286
    reading file, 286
    subfiles, 286
  files, establishing easy-to-research, 282-284
  introduction, 282
  Letter of Transmittal, Sample, 309-311
  trends, tracking, 288-293
    Change Order Summary Sheet, 289-293
    introduction, 288-289
Change Order General Conditions Checklist and Estimation Sheet, 195, 196-197
Change order prices, substantiating, 224-240 (see also Prices of change order, substantiating)
Change order proposal, preparation, and presentation, 179-278
  designing and constructing, 181-223
  project records, using to substantiate change orders and claims, 241-278
  substantiating change order prices, 224-240
Change order rejections, reducing number of, 346-368
  Change Order Research Checklist, 82-85
Change Order Summary Sheet, 81, 289-293
Change Order Telephone Quotation Form, 199, 205-206
Change orders, common sources for, 99-178 (see also Potential change orders, finding)
  discovery checklist, 169-178
Change orders exposed, 59-97
  diagnosis, 61-66
    categories, 63-66
    clarification or change, 62
    consequential changes, 66
    constructive changes, 64-65
    as normal part of construction process, 62
    owner-acknowledged changes, 64
    reasons for change orders, ten, 63
  process, using to your advantage, 78-97
    equitable adjustment, 94
    establishing change order file, 80-81
    notification, 86-89
    Notification Letter to Owner on Changes, 87=88
    P's of change orders, six, 79
    payment, 95-97
    preparing order, 80-85
    presenting, 92-93
    pricing, 89-92 (see also Pricing change order)
    prospecting for (discovery), 79-80
    research, 81
    research checklist, 82-85
    submission, proposal, 92-93
    work, performing, 93-95
  understanding how they arise, 67-77
    coordination among design disciplines, lack of, 72-73
    defective specifications, 68-71 (see also Defective specifications)
    improved information, 75
    improvements in workmanship, time or cost, 75-76
    incomplete design, 73
    "intent" vs. "included," 77
    introduction, 68
    latent conditions (defects), 73-74
    nonapplicable boilerplate, 77
    nondisclosure, 72
    owner changes, 74-75
    proprietary specifications, 76
    restrictions, illegal, 76
Changes, contingency for as owner's responsibility, 16
Changing building industry, 4-5
Claims consciousness, 5
Claims, preparing for winning, 346-368
  arbitration/litigation/mediation, 355-362 (see also under each heading)
    arbitration, 355-358
    attorney, need for, 355, 358-360, 362-364
    conclusions, 362
    introduction, 355
    litigation, 358-360
    mediation, 360-362
  attorney, finding, 362-364
    introduction, 362-363
    "lion" or "pussycat," selecting, 363-364
  claims checklist, 366-368
  consultants, selecting, 364-365
  introduction, 347-348
    action, decisive, 348
    construction contracts, 347
    documentation, 347-348
    resolution options, awareness of, 348
  rejections, turning around, 348-354
    change amount vs. litigation expense, 351
    "good" or "bad" faith, 349-350
    introduction, 348-349
    letter to owner, sample, confirming special meeting, 353-354
    meetings at highest levels, 351-353
    "nothing to lose" attitude, 350-351
Code compliance as architect's responsibility, 25
Column and beam location as potential reason for change order, 128-130
Concessions, 325-326
Conduit Clause, 46-48 (see also Pass-Through Clause)
Consequential changes, 66
Consequential costs (damages), 183, 188
Construction Change Directive, 340
Construction Claims Checklist, 366-368

INDEX

Construction contracts, applying without resistance, 44-54 (*see also* Strategies)
Construction law, 37-38
Construction management, arrangement, pure, 10-12
Construction management with guaranteed maximum price (GMP), 12-13
Construction schedules, using to substantiate change order pricing, 258-260 (*see also* Project records, using to substantiate change orders and claims)
Construction Scheduling Simplified, 95, 120, 209
Construction Specifications Institute, 25, 68
Constructive change order, 326-327
    equitable adjustment, 326
Constructive changes, 64-65
Consultants, selecting, 364-365
Contingency, 325
Contract and bid documents, 111-121
    award date, extending, 111-113
    checklist, 171-173
    contract time, 119-121
        MOST (Management Operation System Technique), 120
    named subcontracts, 113-115
    price/bid allowance, 119
    sample letters:
        to owner, regarding obligation to determine responsibility for questionable work, 115-116
        to subcontractor, regarding owner's decision directing work, 117-118
Contract change order, sample, 369-371
Contract vs. contact, 6-35
    gray areas, caution about, 35
    introduction, 7
    pass-through principle, 35
    responsibilities, clarifying to guarantee accurate assessments, 13-35
        architect, categories of responsibilities of, 22-28
        contractor, general, categories of responsibility of, 29-35
        contractor's expectations, 14
        owner, categories of responsibility of, 14-21
        owner's expectations, 14
        reviewing documents, 13
    structures, ending confusion about, 7-13
        construction management, pure, 10-12
        construction management with guaranteed maximum price (GMP), 12-13
        contract and *contact*, distinction between, 7-8
        design-build, 9-10
        "fast-tracking" construction environment, 11
        general contracting, traditional, 8-9
        specifications, importance of for each job, 7
Contract documents, 38-41
    four C's of contracts, 40-41
    components, 39-40
    General Conditions, 38-40
    Plans and Specifications, 40
    The Standard Form of Agreement Between Owner and Contractor (AIA No. 101). 38
Contract interpretation, rules of, 41-43 (*see also* Strategies, proven for applying construction contracts)

Contract law concepts, 37-38
Contractor, categories of responsibility for, 29-35
    contract payments, 29. 34
    coordination of all parts of work; 29, 33
    duty to inquire, 29-30
    errors, glaring, correction of, 29, 32-33
    insurance, maintaining adequate, 29, 34
    laying out work, 29, 30-31
    planning and scheduling of work, 29, 30
    reasonable review, 29, 30
    safety standards, adherence to, 29, 34-35
    shop drawings, processing, 29, 33-34
    supervision, direction and installation of work, 29, 31
    warranty of clear title, 29, 35
    workmanship, adequate, 29, 31-32
        codes and standard practices, 32
        manufacturer's instructions, 32
        referenced standards, 32
        technical specification, 31
Contractor's move to power, 3-5
    claims consciousness, 5
"Convenience" specifications, 325
Cooperation, achieving, as owner's responsibility, 21
Correspondence, filing, 282-284, 287-288
Cost perceptions in negotiations, 328
Costs, application of, to change order, 183-190
    objectives, three, 188
    practical management of, 189-190
    public projects, easier application to, 189
Credits in negotiations, 328-329
Cut-and-paste preparation of specifications, 68-69

Dates in correspondence, importance of establishing, 244-245
Daily Field Report, 245-250 (*see also* Project records, using to substantiate change orders and claims)
Deadlines in negotiations, 329
Deadlock in negotiations, 329
Defective specifications, 68-71
    cut-and-paste preparation, 68-69
    impossibilities, 71
    inconsistencies, 68, 70-71
        precedence clause, 71
    old and outdated, 68, 70
    reasons for, 68
    silly, 68, 69-70
Defects in contracts, 73-74
Delays in negotiations, excusable, 333-334
Deliberate errors in negotiation, 329-330
Design-build contracting relationship, 9-10
Design change telltales as potential reason for change order, 139-131
Design discipline interfaces, 131
Design professionals, owner's responsibility for, 21
Designing and constructing effective change order proposals, 181-223
    components, 183-190
        costs, three, 183-190
            consequential costs (damages), 183, 188
            direct costs, 183, 184

Designing and constructing effective change order proposals (cont.)
  Direct Project Management and Administrative Costs Form, 185, 186-187
    indirect costs, 183, 184-185
    indirect costs, transforming in to direct, 185
    overhead, 184, 185, 186, 188
  developing, 190-212
    Change Order General Conditions Checklist and Estimate Sheet, 195, 196-197
    Change Order Telephone Quotation Form, 199, 205-206
    component prices, assembling, 195
    identification/notification, 191-193
    schedule impact, determining, 209-212
    subcontractor prices, assembling, 198-199
    telephone quotations, 199
    time impact analysis, 209-212
  finalizing, 212-223
    approval action, requiring, 222-223
    change order components, representing, 219-221
    contract time, presenting effects on, 221-222
    default clause, 222
    format and timing, 213-214
    introduction, 212-213
    price, presenting total, 221
    sample change order proposal cover letter, 214-216
    terms and conditions, additional, 223
  letters, sample:
    to owner, regarding cost escalation due to untimely action, 214, 217-218
    to owner, regarding pending change order, 192, 193-194
    to subcontractor regarding change order price by default, 199, 203-204
    to subcontractor regarding confirmation of telephone quotation, 199, 205, 207-208
    to subcontractor—request for change order quotation, 199-202
Detail in negotiations, level of, 330
Details and dimension strings, multiple, 154-155
Direct costs, 183, 184
Direct Project Management and Administrative Costs Form, 185, 186-187
Discipline in negotiation, 330-331
Discovery in change orders, 79-80
  in litigation, 359
  none in arbitration, 357
Discretion in evaluating low bidders as owner's responsibility, 14-15
Dispute Clause, 48-49
Dispute resolution, 313-368
  negotiation, winning in, 315-345
Duplication of design as potential reason for change order, 132-135
  details and dimension strings, multiple, 1514-155
  letters to owner, sample, 135-138

Easements, 106-107
  owner's responsibility to secure and pay for, 14, 17-18
Eighty–Twenty rule, 331

Elaboration, 331
Electrical equipment, 145-147 (*see also* Mechanical equipment)
Electronic storage of files, 283
Empathy in negotiations, 331
Environmental Protection Agency (EPA), 18
Equitable adjustment, 94, 183, 326, 332
Errors, deliberate, in negotiation, 329-330
Evaluation of work as architect's responsibility, 27-28
Exceptions in negotiations, 333
Excusable delays in negotiations, 333-334
Existing condition, change in, 127-128
Experts, use of in negotiations, 334-335

Face-saving in negotiations, 335
"Fast-tracking" construction environment, 11
"Fat" specifications, 139-140
Field Payroll Report Form, 251, 253
Files of change orders, administrative control of, 281-311 (*see also* Change order and file administration)
Finish schedule vs. specification index as potential reason for change order, 140
Flow-Down Clause, 46-48 (*see also* Pass-Through Clause)
Forms, standardized, use of in negotiations, 335-336
Funding work as owner's responsibility, 14, 15-16

General Conditions of Contract, 4
General contracting relationship, traditional, 8-9
Gentlemen's agreements, 318
"Good faith" rejections, 349-350
Graphics, use of, 339-340
Green Book, 92

Historical cost records, use of to substantiate change order pricing, 238-239

Improved information as cause for change orders, 75
Improvements in workmanship, time or cost as reason for change orders, 75-76
Inadequate level of detail as potential reason for change order, 140-142
Incomplete design in specifications, 73
Inconsistencies in specifications, 68, 70-71
Indirect costs, 183, 184-185
  direct costs, transforming into, 185
Industry and contract environments, 1-57
  contract vs. contact, 6-35
  contractor's move to power, 3-5
  strategies, proven, for applying construction contracts, 36-57
Industry sources to substantiate change order pricing, 239
Inland Wetlands Commission, 18, 107
Inquiry, contractor's responsibility for, 29-30
"Intent" vs. "included" as cause for change order, 77
Interpretation of construction law, 38

INDEX

Interpretation of documents as architect's responsibility, 20-21, 25-26
Invoices—records of direct payment, use of to substantiate change order pricing, 239-240

Job meetings, using to establish dates, scopes, and responsibilities, 261-270 (*see also* Project records)
    Job Meeting Minutes Form, 268-270
Job meeting minutes, importance of in negotiations, 335

Late bids, acceptance of, 15
Latent conditions in contract, 73-74
Letter of Transmittal, Sample, 309
Letter wars, 336
Letters, sample: *see also* Change order proposal, preparation, and presentation
Light fixture locations, 142-144
Litigation, 358-360
    discovery, 359
    evidence, 359-360
    expense, 359
    fact, trier of, 358-359
    formal proceedings, 358
    judgment, 360
    open to public, 359
    requirement, 358
    time, 359
Lump-sum price submissions, 227-229

Match lines, plan orientations and, 144-145
Mechanical equipment, 145-147
    letter to architect, sample, 150-151
    letter to owner, sample, 152-153
    letter to subcontractors, sample, 148-149
Mediation, 320-321, 360-362
    agreement for, 360
    award, 362
    discovery, 361
    evidence, 361-362
    expense, 361
    mediators, 361
    privacy, 361
    proceedings, 360-361
    time, 361
Meetings at highest levels to turn around rejections, 351-353
Military regulations as reason for change order, 76
Minutes of job meeting, importance of, 261-264
Monthly Administrative Time Sheet, 252, 254
MOST (Management Operation System Technique), 120

Negotiation, winning in, 315-345
    acceptance time, 317
    agenda, 317-318
    agreements vs. understanding, 318
    allowances, 318-319
    alternatives, 319
    arbitration, 320-321
    aspiration level, 321
    assumptions, 321
    attorney, need for, 327. 332, 334
    authority, 321-322
    averages, 322
    boilerplate, 322-323
    changes clause, 323, 327
    concessions, 325-326
        "call girl principle," 325
    conclusion, 345
    constructive changes, 326-327
        equitable adjustment, 326
    contingency, 325
    "convenience" specifications, 325
    correlation of contract documents, 327
    cost perceptions, 328
    credits—turning them around, 328-329
    deadlines, 329
    deadlock, 329
    deliberate errors, 329-330
    detail, level of, 330
    discipline, 330-331
    Eighty–Twenty rule, 331
    elaboration, 331
    empathy, 331
    equitable adjustment, 332
    estimates, designer's, 331-332
    exceptions, 333
    excusable delays, 333-334
    experts, use of, 334-335
    face saving, 335
    forms, standardized, 335-336
    general contractor as conduit, 324
    gentlemen's agreements, 318
    hard work. 344
    introduction, 317
    job meeting minutes, 335
    legitimacy, power of, 335-336
    letter wars, 336
    lost notes, 337
    mediation, 320-321
    negotiator, changing, 323-324
    "nonnegotiable" demands, 337
    objections, 338
    off-the-record discussions, 338
    oral promises, 341
    patience, 339
    personal attack, 339
    presentations, 339
    proceed orders, 340
    promises, 341
    questions, 341-342
    quick deals, 342
    "reasonable" review, 342
    reopening change proposals, 342
    splitting difference, 343
    statistics, 343
    by telephone, 343-344
    unit prices, 344
    value of hard work performed, 345
N.I.C. (Not in Contract) items, 146-147 (*see also* Mechanical equipment)

Named subcontracts, 113-115
Nonapplicable boilerplate as reason for change orders, 77
Nondisclosure, change orders and, 72
"Nonnegotiable" demands, 337
"Nothing to lose" attitude in turning around rejection, 350-351
Notification dates in correspondence, importance of establishing, 244-245
Notification letters to owner on changes, 87-88
  components, 89

Objections to change order proposals, 338
Occupational Safety and Health Administration (OSHA), 18
Off-the-record discussions, 338
Oral promises, 341
Ordinances, city, as reason for change orders, 76
Outdated specifications, 68, 70
Overhead, 184, 185, 186, 188, 195
Oversight, 322
Owner, categories of contractual responsibilities of, 14-21
  bidding system integrity, upholding, 14, 15
  clarifications and changes, offering prompt action on, 20
  cooperation, achieving, 21
  design professionals, ultimate responsibility for, 21
  discretion, use of in evaluating low bidders, 14-15
  easements and authorizations, securing and paying for, 14, 17-18
    agency approvals, 18
    site access, 17-18
  "final" interpretation of documents, offering, 20-21
  funding work, 14, 15-16
    conformance to rates/amounts related to job progress, 16
    contingency for changes, 16
    payables, timing of, 16
  plans and specifications, warranting adequacy of, 19
  suitability of furnished materials, warranting, 19
  superior knowledge, obligation to disclose, 19-20
  for surveys of physical characteristics of site, 14, 16-17
    baselines and benchmarks, 17
    material composition, 16-17
    property lines, 16
    utility locations, 17
Owner-acknowledged changes, 64
Owner changes as reason for change order, 74-75
Owner-furnished equipment, 146-147 (*see also* Mechanical equipment)
Owner-selected subcontractors, 113-115
Owner's company policy as reason for change orders, 76

P's of change orders, six, 79
Pass-Through clause, 35, 46-48, 324
  application, 46-47
  description, 46
  exceptions, 47-48
  sample, 48
Patience in negotiations, 339
Payables, timing of as owner's responsibility, 16
Payroll records, use of to substantiate change order pricing, 251-254
Performance and procedure specifications, both, 155-156
Personal attacks, danger of, 339
Photographs, use of to substantiate change order pricing, 255-258 (*see also* Project records, using to substantatiate change orders and claims)
Plans and specifications:
  ceiling spaces (conflict), 123-126
    letter to subcontractors regarding work coordination, sample, 125-126
  checklist, 173-178
  column and beam location, 128-130
  design change telltales, 130-131
  design discipline interfaces, 131
  detail and dimension strings, multiple, 154-155
  duplication of design, 132-135
    letters to owner, sample, 135-138
  electrical equipment, 145-147
  existing condition, changed, 127-128
  "Fat" Specifications, 139-140
  finish schedule vs. specification index, 140
  inadequate level of detail, 140-142
  "as indicated," 121-122
  light fixture locations, 142-144
  match lines and plan orientations, 144-145
  mechanical equipment, 145-147
    letter to architect, sample, 150-151
    letter to owner, sample, 152-153
    letter to subcontractors, sample, 148-149
  N.I.C. (Not in Contract) equipment, 146-147
  owner's responsibility to warrant adequacy of, 19
  owner-furnished equipment, 146-147
  performance and procedure specifications, both, 155-156
  proceeding under protest, 156
  production and coordination of as architect's responsibilities, 22-23
  proprietary restrictions (public), 156-158
    letters to owner, sample, 158-161
  "see specs," 121-122
  as source of potential change orders, 121-162
  specification section "scopes," 162
Potential change orders, finding, 99-178
  change order discovery checklist, 169-178
  contract and bid documents, 111-121, 171-173 (*see also* Contract and bid documents)
  introduction, 103
  plans and specifications, 121-162, 173-178 (*see also* Plans and specifications)
  predesign, 103-111, 170-171 (*see also* Predesign)
  site, 162-168, 178 (*see also* Site as cause for potential change orders)
Potholing, 108
Practicing Law Institute, 15
Precedence clause, 71
Predesign as potential source of change order, 103-111
  boring (subsurface) data, 104-105

INDEX

building code compliance, 105-106
building permit, 109-110
checklist, 170-171
easements, 106-107
interference from utilities not properly shown, 108-109
plan approvals, 109-110
rights of way, 106-107
special agency approvals, 107-108
utilities, temporary, availability of within contract limit lines, 110-111
Presentations, 339-340
Prices of change order, substantiating, 224-240
cost breakdowns, detailed, 231-233
historical cost records, 238-239
industry sources, 239
introduction, 225-227
justification of entitlement, 225
supporting information, importance of, 226-227
invoices—records of direct payment, 239-240
letters, sample:
to subcontractor regarding improper proposal submission, 228-230
lump-sum prices, 227-229
Schedule of Values, 240
time and material (T&M), 233-236
letter, sample to subcontractors regarding T&M requirement notification, 233-234, 235-236
unit prices, 237-238
values, schedule of, 240
Pricing change order, 89-92
methodology, 92
T & M (time and material) option, 91
tone, selecting proper, 92
Proceed orders, 340
Project records, using to substantiate change orders and claims, 241-278
as active working files, 243
administrative staff expenses, 251-252, 255
Monthly Administrative Time Sheet, 252, 254
archives, 244
construction schedules, 258-260
Adjusted Schedule, 258-259
As-Built Schedule, 258-259
As-Planned Schedule, 258-259
presentable evidence, six requirements for, 259-260
daily field reports, 245-250
sample for, 247-250
dates in correspondence, establishing, 244-245
importance of, 243
introduction, 243-244
item completion and close-out, 244
job meetings, using to establish dates, scopes, and responsibilities, 261-270
guidelines, 262-264
Job Meeting Minutes Form, 268-270
minutes, importance of, 261-264
sample letter to subcontractors regarding lack of job meeting attendance, 263, 266-267
sample letter to subcontractors regarding mandatory job meeting attendance, 264-265
payroll records, 251-254
Field Payroll Report Form, 251, 253
photographs, 255-258
layout requirements, 256
"precondition" photos, 255
sample layout form, 256-258
shop drawings and approval submittals, 271-274
approval response times, 273
approval responsibility, 271-273
and architect's responsibility, 272
contractor, absolute responsibility of, 273-274
differing conditions, treatment of, 273
time and material tickets, 274-278
introduction, 274
sample letter to owner regarding acknowledgment of work actually performed, 274, 275-276
sample T&M form, 277-278
Promises, 341
Promptness as architect's responsibility, 27
Property line restrictions as cause of change order, 76
Proprietary restrictions (public), 156-158
letters to owner, sample, 158-161
Proprietary specifications as cause of change order, 76
Prospecting for change orders, 79-80
Public bidding system, owner's responsibility to uphold, 14, 15

Questions in negotiations, 341-342
Quick deals, 342

"Reasonable" as dangerous word, 342
"Reasonable expectations":
concept of, 37
as standard of interpretation of contract, 41-42
Referenced Standard, 24
Rejections of change order, turning around, 348-354 (see also When changes become claims)
Responsibilities, contractual, clarifying, 13-35
Restrictions, illegal, as cause of change orders, 76
Rights of way, 106-107

Schedule of Values, 240
Schedules, as-planned and adjusted, 209-212
Shop drawing review as contractor's responsibility, 33-34
Shop drawings and approval submittals, 271-274 (see also Project records, using to substantiate change orders and claims)
Shop drawings and submittals procedure, 294-308 (see also Change order and file administration)
Silly specifications as reason for defective specifications, 68, 69-70
Site as cause for potential change orders, 162-168
checklist, 178
grades, elevations, and contours, 163-165
introduction, 162
letters to owner, sample, 165-168
Specific design as architect's responsibility, 23-24

Specification section "scopes," 162
Specifications:
   importance of for each job, 7
   owner's responsibility to warrant adequacy of, 19
   production and coordination of as architect's responsibilities, 22-23
Splitting difference in negotiations, 343
Statistics, 343
Strategies, proven, for applying construction contracts, 36-57
   applying contracts to secure power positions, 37-38
   applying construction contracts without resistance, 44-54
      authority (formal/ constructive), 49-50
      Change Clauses, 44-46 (*see also* Change Clauses)
      Costs, three, 54
      Dispute Clause, 48-49
      equitable adjustment, 53-54
      "general scope" of work, 50-51
      "intent" vs. "indication," 51-52
      introduction, 44
      Pass-Through Clause, 46-48 (*see also* Pass-through Clause)
      "performance" and "procedure" specifications, 52-53
      "reasonable review," 51
   construction law interpretation, 38
   contract documents, 38-41 (*see also* Contract and bid documents)
   contract interpretation, rules of, 41-43
      adhesion contracts, 42
      ambiguities resolved against drafter, 42
      of application, 42
      introduction, 41
      right to choose interpretation, 43
      specific vs. general, 43
      standards: reasonable expectations, 41-42
      usage of trade custom, 43
   contract law concept, 37-38
      "reasonable expectations," 37
   interpretation, strategic, 37-38
   references, 54-57
Structures of contract, ending confusion about, 7-13 (*see also* Contract vs. contact)
Subcontractors:
   owner-selected, 113-115
   prices for, assembling, 198-199
   sample letters to:
      confirming telephone quotation, 199, 205, 207-208
      regarding change order price by default, 199, 203-204
      request for change order quotation, 199-202
Submittal process, 294-308 (*see also* Change order and file administration)
Submittal review and approval as architect's responsibility, 26-27
Surveys of physical characteristics of site, owner's responsibility for, 14, 16-17 (*see also* Owner)

Telephone negotiations, 343-344
Telephone quotations, 199
   Form, 199, 205-206
Time and material (T&M) pricing, 233-236
Time and material tickets, 274
Timeliness as architect's responsibility, 27
Trade custom, 43

"Under protest," 156
Uniform Commercial Code, 43
Unit prices, 237-238, 344
"Usage of trade," 43
Utilities, interference from, 108-109
   temporary, availability of, 110-111

Value in negotiations of hard work performed, 345

Work, evaluation of as architect's responsibility, 27-28
Work phases, contractor's responsibility for, 29, 30-31
Work review, contractor's responsibility for, 29, 30
Workability of design as architect's responsibility, 24-25
Working Procedure, 4
Writing letters, 336

Zoning regulations as reason for change orders, 76